The Froehlich / Kent

ENCYCLOPEDIA OF TELECOMMUNICATIONS

VOLUME 7

The Froehlich/Kent
ENCYCLOPEDIA OF TELECOMMUNICATIONS

Editor-in-Chief

Fritz E. Froehlich, Ph.D.

Professor of Telecommunications
University of Pittsburgh
Pittsburgh, Pennsylvania

Co-Editor

Allen Kent

Distinguished Service Professor of Information Science
University of Pittsburgh
Pittsburgh, Pennsylvania

Administrative Editor

Carolyn M. Hall

Pittsburgh, Pennsylvania

VOLUME 7

ELECTRICAL FILTERS: FUNDAMENTALS AND SYSTEM APPLICATIONS to FEDERAL COMMUNICATIONS COMMISSION OF THE UNITED STATES

 CRC Press
Taylor & Francis Group
Boca Raton London New York

CRC Press is an imprint of the
Taylor & Francis Group, an **informa** business

First published 1994 by Marcel Dekker, Inc.

Published 2021 by CRC Press
Taylor & Francis Group
6000 Broken Sound Parkway NW, Suite 300
Boca Raton, FL 33487-2742

© 1994 by Taylor & Francis Group, LLC
CRC Press is an imprint of Taylor & Francis Group, an Informa business

No claim to original U.S. Government works

ISBN 13: 978-0-8247-2905-9 (hbk)

Visit the Taylor & Francis Web site at
http://www.taylorandfrancis.com

and the CRC Press Web site at
http://www.crcpress.com

Library of Congress Cataloging-in-Publication Data

The Froehlich/Kent Encyclopedia of Telecommunications / editor-in-chief, Fritz E.
Froehlich ; co-editor, Allen Kent.
    p. cm.
   Includes bibliographical references and indexes.
   ISBN 0-8247-2902-1 (v. 1 : alk. paper)
   1. Telecommunication—Encyclopedias. I. Froehlich, Fritz E.,
Kent, Allen.
  TK5102.E646 1990   90-3966
   384′.03—dc20   CIP

LIBRARY OF CONGRESS CATALOG CARD NUMBER: 90-3966

CONTENTS OF VOLUME 7

Contributors to Volume 7   v

Electrical Filters: Fundamentals and System Applications   1
  Mebenin Awipi

Electrical Telecommunications Cables: Fundamentals and History   57
  Merle C. Biskeborn

Electromagnetic Signal Transmission   147
  Warren L. Flock

Electron Beam Displays (Cathode-Ray Tubes)   197
  Carlo Infante

Electronic Displays—Human Factors   243
  Gerald M. Murch

Electronic Mail   299
  Eben Lee Kent

Electronic Switching. *See* **Digital Switching Systems**
  Volume 6, pages 259–287

Encryption in Telecommunications. *See* **Cryptology**
  Volume 4, pages 501–517

Engineering for Telecommunications   315
  C. William Anderson

Entropy and Information Theory   333
  Suguru Arimoto

Equalization, Adaptive. *See* **Adaptive Equalization**
  Volume 1, pages 97–103; **Electrical Filters: Fundamentals and**
  System Applications   *Volume 7, pages 1–55*

Equalization of Telecommunication Channels   345
  Hikmet Sari

Espenschied, Lloyd   361
  Marylin K. Sheddan

Evolution of Cellular Switch Networking   367
  Jack M. Scanlon and Martin H. Singer

Exchange Carriers Standards Association   379
  George L. Edwards

Facsimile Image Coding. *See* **Coding of Facsimile Images**
  Volume 3, pages 33–113

Facsimile Standards   387
    Stephen J. Urban

Fading Radio Channels. *See* **Communication over Fading Radio
  Channels**   *Volume 3, pages 261–303*

Fast, High-Performance Local-Area Networks   421
    Jon W. Mark

Fast Packet Network: Data, Image, and Voice Signal Recovery   453
    Farokh A. Marvasti

The Federal Communications Commission of the United States   481
    Byron F. Marchant

CONTRIBUTORS TO VOLUME 7

C. William Anderson   Chief Engineer (Retired), New England Telephone Company, Boston, Massachusetts: *Engineering for Telecommunications*

Suguru Arimoto, Dr. Eng.   Faculty of Engineering, University of Tokyo, Tokyo, Japan: *Entropy and Information Theory*

Mebenin Awipi, Engr. Sc.D. (Columbia)   Associate Professor of Electrical Engineering, Tennessee State University, Nashville, Tennessee; formerly Member of Technical Staff, Bell Telephone Laboratories: *Electrical Filters: Fundamentals and System Applications*

Merle C. Biskeborn, D.Eng.(hon.)   AT&T Bell Laboratories (Retired), Murray Hill, New Jersey; Phelps Dodge Communications Company (Retired), White Plains, New York: *Electrical Telecommunications Cables: Fundamentals and History*

George L. Edwards   President, Exchange Carriers Standards Association, Washington, D.C.: *Exchange Carriers Standards Association*

Warren L. Flock, Ph.D.   Professor and Co-Director, NASA Propagation Information Center, Department of Electrical and Computer Engineering, University of Colorado, Boulder, Colorado: *Electromagnetic Signal Transmission*

Carlo Infante, Dottore in Fisica   CBI Technology Consultants, Scottsdale, Arizona: *Electron Beam Displays (Cathode-Ray Tubes)*

Eben Lee Kent   Product Manager, CompuServe, Inc., Columbus, Ohio: *Electronic Mail*

Byron F. Marchant, J.D.   Legal Advisor, Office of Commissioner Andrew C. Barrett, Federal Communications Commission, Washington, D.C.: *The Federal Communications Commission of the United States*

Jon W. Mark, Ph.D.   Professor, Department of Electrical and Computer Engineering, University of Waterloo, Waterloo, Ontario, Canada: *Fast, High-Performance Local-Area Networks*

Farokh A. Marvasti, Ph.D.   Head, Communications Signal Processing Lab, Communications Research Group, King's College London, University of London, London, U.K.: *Fast Packet Network: Data, Image, and Voice Signal Recovery*

Gerald M. Murch, Ph.D.   Director of Imaging Systems, Apple Computer, Inc., Cupertino, California: *Electronic Displays—Human Factors*

Hikmet Sari, Ph.D.   Department Head, Transmission Systems, Société Anonyme de Télécommunications, Paris, France: *Equalization of Telecommunication Channels*

Jack M. Scanlon   Senior Vice-President and General Manager, Motorola Cellular Infrastructure Group, Arlington Heights, Illinois: *Evolution of Cellular Switch Networking*

Marylin K. Sheddan   Research Associate, Embry-Riddle Aeronautical University, Daytona Beach, Florida: *Espenschied, Lloyd*

Martin H. Singer, Ph.D.   Senior Director of Business Development and Planning, Motorola Cellular Infrastructure Group, Arlington Heights, Illinois: *Evolution of Cellular Switch Networking*

Stephen J. Urban   Delta Information Systems, Inc., Horsham, Pennsylvania: *Facsimile Standards*

Electrical Filters: Fundamentals and System Applications

Introduction

The generic term *filters* as used in electrical and telecommunications engineering was derived from optical science, in which different colored transparent materials were used to block some light components (colors) and pass others. This was based on the fact that white light is made up of different colored lights with varying wavelengths (or, equivalently, frequencies); filtering or separation of one component from another was based on the spectral content of the component.

The following situations provide opportunities for the application of electrical filters.

1. *To shape the frequency spectrum of a signal.* Often design equations and algorithms depend upon the assumption of either a band-limited spectrum or a white (flat) spectrum; if the original signal spectrum does not satisfy either or both of these assumptions, then a "prewhitening" or "band-limiting" filter is applied before the signal is transmitted for further processing.
2. *To select or separate one signal from others if the spectrum of the desired signal does not overlap with the spectra of the interfering signals.* This is the classical application of electrical filters in telecommunications, and its invention led to the implementation of analog carrier systems in the decades from about 1920 to 1950, using the principle of frequency division multiplexing of a number of channels on a common transmission medium.
3. *Improvement of signal-to-noise ratios.* Where overlap of signal spectra prevents the actual separation of a desired signal from other interfering signals or noise, filters can be applied to improve the signal-to-noise ratio of the desired signal over the interference.

These three situations are illustrated in Fig. 1.

In the following sections of this article, we start with a discussion of means of deriving the frequency content of signals. The spectral density functions of deterministic and random signals are defined via Parseval's theorem and the Wiener–Khintchine equation. Both continuous and discrete-time signals are considered.

Next, the convolution properties of causal signals and linear systems are derived for Fourier, Laplace, and Z-transforms. This provides the definition of a transfer function that enables us to modify the spectral content of all types of signals in desired manners (i.e., to filter signals).

The next section discusses system-level classification of filters into four major types: low pass (LP), high pass (HP), band pass (BP), and band elimination

1

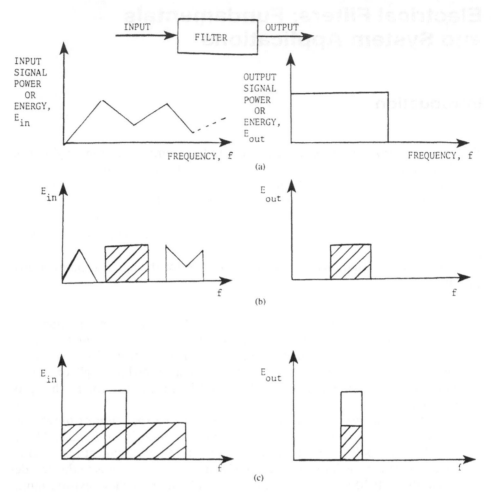

FIG. 1   Situations for application of electrical filters: *a*, prewhitening and bandlimiting; *b*, separation or selection of desired signal from spectrally nonoverlapping interference; *c*, improvement of signal-to-noise ratio.

(BE). Ideal and practical (realizable) characteristics are discussed, as well as scaling and frequency transformation from one type to another.

The article concludes with a summary of key material covered in the text.

Frequency Content of Signals

Introduction to the Frequency Domain

An *electrical filter* is any device, circuit, or system that modifies the frequency contents of electrical signals in a number of desirable ways. Before discussing the frequency domain behavior of devices that can serve as filters, we first discuss the frequency content of signals.

A *signal* is an electrical voltage or current that varies as a function of time; it thus can be made to carry information. Consider a voltage that varies sinusoidally as a function of time:

$$x(t) = A \sin (2\pi f_0 t), \qquad \text{volts (V)}. \qquad (1)$$

In addition to the time-domain plot of this signal (Fig. 2-*a*), we also may describe this signal as having all its power concentrated at the frequency f_0, which equals $1/T$ (Fig. 2-*b*). The function $W_x(f)$ that describes the distribution of the average power of $x(t)$ as a function of frequency is called the power spectral

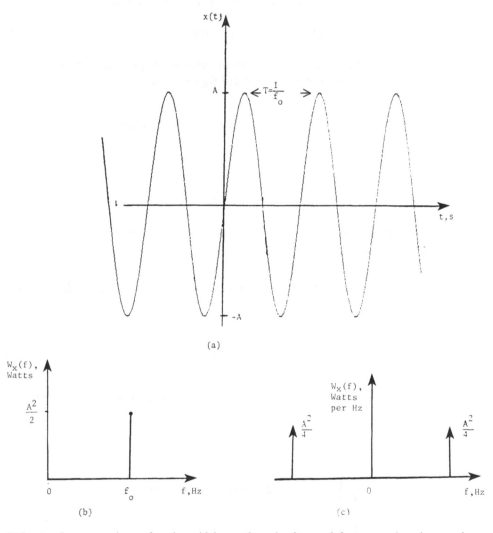

FIG. 2   Representations of a sinusoidal waveform in time and frequency domains: *a*, time domain; *b*, frequency domain; *c*, through use of impulse functions.

density (PSD) function of the signal $x(t)$. A more mathematical representation of the PSD of the sine wave involves the use of impulse functions (Fig. 2-c).

The design of communications systems and other electronic instruments was facilitated by the description of signals in the frequency domain provided by the PSD $W_x(f)$ function much more than it was by the description provided by the time function $x(t)$. Means of determining the PSD functions of various classes of signals are the focus of the next five sections.

Periodic Functions

Beyond the single sinusoidal function, periodic functions represent the next class of functions with significant frequency-domain properties. A *periodic signal* is one that repeats after a finite time interval T (the *period*). That is,

$$x(t + T) = x(t) \text{ for all } t. \tag{2}$$

According to the Fourier series theorem, almost all periodic signals can be represented by an infinite sum of sinusoidal signals. That is,

$$x(t) = \frac{a_0}{2} + \sum_{n=1}^{\infty} (a_n \cos (nw_0t) + b_n \sin (nw_0t)) \tag{3}$$

where

$$w_0 = \frac{2\pi}{T} \text{ (or } f_0 = \frac{w_0}{2\pi} = 1/T) \tag{4}$$

is the fundamental frequency and the Fourier series coefficients a_n and b_n are given by the integrals

$$a_n = \frac{2}{T} \int_{t_0}^{t_0+T} x(t) \cos (nw_0t) \, dt, \qquad a_0 = \frac{2}{T} \int_{t_0}^{t_0+T} x(t) \, dt \tag{5}$$

and

$$b_n = \frac{2}{T} \int_{t_0}^{t_0+T} x(t) \sin (nw_0t) \, dt. \tag{6}$$

For example, the rectangular waveform in Fig. 3 has the Fourier coefficients

$$a_0 = \frac{2}{T} \int_0^{\tau} 1 \, dt = \frac{2\tau}{T} \tag{7}$$

$$a_n = \frac{2}{T} \int_0^{\tau} \cos (nw_0t) \, dt = \frac{1}{n\pi} \sin (nw_0\tau) \tag{8}$$

$$b_n = \frac{2}{T} \int_0^{\tau} \sin (nw_0t) \, dt = \frac{1}{n\pi} [1 - \cos (nw_0\tau)]. \tag{9}$$

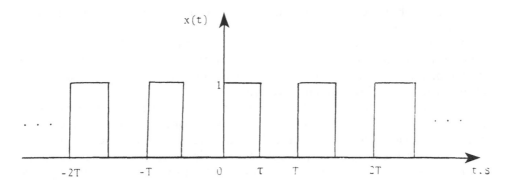

FIG. 3   Rectangular periodic waveform.

For example, let the duty cycle of the pulse train, $d = \tau/T = 1/2$. Then

$$a_0 = 1, \qquad a_n = 0 \text{ for all } n \tag{10}$$

and

$$b_n = \frac{1}{n\pi} (1 - (-1)^n) = \begin{cases} 0, & n \text{ even} \\ \dfrac{2}{n\pi}, & n \text{ odd.} \end{cases} \tag{11}$$

Hence

$$x(t) = \frac{1}{2} + \sum_{n=1,3,5,\ldots}^{\infty} \frac{2}{n\pi} \sin (nw_0 t). \tag{12}$$

Referring to the general Fourier series expansion of Eq. (3), we observe that any periodic function, in general, will have a frequency content consisting of a direct current (DC) term $a_0/2$ (i.e., frequency of zero hertz [Hz] and an infinite number of discrete frequency components, nf_0 from the sine and cosine terms. Sine waves with frequencies that are integral multiples are called *harmonics*. Following the treatment for the single sinusoidal waveform given above, the PSD function of any periodic waveform is as shown in Fig. 4, where $c_n^2 =$

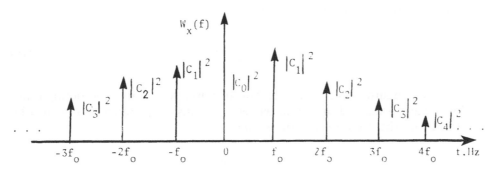

FIG. 4   General discrete spectrum of periodic signal.

$a_n^2 + b_n^2$. As observed for the special case of 50% duty cycle, in specific cases some of the a_n or b_n terms may be zero.

We now view the above result from several perspectives. First, we consider Parseval's theorem, which provides a firmer understanding of the idea of power spectrum, and then a reconstruction of a periodic waveform from a sum of harmonically related sinusoidal waveforms.

Parseval's theorem can be stated for discrete and continuous frequency content. In the first case, consider the periodic waveform $x(t)$ and assume that it represents an electrical current through, or an electrical voltage across, a 1-ohm resistor. Then the quantity $x^2(t)$ represents the instantaneous power being absorbed by the resistor. Since $x(t)$ is periodic, $x^2(t)$ also is periodic with a period T, hence its average value is

$$\overline{P} = \frac{1}{T} \int_{t_0}^{t_0+T} x^2(t)\, dt \tag{13}$$

Now, if $x(t)$ has been expanded in a Fourier series, the quantity $\overline{P}$ also can be computed using the series and the orthogonal properties of the sines and cosines. The result is the discrete version of Parseval's theorem:

$$\frac{1}{T} \int_{t_0}^{t_0+T} x^2(t)\, dt = \frac{a_0^2}{4} + \sum_{n=1}^{\infty} \frac{1}{2}\,(a_n^2 + b_n^2) \tag{14}$$

In other words, the average power $\overline{P}$ can be computed in the time domain by integrating $x^2(t)$ or in the frequency domain by summing the average powers at all discrete frequency components for the periodic signal.

For example, for the symmetric rectangular waveform where $\tau = T/2$

$$\overline{P} = \frac{1}{T} \int_{0}^{\frac{1}{2}T} 1 \cdot dt = \frac{1}{2}\, W.$$

Hence, applying Parseval's theorem,

$$\frac{1}{2} = \frac{1}{4} + \frac{1}{2} \sum_{n=1,3,5,\ldots}^{\infty} \frac{4}{n^2\pi^2}. \tag{15}$$

Rearranging this expression, we obtain

$$\sum_{n=1,3,5,\ldots}^{\infty} \frac{1}{n^2} = \frac{\pi^2}{8} \tag{16}$$

which is a standard result in the algebra of power series. By considering other periodic waveforms and their Fourier series expansions, similar results may be obtained for other power series sums, such as

$$\sum_{n=1}^{\infty} \frac{1}{n^2} = \frac{\pi^2}{6}. \tag{17}$$

The Fourier series expansion is a decomposition process. We also can gain additional insight into the process by constructing a periodic waveform by adding an increasing number of harmonically related sine waves. That is, while the Fourier series involves an infinite number of such sinusoids, the truncated series is also an optimum approximation, in a least-mean-square-error sense, to the value of the waveform at all points of continuity.

Once again, consider the series for the symmetric rectangular waveform

$$x(t) = \frac{1}{2} + \sum_{n=1,3,5,\dots}^{\infty} \frac{2}{n\pi} \sin(nw_0 t). \tag{18}$$

Figures 5 and 6 show in succession the terms $\frac{1}{2}$, $2/\pi \sin w_0 t$, $2/(3\pi) \sin 3w_0 t$, $\dots$, and the partial sums

$$s_0(t) = 1/2, \tag{19}$$

$$s_1(t) = \frac{1}{2} + \frac{2}{\pi} \sin w_0 t, \tag{20}$$

$$s_3(t) = \frac{1}{2} + \frac{2}{\pi} \sin w_0 t + \frac{2}{3\pi} \sin 3w_0 t. \tag{21}$$

As we can see, the more terms we have, the closer $s_n(t)$ approaches the original function $x(t)$. Examples in which up to 100 terms are included in the sum are shown in Fig. 7 (1).

With the frequency content of periodic signals established, we can make a mathematical transformation of the Fourier series to obtain a complex version. Using Euler's formulas

$$\cos\theta = \frac{e^{j\theta} + e^{-j\theta}}{2}, \qquad \sin\theta = \frac{e^{j\theta} - e^{-j\theta}}{2j},$$

the trigonometric Fourier terms

$$a_n \cos\theta + b_n \sin\theta = \frac{a_n(e^{j\theta} + e^{-j\theta})}{2} + b_n\left(\frac{e^{j\theta} - e^{-j\theta}}{2j}\right)$$

$$= \frac{a_n - jb_n}{2} e^{j\theta} + \frac{a_n + jb_n}{2} e^{-j\theta}.$$

Thus, replacing θ by the argument $(nw_0 t)$, the Fourier series becomes

$$x(t) = \sum_{n=-\infty}^{\infty} c_n e^{jnw_0 t} \tag{22}$$

where

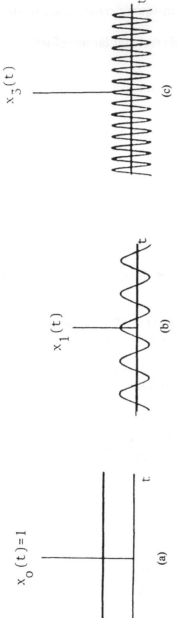

FIG. 5 Harmonic terms of Fourier series expansion: *a*, direct current; *b*, first; *c*, third. (Reprinted with permission of Ref. 1.)

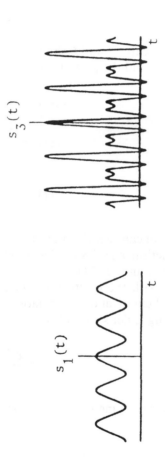

FIG. 6 Partial sums of harmonically related sinusoidal waveforms.

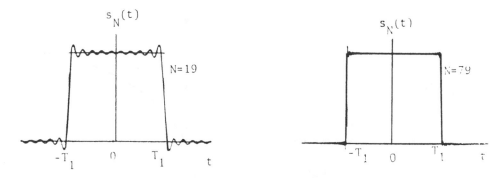

FIG. 7   Partial sums of a large number of Fourier series terms.

$$c_0 = \frac{a_0}{2},$$

$$c_n = \frac{a_n - jb_n}{2}, \tag{23}$$

$$c_{-n} = \frac{a_n + jb_n}{2}. \tag{24}$$

But the set $\{e^{jnw_0 t}\}$ is also orthogonal; that is, if a product of two terms of the set are integrated over an interval of time equal to the period $T = (2\pi)/(w_0)$, starting from any arbitrary instant of time t_0, the result is zero except when the pair is identical:

$$\int_{t_0}^{t_0 + T} e^{jmw_0 t} e^{jnw_0 t} \, dt = \begin{cases} 0, & m \neq n \\ T, & m = -n. \end{cases}$$

It was this orthogonal condition that was used to derive the equations for the Fourier series given in Eqs. (5) and (6). When applied to the exponential Fourier series, the coefficients are given by

$$c_0 = \frac{1}{T} \int_{t_0}^{t_0 + T} x(t) \, dt \tag{25}$$

and

$$c_n = \frac{1}{T} \int_{t_0}^{t_0 + T} x(t) e^{-jnw_0 t} \, dt. \tag{26}$$

The complex form of the Fourier series is closer to the Fourier transform, which is used when dealing with nonperiodic deterministic signals, discussed in the next section.

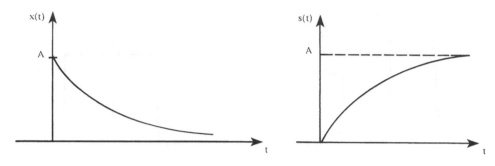

FIG. 8 Exponential pulses.

Nonperiodic Deterministic Signals

Many other information bearing signals are of the nonperiodic deterministic nature, such as the exponential pulses

$$x(t) = Ae^{-at}u(t)$$

or

$$s(t) = A(1 - e^{-at})u(t)$$

shown in Fig. 8, where $u(t)$ is the unit step function defined by

$$u(t) = \begin{cases} 0, & t < 0 \\ 1, & t \geq 0 \end{cases}.$$

Knowing the frequency content of such signals is important for designing physical communications media appropriate for transporting them within acceptable distortion limits. In fact, the rectangular pulse, when filtered with finite-bandwidth circuits, is transformed into exponential pulses with finite rise and decay times (Fig. 9).

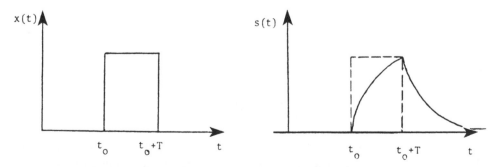

FIG. 9 Transformation of rectangular pulse to exponentials.

The Fourier transform of finite-energy nonperiodic signals may be derived from the complex form of the Fourier series expansion. For the expressions

$$x(t) = \sum_{n=-\infty}^{\infty} c_n e^{jn2\pi f_0 t}, \qquad c_n = \frac{1}{T} \int_{t_0}^{t_0+T} x(t) e^{-j2\pi n f_0 t} \, dt$$

where $f_0 = 1/T$. Let $T \to \infty$ while the discrete nf_0 approaches a continuous-frequency variable f and the summation also becomes an integral. Then, the expressions for $x(t)$ and c_n become

$$x(t) = \int_{-\infty}^{\infty} X(f) e^{j2\pi ft} \, df \tag{27}$$

and

$$X(f) = \int_{-\infty}^{\infty} x(t) e^{-j2\pi ft} \, dt \tag{28}$$

which are called Fourier transform integrals. For example, if $x(t) = e^{-at} u(t)$, then we easily can obtain

$$X(f) = \int_{0}^{\infty} e^{-at} e^{-j2\pi ft} \, dt = \frac{1}{a + j2\pi f}$$

if a is a real positive number. We therefore obtain the Fourier transform pair

$$e^{-at} u(t) \leftrightarrow \frac{1}{a + j2\pi f}. \tag{29}$$

In order to deduce the frequency content of a signal from its Fourier transform, we apply the continuous version of Parseval's theorem, which states that if $x(t)$ and $X(f)$ are a Fourier transform pair, then

$$\int_{-\infty}^{\infty} |x(t)|^2 \, dt = \int_{-\infty}^{\infty} |X(f)|^2 \, df. \tag{30}$$

Once again, if $x(t)$ represents a real electrical voltage across, or current through a 1-ohm resistor, then the square of the absolute value of $x(t)$ becomes the real square of $x(t)$, which is the instantaneous power being absorbed by the resistor. Hence the integral of $x^2(t)$ on the left side of Parseval's theorem represents the total energy delivered by the signal $x(t)$. Thus the quantity $|X(f)|^2$, when integrated over frequency, provides the energy of the signal. We therefore define $|X(f)|^2$ as the energy spectral density function of the signal $x(t)$, $W_x(f)$. For the specific case of the exponential pulse, these quantities are (Fig. 10)

$$W_x(f) = \frac{1}{a^2 + 4\pi^2 f^2} = \frac{1/4\pi^2}{\left(\dfrac{a}{2\pi}\right)^2 + f^2}. \tag{31}$$

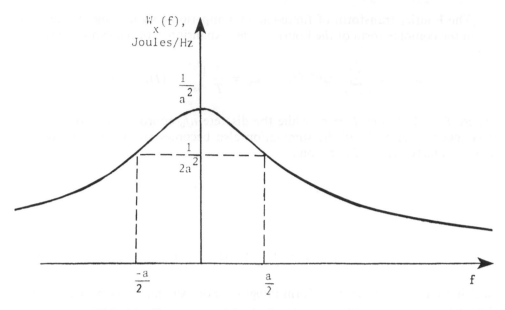

FIG. 10   Energy spectral density function of exponential pulse.

The total energy can be calculated in both the frequency and time domains as follows:

$$E_x = \int_{-\infty}^{\infty} W_x(f)\, df = \frac{1}{4\pi^2} \cdot \frac{2\pi}{a}\, \tan^{-1}\left(\frac{2\pi f}{a}\right) \Big|_{-\infty}^{\infty}$$

$$= \frac{1}{2a}\ \text{joules.}$$

Also,

$$E_x = \int_0^{\infty} x^2(t)\, dt = \int_0^{\infty} e^{-2at}\, dt = \frac{1}{2a}\ \text{joules.}$$

As indicated by the derivation, the main difference between the frequency contents of periodic and nonperiodic signals is that the former are discrete in frequency, while the latter are continuous in frequency. This difference is carried over to the cases in which time is discrete, as discussed in the next section.

The Fourier transform has a number of useful properties (see Table 1). In Table 2, we list a number of common Fourier transform pairs.

Using the frequency translation and linearity properties and the Fourier transform of a constant signal (DC), we now can obtain the Fourier transform of a periodic signal, and thus the impulse form of the spectrum alluded to in the previous section. That is, for

TABLE 1   Selected Properties of the Fourier Transform

1. Linearity: $ax_1(t) + bx_2(t) \leftrightarrow aX_1(f) + bX_2(f)$
2. Time translation: $x(t - t_0) \leftrightarrow e^{-j2\pi f t_0}X(f)$
3. Frequency translation: $e^{j2\pi f_0 t}x(t) \leftrightarrow X(f - f_0)$
4. Time/frequency scaling: $x(at) \leftrightarrow \dfrac{1}{|a|} X(f/a)$
5. Modulation property: $x(t)\cos2\pi f_0 t \leftrightarrow \dfrac{1}{2}[X(f - f_0) + X(f + f_0)]$
6. Convolution:

    $x(t) * h(t) \leftrightarrow X(f)H(f)$
    $x(t)h(t) \leftrightarrow X(f) * H(f)$

7. Duality property: If $x(t) \leftrightarrow X(f)$, then $X(t) \leftrightarrow x(-f)$
8. Differentiation: $\dfrac{dx(t)}{dt} \leftrightarrow (j2\pi f)X(f)$

$$x(t) = \sum_{n=-\infty}^{\infty} c_n e^{j2\pi n f_0 t},$$

$$X(f) = \sum_{n=-\infty}^{\infty} c_n \delta(f - nf_0) \tag{32}$$

where the impulse is a special function with argument λ defined by the mathematical properties

$$\delta(\lambda) = 0, \qquad \lambda \neq 0$$

and

$$\int_{-\infty}^{\infty} \delta(\lambda)d\lambda = 1.$$

Physically, the impulse function is an infinitesimally narrow and infinitely tall pulse with a finite area interpreted as a weight.

Hence, the PSD function for the periodic signal is

$$W_x(f) = \sum_{n=-\infty}^{\infty} |c_n|^2 \, \delta(f - nf_0) \tag{33}$$

as plotted in Fig. 4.

We also observe that using Parseval's theorem to calculate the total energy can lead to the evaluation of some definite integrals, just as the discrete case led to the evaluation of algebraic series. For example, the Fourier transform pair (Fig. 11)

TABLE 2   Common Fourier Transform Pairs

Time Domain	Frequency Domain
1	$\delta(f)$
$\delta(t)$	1
$u(t)$	$\dfrac{1}{2}\,\delta(f) + \dfrac{1}{j2\pi f}$
$e^{-at}u(t)$	$\dfrac{1}{a + j2\pi f}$
$\cos 2\pi f_0 t$	$\dfrac{1}{2}\,[\delta(f - f_0) + \delta(f + f_0)]$
$\sin 2\pi f_0 t$	$\dfrac{1}{2j}\,[\delta(f - f_0) - \delta(f + f_0)]$
$\dfrac{\sin(at)}{\pi t}$	$\begin{cases} 1, & \|f\| < a/2\pi \\ 0, & \text{otherwise.} \end{cases}$

$$p(t) = u\left(t + \frac{a}{2}\right) - u\left(t - \frac{a}{2}\right) \leftrightarrow P(f) = \frac{\sin \pi a f}{\pi f}$$

provides the result

$$\int_{-\infty}^{\infty} \frac{\sin^2 \pi a f}{(\pi f)^2}\, df = 2 \int_{0}^{\infty} \left(\frac{\sin \pi a f}{\pi f}\right)^2 df = \int_{-a/2}^{a/2} 1^2\, dt = a.$$

In standard calculus notation,

$$\int_{0}^{\infty} \frac{\sin^2 \pi a x}{(\pi x)^2}\, dx = a/2$$

or

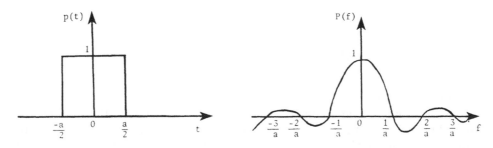

FIG. 11   Unit pulse and its Fourier transform.

$$\int_0^\infty \frac{\sin^2 y}{y^2}\, dy = \pi/2 \tag{34}$$

as listed in tables of integrals (2).

Discrete-Time Deterministic Signals

In the wake of the revolution in digital electronics, all branches of electrical engineering are dealing more and more with discrete-time signals, which can be converted to digital signals by amplitude quantization. Thus, even starting out with such analog signals as speech, there are often technological advantages in processing, including filtering, if such analog signals first are converted to digital signals. We derive the frequency content of discrete-time deterministic signals by methods analogous to those given in the last two sections, that is, by Fourier methods.

Discrete-time signals arise routinely, as in radar returns indicating the position of a target $x(kT)$, or by sampling a continuous-time signal $x(t)$. The frequency content of a sampled signal $x(kT)$ is related to the frequency content of the underlying signal $x(t)$, while also exhibiting characteristics of all discrete-time signals. We therefore consider the sampling process first.

Ideally, the process of sampling an analog signal at uniform intervals can be modeled by multiplying the signal with an impulse train (see Fig. 12). The periodic impulse train may be represented by a time-domain expression, or by the Fourier series expansion, as

$$p(t) = \sum_{k=-\infty}^{\infty} \delta(t - kT) \tag{35}$$

or

$$p(t) = \sum_{k=-\infty}^{\infty} c_n e^{j2\pi n f_0 t}, \tag{36}$$

where $f_0 = (1)/(T)$ and

$$c_n = \frac{1}{T} \int_{t_0}^{t_0/T} p(t) e^{-j2\pi n f_0 t}\, dt = \frac{1}{T}. \tag{37}$$

The sampled signal $y(t)$ now can be written as

$$y(t) = x(t)p(t) = \sum_{k=-\infty}^{\infty} x(kT)\, \delta(t-kT) \tag{38}$$

or

$$y(t) = \frac{1}{T} \sum_{n=-\infty}^{\infty} x(t) e^{j2\pi n f_0 t}. \tag{39}$$

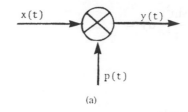

(a)

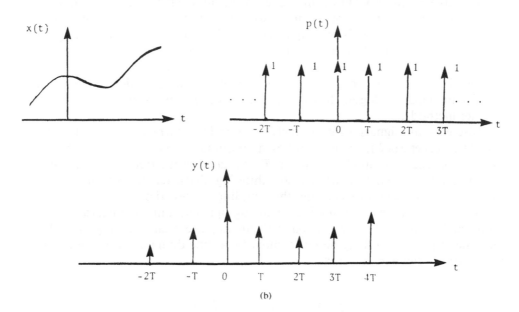

(b)

FIG. 12   Sampling process: a, system model; b, related waveforms.

Using the linearity and frequency translation properties, the Fourier transform of $y(t)$ can be obtained as

$$Y(f) = \frac{1}{T} \sum_{n=-\infty}^{\infty} X(f - nf_0). \tag{40}$$

According to the sampling theorem, if $|X(f)|$ is band limited to B Hz and the sampling frequency $f_0 = 1/T$ is at least two times B, then the translated versions of $|X(f)|$ do not overlap, and an undistorted version of the analog signal $x(t)$ can be recovered from the sampled signal $y(t) = \{x(kT)\}$ via the interpolation formula

$$x(t) = \sum_{k=-\infty}^{\infty} x(kT) \frac{\sin (2\pi f_0 t - k\pi)}{(2\pi f_0 t - k\pi)} \tag{41}$$

where $f_0 = 1/T = 2B$.

For the case $f_0 \geq 2B$, the relation between the spectra of $x(t)$ and $y(t)$ is illustrated in Fig. 13. Further, the repetitive nature of $|Y(f)|$ is typical of all discrete-time signals $\{y(kT)\}$, that is, the spectrum of $\{y(kT)\}$ typically is periodic with period $f_0 = 1/T$.

Let us now consider the problem of determining the frequency spectrum of a discrete-time signal in the situation in which the underlying continuous-time signal is unknown, that is, we only have the signal

$$x(kT), \quad k = \ldots, -1, 0, 1, 2, \ldots$$

First, let us consider the case for periodic signals, that is,

$$x(kT) = x((k + N)T) \text{ for some } N.$$

Thus $x(kT)$ is periodic with period NT. The signal $x(kT)$ then can be expanded in a Fourier series

$$x(kT) = \sum_{n=n_0}^{n_0+N-1} c_n e^{jk2\pi f_0 nT} \tag{42}$$

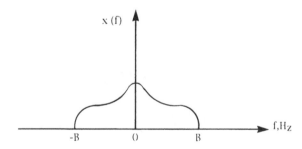

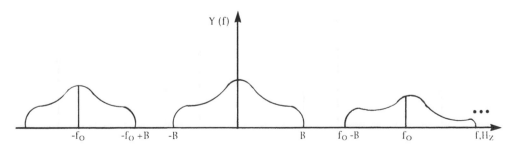

FIG. 13   Typical case of spectra of $x(t)$ and its sampled version $[x(kT)]$ where $f_0 > 2B$.

where $f_0 = 1/NT$ and

$$c_n = \frac{1}{NT} \sum_{k=k_0}^{k_0+N-1} x(kT)e^{-jk2\pi n f_0 T}. \tag{43}$$

For example, consider the periodic signal in Fig. 14. Here $x(kT)$ is periodic with period $4T$. Hence, its Fourier series expansion is

$$x(kT) = \sum_{n=0}^{3} c_n e^{jk2\pi n f_0 T}$$

where $f_0 = 1/4T$ and

$$c_n = \frac{1}{4} \sum_{k=0}^{3} x(kT)e^{-jk2\pi n f_0 T}$$

$$= \frac{1}{4}[x(0) + x(T)e^{-j2\pi n f_0 T} + x(2T)e^{-j4\pi n f_0 T} + x(3T)e^{-j6\pi n f_0 T}].$$

Since $x(2T) = x(3T) = 0$,

$$c_n = \frac{1}{4}[x(0) + x(T)e^{-j2\pi n f_0 T}]$$

that is,

$$c_0 = \frac{1}{4}[2 + 1] = \frac{3}{4}$$

$$c_1 = \frac{1}{4}[2 + e^{-j2\pi f_0 T}]$$

$$c_2 = \frac{1}{4}[2 + e^{-j4\pi f_0 T}]$$

$$c_3 = \frac{1}{4}[2 + e^{-j6\pi f_0 T}].$$

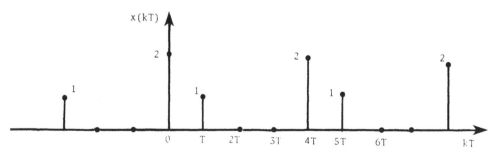

FIG. 14   Example of periodic discrete-time signal.

Further, let $T = 1s$. Then

$$c_0 = 3/4$$

$$c_1 = \frac{1}{4} [2 + e^{-j2\pi \cdot 1/4}] = \frac{1}{4} [2 - j]$$

$$c_2 = \frac{1}{4} [2 + e^{-j\pi}] = \frac{1}{4}$$

$$c_3 = \frac{1}{4} [2 + e^{-j3\pi/2}] = \frac{1}{4} [2 + j].$$

With these Fourier coefficients,

$$x(kT) = \frac{3}{4} + \frac{1}{4} (2 - j)e^{jk\pi/2} + \frac{1}{4} e^{jk\pi}$$

$$+ \frac{1}{4} (2 + j)e^{jk3\pi/2}. \tag{44}$$

For example,

$$x(0) = \frac{3}{4} + \frac{1}{4} (2 - j) + \frac{1}{4} + \frac{1}{4} (2 + j) = 2$$

$$x(T) = \frac{3}{4} + \frac{1}{4} (2 - j)e^{j\pi/2} + \frac{1}{4} e^{j\pi} + \frac{1}{4} (2 + j)e^{j3\pi/2}$$

$$= 1$$

and

$$x(2T) = \frac{3}{4} + \frac{1}{4} (2 - j)e^{j\pi} + \frac{1}{4} e^{j2\pi} + \frac{1}{4} (2 + j)e^{j3\pi}$$

which equals 0, as does $x(3T)$. Thus we have verified the complete equivalence of the signal and its Fourier series expansion.

Now if the periodic signal $x(kT)$ is expanded in a discrete Fourier series, then we can apply another version of Parseval's theorem (1):

$$\frac{1}{NT} \sum_{n=n_0}^{n_0+N-1} |x(nT)|^2 = \sum_{k=k_0}^{k_0+N-1} |c_k|^2. \tag{45}$$

In other words, the average power of the signal $x(kT)$ is distributed in the frequency domain at discrete frequencies nf_0 with average power of each component at $|c_k|^2$.

The frequency spectrum of the periodic signal shown in Fig. 14 is shown in Fig. 15, where

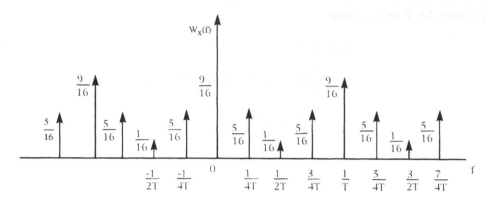

FIG. 15   Frequency spectrum of periodic discrete-time signal where $f_0 = 1/4T$ = fundamental frequency of $[2x(kT)]$, and $f_s = 1/T$ = sampling frequency.

$$|c_0|^2 = 9/16, \qquad |c_1|^2 = 5/16 = |c_3|^2, \text{ and} \qquad |c_2|^2 = 1/16.$$

The frequency content of a discrete-time periodic signal is discrete at harmonics of the fundamental frequency f_0 and also repeats at harmonics of the sampling frequency fs, where, in general, $fs = Nf_0$ when the signal is periodic with period NT.

We next consider the case of nonperiodic discrete-time deterministic signals with finite total energy. The discrete Fourier transform theorem is applicable to such cases where

$$\sum_{k=-\infty}^{\infty} |x(kT)|^2 < \infty.$$

We have the Fourier transform pair $x(kT) \leftrightarrow X(f)$ where

$$X(f) = \sum_{k=0}^{\infty} x(kT)e^{-j2\pi kfT}$$

and

$$x(kT) = \frac{1}{T} \oint_c X(f)e^{j2\pi kfT}df.$$

For example, if the signal $x(kT)$ is the pulse shown in Fig. 16, the Fourier transform is

$$X(f) = \sum_{k=0}^{2} x(kT)e^{-j2\pi kfT}$$

that is,

$$X(f) = 1 + e^{-j2\pi fT} + e^{-j4\pi fT}.$$

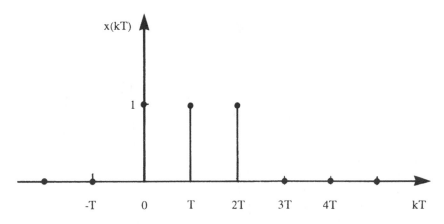

FIG. 16   Discrete-time pulse.

In this case, where $x(kT)$ is non-zero only for values of time $\leq 2T$, we observe $X(f)$ is periodic with period $f_0 = 1/(2T)$. In general, if the record length is NT, the Fourier transform is periodic in frequency with period $f_0 = 1/NT$. We therefore can sample the spectrum $X(f)$ N times in the interval $[0, 1/T]$ and obtain a discrete version of $X(f)$, which we denote by $X(m\Omega)$, where $\Omega = 1/NT$. Then both the Fourier transform and its inverse become summations and we have the discrete Fourier transform pair

$$X(m\Omega) = \frac{1}{N} \sum_{n=0}^{N-1} x(nT)e^{-j2\pi kn/N}, \qquad m = 0,1, \ldots, N-1 \qquad (46)$$

and

$$x(nT) = \sum_{m=0}^{N-1} X(m\Omega)e^{j2\pi km/N}, \qquad n = 0,1, \ldots, N-1. \qquad (47)$$

The frequency spectrum of the discrete-time signal $x(kT)$ can be derived from the final version of Parseval's theorem:

$$\sum_{n=-\infty}^{\infty} |x(nT)|^2 = \int_c |X(f)|^2 \, df. \qquad (48)$$

Hence $|X(f)|^2$ represents the energy spectral density function of the discrete-time signal as it does for the continuous-time signal. For the pulse of Fig. 16, the spectrum is thus

$$W_x(f) = |X(f)|^2 = 3 + 4 \cos 2\pi fT + 2 \cos 4\pi fT \qquad (49)$$

(see Fig. 17).

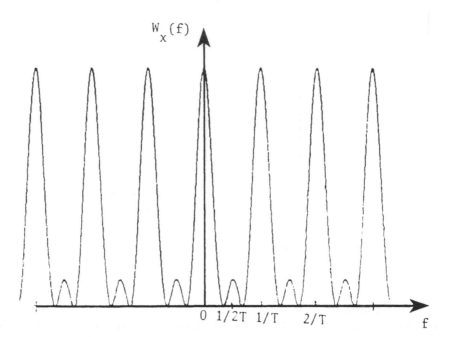

FIG. 17   Spectrum of discrete-time pulse.

Random Signals

The signals we discussed to this point are deterministic in the sense that their future values are completely predictable. For example, the values of the signal

$$x(t) = 10e^{-2t}u(t)$$

can be calculated for each and all values of time.

In communications systems, we also deal with signals that are random in some aspect. The values of such signals at any instant in time constitute a random variable. The analytic treatment of random signals starts with the theory of probability and random variables (RVs).

The probability of a random event is a real number between 0 and 1, obeying a number of axioms. For example, in tossing a fair coin, the probability of either a head or a tail is 1/2. This can be supported partially by the empirical, relative frequency definition of the probability of occurrence of an event that is one possible outcome of an experiment that can be repeated a large number of times. If n is the number of times an event E occurs out of N trials of the experiment, then

$$\text{probability of } E = \lim_{N \to \infty} \left(\frac{n}{N}\right)$$

An RV is a function of the possible outcomes of a probabilistic experiment or phenomenon. That is, a value is assigned for all possible outcomes. For example, consider the experiment of tossing 2 fair coins simultaneously and define a random variable X in accordance with the table below. Because of our knowledge of the underlying experiment, we can derive the probabilities of the random variable X assuming all possible values, that is,

$$\text{prob } (x = k) = p_x(k) = \begin{cases} 1/4, & \text{for } k = 4 \\ 1/2, & k = 2 \\ 1/4, & k = 1 \\ 0, & \text{all other values.} \end{cases}$$

Possible Outcomes	Value of Random Variable X
HH	4
HT or TH	2
TT	1

For this discrete-value random variable, the quantity $p_x(k)$ represents its probability density function (pdf). For a continuous value random variable, a continuous-value function $f_X(x)$ is specified as the pdf such that

$$\text{prob } (x_1 \leq X \leq x_2) = \int_{x_1}^{x_2} f_X(x) \, dx$$

and

$$\int_{-\infty}^{\infty} f_X(x) \, dx = 1.$$

Some common probability density functions are the exponential distribution

$$f_X(x) = \lambda e^{-\lambda x} u(x)$$

and the normal distribution

$$f_X(x) = \frac{1}{\sigma\sqrt{2\pi}} \exp \left[(x - m)^2/2\delta^2\right].$$

All statistical information about an RV can be derived once its pdf is specified. Such statistics include the expected or mean value $\overline{X}$, the mean square value $\overline{X^2}$, and the variance $(X - \overline{X})^2$ denoted by σ_x^2.

For the discrete RV defined in the informal table above, these statistics are

$$m = E\{X\} = \overline{X} = \sum_{k=1,2,4} k p_x(k) = 2.25$$

$$E\{X^2\} = \sum_{k} k^2 p_x(k) = 6.25$$

and

$$\sigma_x^2 = E\{(X - m)^2\} = \sum_k (k - m)^2 p_x(k) = 1.1875.$$

Similarly, for an exponentially distributed continuous random variable with pdf

$$f_Y(y) = \lambda e^{-\lambda y} u(y),$$

$$E\{Y\} = \overline{Y} = \int_{-\infty}^{\infty} y f_y(y) dy = \lambda \int_0^{\infty} y e^{-\lambda y} dy = \frac{1}{\lambda},$$

$$E\{Y^2\} = \lambda \int_0^{\infty} y^2 e^{-\lambda y} dy = \frac{2}{\lambda^2},$$

and

$$\sigma_y^2 = E\{(Y - m_Y)^2\} = 1/\lambda^2.$$

The random signals, also called *random processes*, are functions of time and some random phenomenon such that the value at any instant of time is an RV. An example of a random signal is the function

$$x(t) = A \cos (w_0 t + \theta), \tag{50}$$

where θ is an RV uniformly distributed in the interval $(0, 2\pi)$. This distribution and a few sample functions of the random process $x(t)$ are shown in Fig. 18. For example, the random variable $Y = x(0)$ can be shown to have a pdf

$$f_Y(y) = \begin{cases} \dfrac{1}{\pi \sqrt{A^2 - y^2}}, & |y| < A \\ 0, & \text{otherwise.} \end{cases} \tag{51}$$

The statistic of a random process that relates to its frequency content is the autocorrelation function

$$R_x(t_1, t_2) = E\{x(t_1)x(t_2)\}. \tag{52}$$

If the random process is *wide-sense stationary* (WSS) (3), then the autocorrelation is an even function of $\tau = t_1 - t_2$ only, that is,

$$R_x(\tau) = E\{x(t)x(t - \tau)\}. \tag{53}$$

If $x(t)$ is not WSS, then we take the time average

$$R_x(\tau) = <R_x(t, \tau)> = <E\{x(t)x(t - \tau)\}>. \tag{54}$$

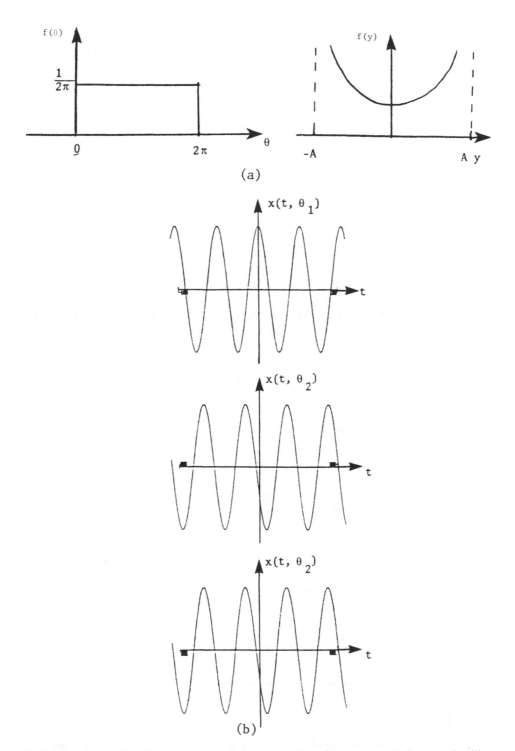

FIG. 18 Probability density and sample functions of random process $x(t,\theta)$: a, probability density functions of phase angle and $x(\theta)$; b, sample functions of random process.

In either case, the Wiener–Khinchine theorem states that the PSD function of the random process $x(t)$ is the Fourier transform of the autocorrelation function $R_x(\tau)$. For example, for the random-phase process $x(t, \theta)$, it can be shown that

$$R_x(\tau) = \frac{A^2}{2} \cos (w_0\tau) \tag{55}$$

and, hence,

$$W_x(f) = \frac{A^2}{4} (\delta(f - f_0) + \delta(f + f_0)) \tag{56}$$

where

$$f_0 = \frac{w_0}{2\pi}.$$

For system applications, a random process usually is specified by its autocorrelation function or its PSD function. A common example is the Gaussian white-noise process for which

$$R_n(\tau) = A\delta(\tau)$$

and

$$W_n(f) = A, \quad -\infty < f < \infty.$$

A more realistic approximation is the band-limited case (Fig. 19):

$$W_n(f) = \begin{cases} N_0, & |f| < B. \\ 0, & \text{otherwise} \end{cases} \tag{57}$$

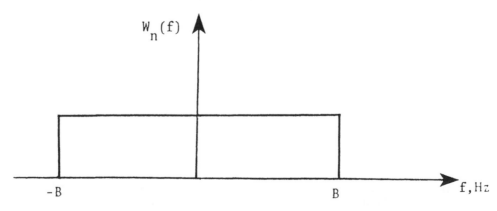

FIG. 19   Spectral density function of band-limited white noise.

Speech, Video, and Data Signals

The electrical signals that represent speech and television pictures that commonly are transmitted in analog form through communications systems are too complex to be expressed as analytical functions using polynomials, or exponential or sinusoidal functions in the time domain. Data signals or digital signals derived from analog waveforms by analog-to-digital conversion are transmitted by assigning distinct electrical waveforms to the different symbols of an alphabet, typically the binary digits 0 and 1. Digital signals therefore are most amenable to analysis using the frequency-content-producing methods described in the previous sections.

The frequency content of the speech signal largely has been derived experimentally and averaged statistically. A picture of the electrical waveform generated by a microphone from the acoustic signal produced by a male subject speaking the phrase, "a speech waveform," is shown in Fig. 20 (4). Using a time scale of one millisecond (ms) per inch, we can estimate frequencies of about 5 kilohertz (kHz) in those portions of the waveform corresponding to fricatives ("f," "s") and about 1–2 kHz for the portions corresponding to vowels ("a," "e," "o," etc.). On average, an amplitude-normalized sketch of the frequency content of human speech is shown in Fig. 21. When song and music are included, the spectrum can extend to 15–20 kHz, which is about the audible limit for humans. The standard bandwidth for a speech communications channel has been set for from 100 Hz to 3.5 kHz or 4 kHz.

The electrical video signal for television is derived by scanning the scene being televised and converting the variations in light intensities and color hues into electrical waveforms, and then adding synchronizing pulses (horizontal and vertical) to indicate the end of a scan trace and the end of a frame of scans. In U.S. commercial broadcast television, there are 525 lines/frame and 60 frames/second. Combining the scanning requirements with picture resolution requirements, the bandwidth of the TV signal is found to be 8 megahertz (MHz). However, for spectrum conservation reasons, the bandwidth allocated for

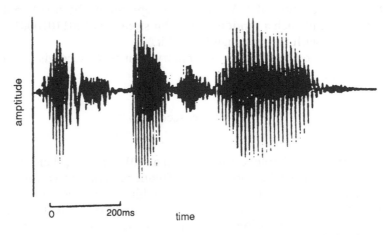

FIG. 20   Waveform generated by human speech. (Reprinted with permission from Ref. 4.)

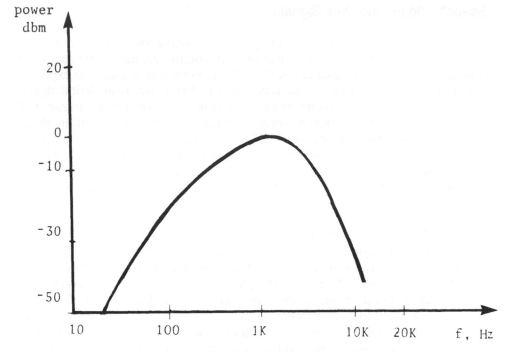

FIG. 21   Normalized average frequency spectrum of human speech.

broadcast TV is 4.1 MHz and bandwidth reduction techniques (e.g., interlaced scanning) are applied to the basic signal before it is broadcast. A typical segment of a television waveform and the frequency composition of a broadcast TV signal are shown in Fig. 22.

The data signal is most amenable to analytical study. For example, consider the most common digital baseband signaling scheme, the unipolar non-return-to-zero (NRZ) scheme. Here, to transmit the binary alphabets 0 and 1, we represent a 1 with a pulse of 5 V and a 0 is represented by a 0-V pulse. The waveform to transmit such a random binary bit stream as 10011010001 is shown in Fig. 23, starting with the least significant bit.

If the time allowed to transmit one binary digit is T_b seconds, then the bit transmission rate is

$$R = \frac{1}{T_b} \text{ bits per second (b/s).}$$

If the bit stream were a simple test signal such as 101010 . . . , then the resulting waveform would be periodic and hence its frequency content could be derived from a Fourier series expansion. For a random bit stream generated by a pulse code modulation (PCM) code of an analog waveform or an ASCII (American Standard Code for Information Interchange) code of alphanumeric text, we assume the waveform is a random process

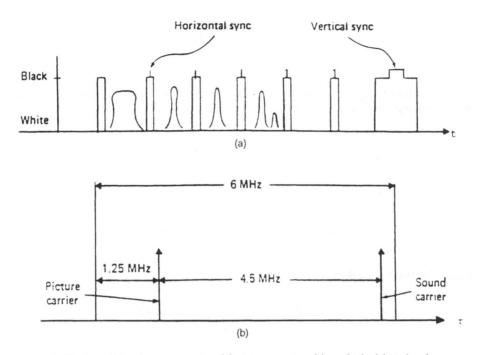

FIG. 22   Waveform segment and frequency composition of television signal.

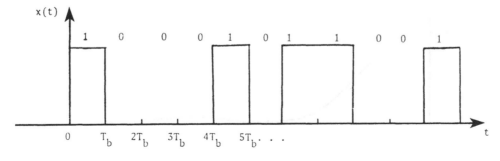

FIG. 23   Waveform for unipolar non-return-to-zero transmission of a random bit stream.

$$x(t) = \sum_{k=-\infty}^{\infty} a_k p(t) \tag{58}$$

where $a_k = 0$ or 1 with equal probabilities of $1/2$, and $p(t)$ is the rectangular pulse of duration T_b and amplitude A. It can be shown that the autocorrelation of this process, averaged over time, is (5)

$$R_x(\tau) = \frac{1}{T_b}\left(\frac{A^2}{2} + \frac{A^2}{4} \text{ tri }\left(\frac{\tau}{T_b}\right)\right) \tag{59}$$

where tri is the triangular function defined as

$$\text{tri }(\tau/T_b) = \begin{cases} 1 - \tau/T_b |\tau| \leq T_b \\ 0 \qquad\quad |\tau| > T_b \end{cases}.$$

Hence, its PSD function is

$$W_x(f) = \frac{A^2}{2}\delta(f) + \frac{A^2}{4}T_b \frac{\sin^2(\pi f T_b)}{(\pi f T_b)^2}. \tag{60}$$

This PSD is shown in Fig. 24.

If the bandwidth to the first null is taken as the approximate bandwidth of the signal, then it equals $R = 1/T_b$ and the bandwidth efficiency

$$= \frac{\text{bit rate}}{\text{bandwidth}} = \frac{R}{B} = 1$$

for this signaling scheme.

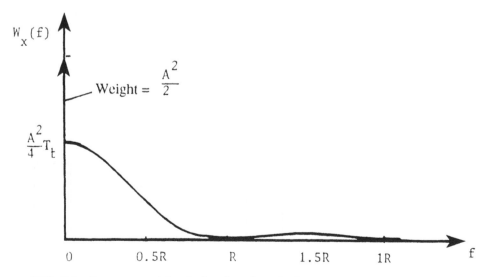

FIG. 24   Power spectral density function of a unipolar non-return-to-zero waveform.

There are several other signaling formats, both baseband and bandpass (BP) depending upon the waveforms $s_0(t)$ and $s_1(t)$ selected to represent binary digits 0 and 1 in a per-bit interval T_b. Among the baseband formats are the polar scheme, in which $s_0(t) = -A$ and $s_1(t) = A$, and the bipolar scheme in which $s_0(t) = 0$ and $s_1(t) = \pm A$ for alternating binary 1s.

We also have return-to-zero schemes in which the pulse returns to zero during the per-bit interval. Typical waveforms for several of these schemes are shown in Fig. 25 and the corresponding PSD functions are shown in Fig. 26.

We also can define several BP signaling schemes:

Amplitude shift keying (ASK): $s_0(t) = 0$, $s_1(t) = A \cos 2\pi f_0 t$

Frequency shift keying (FSK): $s_0(t) = A \cos 2\pi f_0 t$, $s_1(t) = A \cos 2\pi f_1 t$

Phase shift keying (PSK): $s_0(t) = A \cos(2\pi f_0 t + \theta_0)$, $s_1(t) = A \cos(2\pi f_0 t + \theta_1)$. For example, if $\theta_0 - \theta_1 = 180$ then $s_0(t) = -s_1(t)$.

The relative PSD functions of these schemes are shown in Fig. 27.

Frequency Response of Linear Systems

Every electrical circuit that includes an energy storage device such as an inductor (L) or capacitor (C) or any structure that simulates an inductor or capacitor, responds differently to signals at different frequencies or responds differently to the different frequency components of any signal. Hence, all can be constructed to perform as filters. Such devices or structures include wire pairs

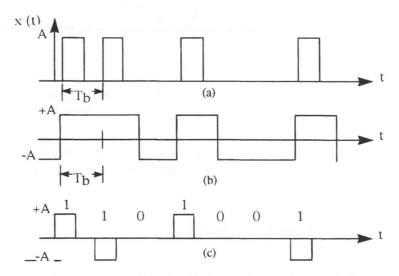

FIG. 25   Waveforms for several digital baseband transmission codes: *a*, unipolar non-return-to-zero; *b*, polar non-return-to-zero; *c*, bipolar return-to-zero (RZ).

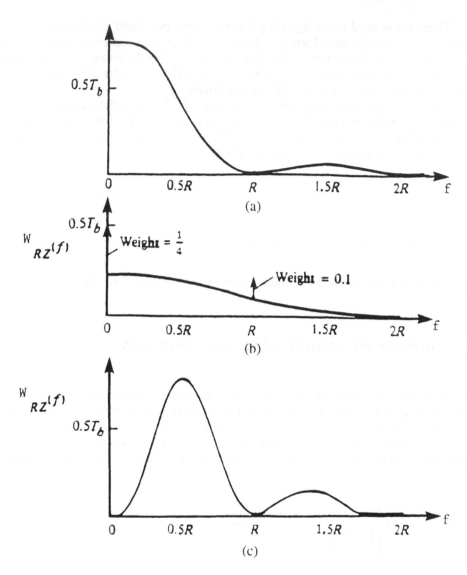

FIG. 26   Power spectral density functions for several digital baseband transmission waveforms: *a*, polar non-return-to-zero; *b*, unipolar return-to-zero; *c*, bipolar return-to-zero.

(transmission lines), coaxial cables, waveguides, crystals, antennas, the atmosphere, and other media, as well as the common resistor-inductor-capacitor (RLC) circuits.

Linear Resistor-Inductor-Capacitor Circuits in Time and Frequency Domains

Consider the simple resistor-capacitor (RC) circuit (Fig. 28). In the time domain, a differential equation relating the output to the input can be obtained. With

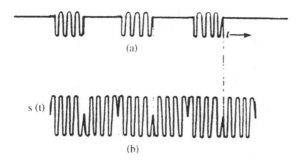

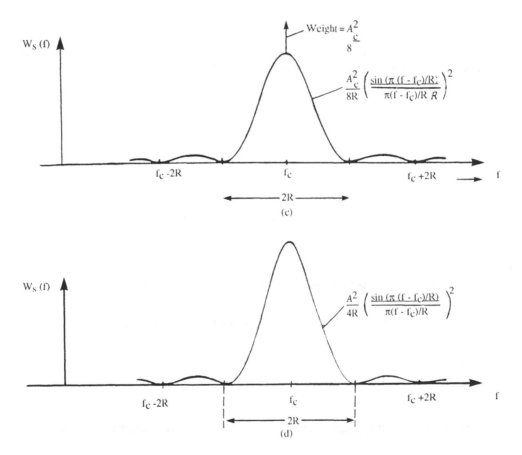

FIG. 27   Power spectral density functions of several band-pass (BP) digital transmission wave-forms: *a*, on–off keying signal; *b*, binary phase-shift keying signal; *c*, power spectral density function of on–off keying; *d*, power spectral density function of binary phase-shift keying.

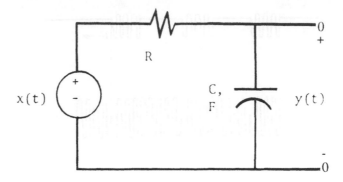

FIG. 28   Resistor-capacitor circuit.

the input voltage source $x(t)$ and the output $y(t)$ being the voltage across the capacitor, equate two expressions for the capacitor current:

$$\frac{x(t) - y(t)}{R} = C\frac{dy}{dt}$$

or

$$RC\frac{dy}{dt} + y(t) = x(t). \qquad (61)$$

If the input $x(t)$ is a sinusoidal wave at frequency f_0 Hz, that is,

$$x(t) = A \cos 2\pi f_0 t,$$

then the steady-state portion of the response (t) may be shown to be

$$y(t) = \frac{A}{\sqrt{1 + w_0^2\tau^2}} \cos (2\pi f_0 t - \tan^{-1}(w_0\tau)) \qquad (62)$$

where
$$w_0 = 2\pi f_0$$
$$\tau = RC$$

The ratio of the amplitude of the response $y(t)$ to the amplitude of the input $x(t)$ is

$$H(w_0,\tau) = \frac{1}{\sqrt{1 + w_0^2\tau^2}}. \qquad (63)$$

We observe that if the frequency of the input is low, ($w_0\tau \ll 1$), then the amplitude of the output sinusoidal waveform almost is equal to that of the input. But, if $w_0\tau \gg 1$, then the output amplitude is very small compared to

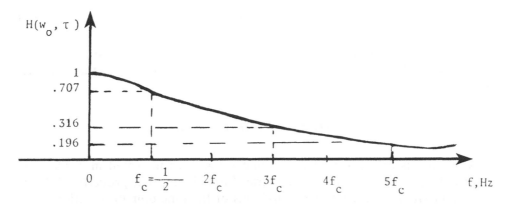

FIG. 29   Amplitude response of resistor-capacitor circuit as a function of frequency.

that of the input. The function $H(W_0, \tau)$ is portrayed in Fig. 29 and shows the relative favoring of low-frequency signals over higher frequency signals.

Another method of analyzing linear systems is by the impulse response. The impulse function $\delta(t - \tau)$ is defined as infinitely narrow, yet with a finite area:

$$\delta(t - \tau) = 0 \quad \text{if } t \neq \tau$$

and

$$\int_{-\infty}^{\infty} k\delta(t - \tau)dt = k.$$

Further, the impulse has a sifting property such that

$$\int_{-\infty}^{\infty} f(t)\delta(t - \tau)\ dt = f(\tau). \tag{64}$$

Thus, any function $f(t)$ can be represented as a weighted sum of impulses

$$f(t) = \int_{-\infty}^{\infty} f(\tau)\delta(t - \tau)\ d\tau. \tag{65}$$

Hence, if the response of a linear system to an impulse input $\delta(t - \tau)$ is $h(t,\tau)$, then the response to any other input $x(t)$ expressed as a weighted sum of impulses, yields a similarly weighted sum of the responses to a single impulse. Thus, in general, for a linear system (see Fig. 30).

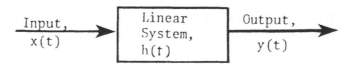

FIG. 30   Linear system characterized by its impulse response.

$$y(t) = \int_{-\infty}^{\infty} x(\tau)h(t,\tau)d\tau. \tag{66}$$

If the system is linear time invariant (LTI) and causal such that an input $\delta(t - \tau)$ yields $h(t - \tau)$ that is zero for $t < \tau$, then, for inputs $x(t)$ that are zero for $t < 0$,

$$y(t) = \int_{0}^{t} x(\tau)h(t - \tau)d\tau \tag{67}$$

This relation is called a *convolution integral* and, as indicated in Table 1, is among the properties of the Fourier transform. Using that property, if $Y(f)$, $X(f)$, and $H(f)$ are the Fourier transforms of the time-domain functions $y(t)$, $x(t)$, and $h(t)$, respectively, then $Y(f) = H(f)X(f)$.

The quantity $H(f)$, which is the ratio of the transform of the output of the LTI system to the transform of the input, is called the *transfer function*, that is,

$$H(f) = \frac{Y(f)}{X(f)}. \tag{68}$$

In general, $H(f)$ is a complex function, hence it can be expressed in polar form as

$$H(f) = A(f)e^{j\phi(f)}. \tag{69}$$

Hence, since $Y(f) = H(f)X(f)$, the magnitude of $Y(f)$ is given by

$$A_Y(f) = |Y(f)| = |H(f)||X(f)| = A_H(f) \cdot A_X(f). \tag{70}$$

Thus, the ratio of the amplitude of the response to the amplitude of the input is the magnitude of the transfer function.

Further, for those functions with well-defined Fourier transforms, we have shown that the energy spectral density function is given by

$$W_x(f) = |X(f)|^2. \tag{71}$$

Hence, for a linear system, the energy spectral density function of the output is given by

$$W_y(f) = |H(f)|^2 W_x(f). \tag{72}$$

Thus, by specifying the function $A(f) = |H(f)|$, we can modify the frequency content of the input of a linear system to any shape desired at the output, within certain physical limitations.

For example, for the RC circuit considered above, the amplitude transfer function is

$$A(f) = \frac{1}{\sqrt{1 + w^2\tau^2}} \tag{73}$$

where $w = 2\pi f$. This is identical to $H(w_0, \tau)$ plotted earlier (Fig. 29). Hence, the amplitude response to sinusoidal waveforms of various frequencies observed above is applicable to any other input waveform with a known frequency content.

Thus, let the input be a periodic rectangular pulse as discussed in the section concerning periodic functions (see Fig. 31-a). The frequency contents of the input and output for the special case $T = RC$ also are shown in Figs. 31-b to 31-d.

If the input were the exponential pulse

$$x(t) = e^{-t/T}u(t)$$

with $\tau = RC$, the corresponding frequency contents of the input and output are shown in Fig. 32. The frequency content of random signals can be modified in a similar manner. Suppose the input to the RC filter consists of a deterministic signal

$$x(t) = A \cos 2\pi f_0 t$$

and a band-limited random noise with the PSD function

$$W_n(f) = N_0, \ |f| < B.$$

Then the average signal power at the input is

$$P_{s_i} = \frac{1}{2} A^2 \ \text{watts}$$

and the noise power is

$$P_{n_i} = \int_{-B}^{B} N_0 df = 2N_0 B \ \text{watts}.$$

Then the signal-to-noise ratio at the input is

$$\left(\frac{S}{N}\right)_i = \frac{P_{s_i}}{P_{n_i}} = \frac{A^2}{4N_0 B}.$$

At the output of the RC filter with $\tau = RC = 1$ and $f_0 = 1/(2\pi\tau)$, the signal power is now

$$P_{s_0} = \frac{1}{4} A^2$$

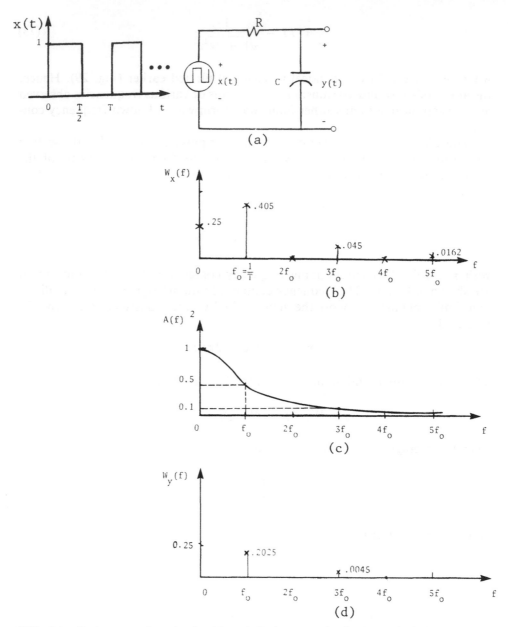

FIG. 31   Resistor-capacitor circuit with periodic input: *a*, circuit with periodic input; *b*, frequency content of input signal; *c*, amplitude characteristic of circuit transfer function; *d*, frequency content of output signal.

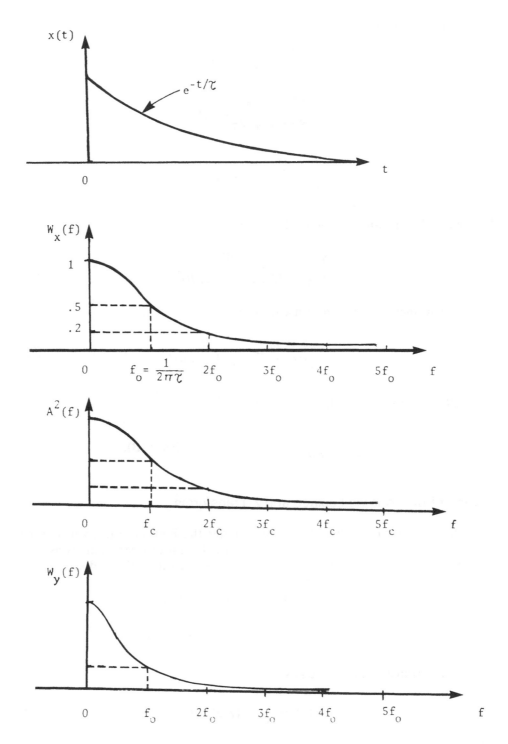

FIG. 32   Input and output frequency spectra for resistor-capacitor circuit with exponential pulse input.

and the noise power is

$$P_{n_0} = \int_{-B}^{B} |H(f)|^2 W_n(f) \, df$$

$$= \int_{-B}^{B} \frac{N_0}{1 + 4\pi^2 f^2} \, df$$

or

$$P_{n_0} = \frac{N_0}{\pi} \tan^{-1}(2\pi B).$$

Hence the signal-to-noise ratio at the output is

$$\left(\frac{S}{N}\right)_0 = \frac{\pi A^2}{4 N_0 \tan^{-1}(2\pi B)}. \tag{74}$$

The improvement in signal-to-noise ratios is

$$\eta = \frac{(S/N)_0}{(S/N)_i} = \frac{\pi B}{\tan^{-1}(2\pi B)}. \tag{75}$$

Typically, the noise bandwidth $B = 10 f_0 = \dfrac{10}{2\pi}$, then

$$\eta = \frac{5}{\tan^{-1}(10)} = 3.4 \text{ or } 5.3 \text{ decibels (dB)}.$$

Transfer Functions via Laplace and Z-Transforms

The idea of a transfer function defined under the Fourier transform can be extended to the other transforms commonly used in linear systems analysis. For continuous-time systems, we use the Laplace transform and for discrete-time systems, we use the Z-transform.

For signals $x(t)$ that are zero for $t < 0$ and for which

$$\int_0^\infty |x(t) e^{-\sigma t}| \, dt < \infty \quad \text{for some real } \sigma, \tag{76}$$

the Laplace transform is defined as

$$X(s) = \int_0^\infty x(t) e^{-st} dt \tag{77}$$

where $s = \sigma + jw$ is a complex frequency variable. For any function possessing both Fourier and Laplace transforms, the Fourier transform $X(f)$ can be obtained from the Laplace transform by setting $s = j2\pi f$.

Thus, for $x(t) = e^{-at}u(t)$, the Laplace transform is

$$X(s) = \frac{1}{a + s} \tag{78}$$

and by setting $s = j2\pi f$, we have the Fourier transform of the same signal $X(f) = 1/(a + j2\pi f)$ as derived in the section, "Nonperiodic Deterministic Signals."

The Laplace transform is advantageous in system analysis in that more time-domain functions are transformable and, further, both transient and steady-state responses can be derived. It therefore is useful in stability analysis of feedback control systems, but does not provide additional information on the frequency content of communications signals than that provided by the Fourier transform.

The Laplace transform also satisfies the convolution property: if $y(t) = x(t)*h(t)$, then $Y(s) = H(s)X(s)$ and the transfer function is

$$H(s) = \frac{Y(s)}{X(s)}.$$

For the RC filter,

$$H(s) = \frac{1}{1 + s\tau}, \qquad \text{where } \tau = RC.$$

This yields the transfer function $H(f)$ by setting $s = j2\pi f$.

For discrete-time signals and systems, we use the Z-transform, which is derived from the Laplace transform by setting $Z = e^{sT}$. For a sequence or discrete-time signal $x(kT)$ defined for $k = 0, 1, 2, \ldots$, the Z-transform is

$$X(Z) = \sum_{k=0}^{\infty} x(kT)Z^{-k} \tag{79}$$

For example, for the discrete-time exponential signal $x(kT) = a^k$, $k = 0, 1, \ldots$, and $|a| < 1$,

$$X(Z) = \sum_{k=0}^{\infty} a^k z^{-k} = \sum_{k=0}^{\infty} (aZ^{-1})^k$$

that is,

$$X(Z) = \frac{1}{1 - aZ^{-1}} \text{ provided } |aZ^{-1}| < 1.$$

Similarly, the Z-transform of the unit impulse

$$\delta(kT) = \begin{cases} 1, & k = 0 \\ 0, & \text{otherwise} \end{cases}$$

is $X(Z) = 1$.

A discrete-time system is shown in Fig. 33. If the system is linear and time invariant, it can be characterized by its impulse response $h(kT)$. Then the input and output are related by a convolution sum

$$y(kT) = x(kT)*h(kT) = \sum_{n=0}^{\infty} x(nT)h((k - n)T). \tag{80}$$

Once again, the Z-transform possesses a convolution property such that

$$Y(Z) = H(Z)X(Z) \tag{81}$$

so that a transfer function can be defined as

$$H(Z) = \frac{Y(Z)}{X(Z)}. \tag{82}$$

The transfer function $H(Z)$ can be used to modify the frequency content of the input signal $x(kT)$ by using the amplitude response function.

$$A(f) = |H(Z)|z = e^{+j2\pi fT}. \tag{83}$$

For example, suppose the input/output relation for a discrete-time system is

$$ay((k - 1)T) + y(kT) = x(kT). \tag{84}$$

The transfer function is

$$H(Z) = \frac{1}{1 + aZ^{-1}} \tag{85}$$

and the amplitude characteristic is (see Fig. 34)

$$A(f) = \frac{1}{\sqrt{1 + a^2 + 2a \cos 2\pi fT}}. \tag{86}$$

The amplitude characteristic is periodic with the period equal to the sampling frequency $f_0 = 1/T$. But, the frequency spectrum of the discrete-time input

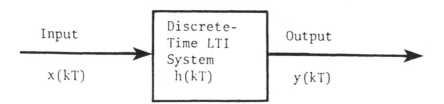

FIG. 33   Discrete-time linear system.

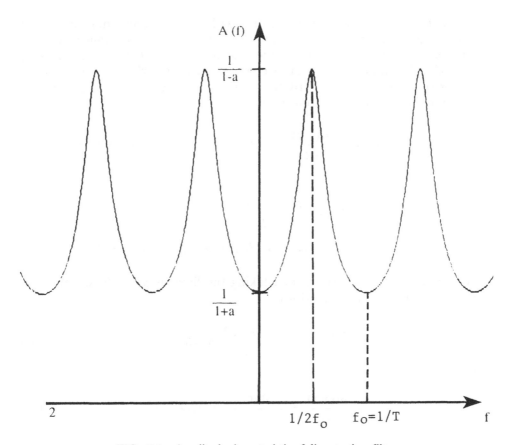

FIG. 34   Amplitude characteristic of discrete-time filter.

$x(kT)$ also is periodic with period f_0. Hence, the filter can be used to modify the spectrum uniquely in any interval of half the sampling frequency. Thus, it is sufficient to specify the desired characteristics only in the interval $(0, 1/2f_0)$. The characteristic sketched in Fig. 34 represents a high pass filter.

Transmission Lines

In the introduction to the section, "Frequency Response of Linear Systems," we stated that all physical structures exert some selective frequency response and can be used for filtering. These structures include such transmission media as wire pairs, thin-film lines on circuitboards, cables, waveguides, crystals, and antennas. For these structures, the lumped circuit theory approximation is not valid, and hence electromagnetic field analysis needs to be applied. For transmission lines, an intermediate approximation using distributed circuit elements may be utilized.

A differential line segment with input and output voltages and currents is shown in Fig. 35. If the line is uniform with certain series inductance and resistance per unit length, and shunt capacitance and conductance per unit length, the total circuit elements for the segment are Ldx, Cdx, Rdx, and Gdx, respectively. The input waveforms are $v(x,t)$ and $i(x,t)$, while the outputs are $v(x + dx,t)$ and $i(x + dx,t)$. Writing Kirchoff's voltage law (KVL) and Kirchoff's current law (KCL) equations, we have

$$-v(x,t) + (L\ dx)\frac{di(x,t)}{dt} + (R\ dx)i(x,t) + v(x + dx,t) = 0$$

and

$$-i(x,t) + (C\ dx)\frac{dv(x + dx,t)}{dt}$$
$$+ (G\ dx)(v(x + dx,t)) + i(x + dx,t) = 0.$$

Rearranging these equations and taking the limit as $dx \rightarrow 0$, we obtain a pair of partial differential equations

$$-\frac{\partial v(x,t)}{\partial x} = \left(L\frac{\partial i(x,t)}{\partial t} + Ri(x,t)\right) \tag{87}$$

and

$$-\frac{\partial i(x,t)}{\partial x} = C\frac{\partial v(s,t)}{\partial t} + Gv(x,t). \tag{88}$$

Considering only the steady-state sinusoidal responses, we can assume the currents and voltages of the form

$$V(x,t) = V(x)e^{jwt} \tag{89}$$

and

$$i(x,t) = I(x)e^{jwt}. \tag{90}$$

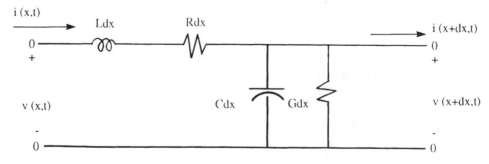

FIG. 35   Distributed circuit model of transmission line segment.

We then obtain ordinary differential equations for the space variations

$$-\frac{dV(x)}{dx} = jwLI(x) + RI(x) \tag{91}$$

$$-\frac{dI(x)}{dx} = jwCV(x) + GV(x). \tag{92}$$

By defining the impedance $Z = R + jwL$ and admittance $Y = G + jwC$, these equations are

$$\frac{dV(x)}{dx} = -ZI \tag{93}$$

$$\frac{dI(x)}{dx} = -YV. \tag{94}$$

Eliminating one variable by a second differentiation, we have

$$\frac{d^2V}{dx^2} = ZYV(x) \tag{95}$$

Let $\gamma^2 = ZY = (R + jwL)(G + jwC)$, then the general solution is

$$V(x) = Ae^{\gamma x} + Be^{-\gamma x} \tag{96}$$

where A and B are constants of integration to be determined by boundary conditions.

Combining the spatial solution with the time harmonic function, we have general expressions for the instantaneous voltage and current at a location of the line

$$v(x,t) = Re\ \{Ae^{(\gamma x + jwt)} + Be^{-(\gamma x + jwt)}\} \tag{97}$$

$$i(x,t) = Re\ \{Ce^{(\gamma x + jwt)} + De^{-(\gamma x + jwt)}\}. \tag{98}$$

The quantity $\gamma(w)$ is called the *propagation constant*, with real and imaginary parts:

$$\gamma(w) = \alpha(w) + j\beta(w) \tag{99}$$

where α is the attenuation constant and β is the phase constant.

With these quantities, the currents and voltages may be written as

$$v(x,t) = Re\ \{Ae^{\alpha x}e^{j(\beta x + wt)} + Be^{-\alpha x}e^{j(-\beta x + wt)}\} \tag{100}$$

$$i(x,t) = Re\ \{Ce^{\alpha x}e^{j(\beta x + wt)} + Be^{-\alpha x}e^{j(-\beta x + wt)}\}. \tag{101}$$

Several interpretations of these expressions are relevant. First, the phase portions $e^{-j(\beta x + wt)}$ represent waves traveling in the positive and negative x directions with velocity $c = w/\beta$. Second, the factors $e^{\pm\alpha x}$ represent the attenuations per unit length of the waves traveling in the positive and negative x directions. Since both α and β are functions of frequency as well as the line parameters R, G, L, and C, the attenuation is frequency dependent and signals at certain frequencies are transmitted selectively. The attenuation constant as a function of frequency for a typical communication-gauge wire pair is shown in Fig. 36 (6).

With different methods of analysis, propagation constants can be derived for such various transmission media as waveguides and the atmosphere. Filters thereby can be implemented for such various ranges of frequencies as microwaves and ultrasonics using such structures as thin films, waveguide cavities, and crystals.

System-Level Classification of Filters

For purposes of system specification, the characteristics of filters have been divided into four major types: LP, HP, BP, and band-stop filters. The ideal amplitude characteristics of these filter types are shown in Fig. 37. As indicated by the partial sum construction of a rectangular pulse from its Fourier components, it will require a very high order transfer function $H(s)$ to obtain discontinuous transitions from passbands to stopbands. In fact, the relation between the order of the transfer function and the transition rate can be demonstrated by considering the first-order function

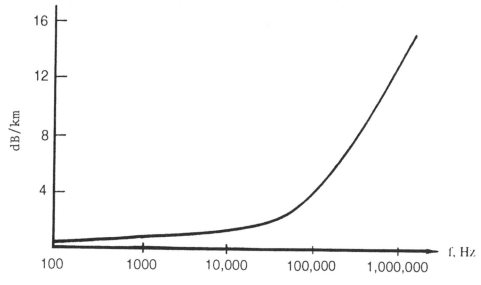

FIG. 36   Attenuation constant of communications wire pair.

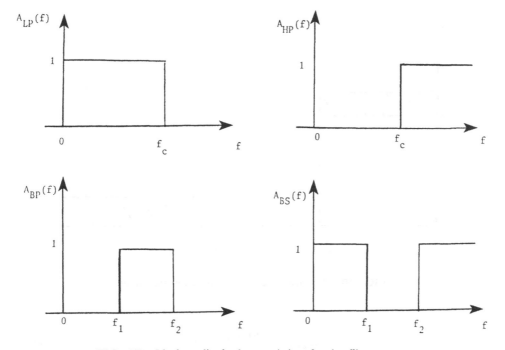

FIG. 37   Ideal amplitude characteristics of major filter types.

$$H(s) = \frac{1}{1 + s}. \tag{102}$$

The corresponding amplitude characteristic is

$$A(f) = \frac{1}{\sqrt{1 + 4\pi^2 f^2}} \tag{103}$$

and as $f \to \infty$, $A(f) \to 1/f$. The asymptotic rate in dB is $\alpha = -20\log_{10}f$ such that as f increases by a factor of 10 (i.e., from any f_0 to $10f_0$), α decreases by 20 dB. Thus, for a first-order transfer function, the transition rate is 20 dB per decade, whether going from a passband to a stopband or vice versa. In general, for an nth order function, the rate is $20n$ dB per decade. We therefore specify filters with this transition rate in mind, yielding a result such as that shown in Fig. 38 for an LP filter. The additional parameters are

$$A_{min} = \text{minimum transmission in passband}$$

$$A_{max} = \text{maximum transmission factor in stopband}$$

$$f_T = \text{lowest stopband frequency.}$$

The required order of $H(s)$ is estimated from

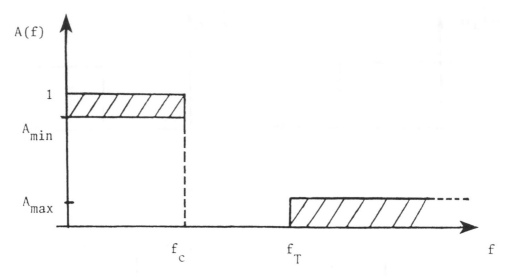

FIG. 38   A realistic specification for a low-pass filter.

$$-20n \geq \frac{20\log_{10}(A_{\max})}{(f_T - f_C)/10} \tag{104}$$

For example, if $A_{\max} = 0.01, f_C = 4$ kHz, and $f_T = 5$ kHz, then $n \geq 20$, which will be a highly complex filter.

Scaling Transformations

Many tables and design aids have been developed for the implementation of electrical filters. In order to minimize the amount of data required, most design aids are provided only for LP filters with a cutoff frequency of 1 radian per second. The generic data for the LP filter can be used to design LP filters with other cutoff frequencies and for all other types of filters by applying the following principles of scaling and frequency transformations.

Suppose a filter has a transfer function $H(s)$ and we make the transformation of the variable

$$s \Rightarrow \frac{s}{K_f}.$$

Then, a critical frequency at $s = 1$ now will occur at $s = K_f$. Thus, all features of the filter at frequencies w_n now will occur at frequencies $K_f w_n$. The result is a frequency-scaled version of the original filter. For example, consider the first-order LP filter with transfer function

$$H(s) = \frac{1}{1 + s}.$$

Its amplitude response is

$$A(w) = \frac{1}{\sqrt{1 + w^2}}. \qquad (105)$$

with a half-power frequency of 1 radian per second.

If we make the transformation $s \rightarrow s/10$, the new transfer function is

$$H_f(s) = \frac{1}{1 + s/10} = \frac{10}{s + 10}$$

with amplitude response

$$A_f(w) = \frac{1}{\sqrt{1 + \left(\dfrac{w}{10}\right)^2}} \qquad (106)$$

The half-power frequency is now 10 radians per second, as expected.

A physical realization of the original transfer function is an RC circuit with a transfer function of

$$H(s) = \frac{1}{1 + sRC}$$

that becomes the required function if we set $R = 1$ ohm and $C = 1F$. The scaled transfer function can be realized by an RC circuit with $R = 1$ ohm and $C = 1/K_f = 1/10 \ F$. For a general RLC implementation, the element values can be scaled to realize the new frequency-scaled transfer function according to the relation $R \rightarrow R$ (unchanged), $C \rightarrow C/K_f$, and $L \rightarrow L/K_f$. We also can scale the impedance level of the circuit, thereby reducing the current level and power dissipated according to the relations $R \rightarrow K_m R$, $C \rightarrow C/K_m$, and $L \rightarrow K_m L$. The magnitude scaling does not change a transfer function that is a ratio of voltages or a ratio of currents.

Frequency Transformations

Frequency and magnitude scaling maintain the type of filter as before (i.e., an LP filter still will be LP, or a BP filter still will be BP). In order to transform a filter from one type to another, we apply one of several frequency transformations. For example, if a transfer function $H(s)$ is LP, then the frequency transformation $s \rightarrow 1/s$ will change it to an HP transfer function. For the first-order case

$$H_{\mathrm{LP}}(s) = \frac{1}{1 + s}.$$

After the transformation $s \rightarrow 1/s$, we have

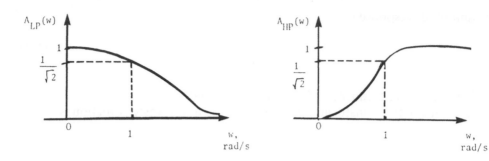

FIG. 39   Amplitude characteristics of first-order low-pass and high-pass filters.

$$H_{\text{HP}}(s) = \frac{1}{1 + (1/s)} = \frac{s}{s + 1}$$

with corresponding amplitude characteristic

$$A_{\text{HP}}(w) = \frac{w}{\sqrt{1 + w^2}}.$$

The amplitude characteristics $A_{\text{LP}}(w)$ and $A_{\text{HP}}(w)$ are shown in Fig. 39. Further, the RLC circuits realizing the equivalent LP and HP filters are shown in Fig. 40. We observe that, in transforming from LP to HP, the capacitor becomes an inductor and vice versa. Table 3 shows other frequency transformations and the equivalent changes of circuit elements. Applying the listed transforms, we observe that the lowest-order BP and BE filters are second order:

$$H_{\text{BP}}(s) = \frac{s}{s^2 + s + 1} \tag{107}$$

$$H_{\text{BE}}(s) = \frac{s^2 + 1}{s^2 + s + 1}. \tag{108}$$

The corresponding amplitude functions are

$$A_{\text{BP}}(w) = \frac{w}{\sqrt{(1 - w^2)^2 + w^2}} \tag{109}$$

and

$$A_{\text{BS}}(w) = \frac{1 - w^2}{\sqrt{(1 - w^2)^2 + w^2}} \tag{110}$$

These are shown in Fig. 41. The normalized half-power frequencies are $w_1 = 0.618$, $w_2 = 1.618$ where $w_1 w_2 = w_0^2 = 1$.

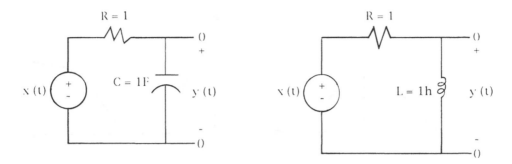

FIG. 40 Resistor-inductor-capacitor circuits realizing first-order low-pass and high-pass filters.

The amplitude characteristic $A(f)$ plotted for various filters indicates how each filter would respond to a steady-state signal. In most cases, the signals turn on and off and it takes some finite time interval to achieve a steady-state response. This time interval is best measured by the rise time of the unit step response of a particular filter. Using the Laplace transform technique of circuit analysis, the step response is given by

$$R(s) = H(s)U(s)$$

where $R(s)$ is the Laplace transform of the step response $r(t)$ and $U(s)$ is the Laplace transform of the unit step function $u(t)$ that equals $1/s$. Hence, $R(s) = (1/s) H(s)$. For example, for the first-order LP filter $H(s) = 1/(s + 1)$,

TABLE 3 Frequency Transformations of Filters

	Low Pass	High Pass	Band Pass	Band Elimination
Low pass		$s \rightarrow 1/s$ $L \rightarrow C$ $C \rightarrow L$	$s \rightarrow s + 1/s$	$s \rightarrow \dfrac{1}{s + 1/s}$
High pass	$s \rightarrow 1/s$ $L \rightarrow C$ $C \rightarrow L$		$s \rightarrow \dfrac{1}{s + 1/s}$	$s \rightarrow s + 1/s$
Band pass	$s + 1/s \rightarrow s$	$s + 1/s \rightarrow 1/s$		$s + 1/s \rightarrow \dfrac{1}{s + 1/s}$
Band Elimination	$s + 1/s \rightarrow 1/s$	$s + 1/s \rightarrow s$	$s + 1/s \rightarrow \dfrac{1}{s + 1/s}$	

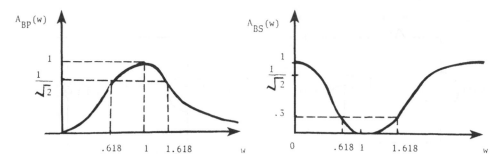

FIG. 41 Amplitude characteristics of second-order band-pass and band-stop filters.

$$R(s) = \frac{1}{s(s+1)} = \frac{1}{s} - \frac{1}{s+1}. \tag{111}$$

By inverse Laplace transformation,

$$r(t) = u(t) - e^{-t}u(t) = (1 - e^{-t})u(t). \tag{112}$$

The unit step responses of the elementary LP, HP, BP, and BE filters are listed in Table 4 and those of the LP, BP, and band-stop (BS) filters are shown in Fig. 42. The settling time for the LP or HP filter is about 2.3 seconds, whereas the settling time of the BP or BS filter is about 11 seconds. When these filters are scaled in frequency, the settling times are reduced proportionally. For example, if the LP filter with transfer function $H(s) = 1/(s+1)$ is scaled in frequency by a factor of 1000, the transfer function becomes

$$H(s) = \frac{1}{(s/1000) + 1} = \frac{1000}{s + 1000},$$

TABLE 4   Unit Step Responses of Low-Pass, High-Pass, Band-Pass, and Band-Elimination Filters

Filter Type	Transfer Function, $H(s)$	Unit Step Response $r(t)$
First-order low pass	$\dfrac{1}{s+1}$	$(1 - e^{-t})u(t)$
First-order high pass	$\dfrac{s}{s+1}$	$e^{-t}u(t)$
Second-order band pass	$\dfrac{s}{s^2 + s + 1}$	$\dfrac{2}{\sqrt{3}}\left[e^{-1/2t}\sin\left(\dfrac{\sqrt{3}}{2}\right)t\right]u(t)$
Second-order band elimination	$\dfrac{s^2 + 1}{s^2 + s + 1}$	$1 - \dfrac{2}{\sqrt{3}}\left[e^{-1/2t}\sin\left(\dfrac{\sqrt{3}}{2}\right)t\right]u(t)$

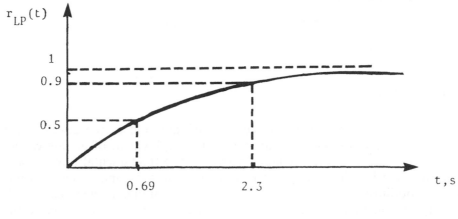

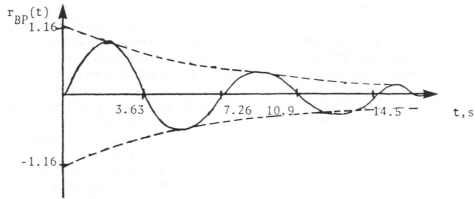

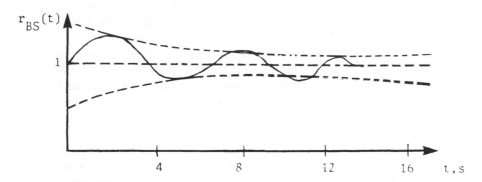

FIG. 42 Unit step responses of low-pass, band-pass, and band-stop filters.

the unit step response is

$$r(t) = (1 - e^{-1000t})u(t),$$

and the settling time is about 2.3 ms.

Conclusion

The design of electrical filters played a significant role in the development of telecommunications systems, particularly the deployment of analog carrier transmission systems in the telephone network. The fundamental physical and mathematical principles that made this engineering accomplishment possible have been outlined in this article. First, we have shown that all electrical signals that vary as functions of time possess frequency contents that describe the distribution of their energies over different bands of frequency. The electrical signals may be such real-world signals as those representing speech or dramatic scenes; these have been converted by appropriate energy conversion devices from their original physical forms into electrical form. For speech, as an example, the transducer involved is a microphone that converts speech from its natural energy form of acoustic pressure waves into a time-varying electrical current or voltage waveform. In addition to such natural signals, we also studied "test" signals mathematically defined using the elementary functions of algebra, trigonometry, and calculus, the properties of which can be related by extension to the properties of the natural signals encountered by telecommunications systems.

The attributes of electrical signals include whether they are analog or digital, continuous time or discrete time, random or deterministic, and periodic or nonperiodic. Evaluating the frequency contents of signals with all possible combinations of these attributes involves variations in the application of the Fourier series or transform. The resulting frequency distributions are discrete or continuous, periodic or nonperiodic, over the frequency variable, as detailed in the article.

The function of an electrical filter is to modify the frequency content of an electrical signal in a desired manner. Examples of technical situations where such modifications are needed are listed in the introductory section. These situations can arise in the design of carrier transmission systems (analog and digital), data communications systems, such signal processing systems as sonar and radar, and in many instrumentation and control systems; these are discussed in articles in this and other volumes of this encyclopedia.

Whereas the performance of electrical filters naturally is measured most often in the frequency domain, some time-domain behavior of filters also is considered. Since the frequencies present in a signal vary from one instant to another, the filter must act on those frequencies in the time interval that they are present. For example, consider a filter designed to attenuate the frequencies of a signal above one kHz; if a 2-kHz segment burst occurs for 20 ms, then the filter must perform its designed function within one or two ms of the occurrence of the burst. The ability of the filter to do so can be estimated from the step response rise time as discussed in the final section.

The electrical circuit elements that primarily are employed in physical implementation of filters are those with terminal responses that are rate dependent.

For example, for a capacitor the current depends upon the time rate of change of the voltage, and for an inductor the voltage depends upon the rate of change of the current. Hence, these are the components, in combination with resistors, amplifiers, and transistors, that are used to construct filters. Such other elements as crystals, delay lines, and waveguides derive their filtering properties by simulating the effects of capacitance and inductance. These physical implementation technologies are covered in a companion article on electrical filters (7).

Bibliography

Ahmed, N., and Natarajan, T., *Discrete-Time Signals and Systems,* Reston Publishing Co., Reston, VA, 1983.

Allen, P. E., and Sanchez-Sinencio, E., *Switched Capacitor Circuits*, Van Nostrand Reinhold, New York, 1984.

Bracewell, R. N., *The Fourier Transform and Its Applications*, 2d ed., McGraw-Hill, New York, 1986.

Haykin, S., *Modern Filters*, Macmillan, New York, 1989.

Jackson, L. B., *Digital Filters and Signal Processing*, Kluwer, Hingham, MA, 1986.

Jackson, L. B., *Signals, Systems, and Transforms*, Addison-Wesley, Reading, MA, 1990.

Lindquist, C. S., *Active Network Design with Signal Filtering Applications*, Steward and Sons, Long Beach, CA, 1977.

Mitra, S. K., and Kurth, C. F. (eds.), *Miniaturized and Integrated Filters*, John Wiley, New York, 1989.

Rhodes, J. D., *Theory of Electrical Filters*, John Wiley, New York, 1976.

Sinha, N. K., *Linear Systems*, John Wiley, New York 1991.

Soliman, S. S., and Srinath, M. D., *Continuous and Discrete Signals and Systems*, Prentice-Hall, Englewood Cliffs, NJ, 1990.

References

1. Openheim, A. V., Willsky, A. S., with Young, I. T., *Signals and Systems*, Prentice-Hall, Englewood Cliffs, NJ, 1983.

2. Beyer, W. H. (ed.), *CRC Standard Mathematical Tables*, 25th ed., CRC Press, Boca Raton, FL, 1978, pp. 377.

3. Papaoulis, A., *Probability, Random Variables, and Stochastic Processes*, 2d ed., McGraw-Hill, New York, 1984, pp. 219–231.

4. Nolan, F., The Nature of Speech. In: *Electronic Speech Recognition* (G. Bristow, ed.), McGraw-Hill, New York, 1986, pp. 19–48.

5. Couch, L. W., *Digital and Analog Communication Systems*, 3d ed., Macmillan, New York, 1990.

6. Roden, M. S., *Analog and Digital Communication Systems*, 3d ed., Prentice-Hall, Englewood Cliffs, NJ, 1991.

7. Moschytz, G. S., Electrical Filters. In: *The Froehlich/Kent Encyclopedia of Telecommunications*, Vol. 6 (F. E. Froehlich and A. Kent, eds.), Marcel Dekker, New York, 1993, pp. 427–543.

MEBENIN AWIPI

Electrical Telecommunications Cables: Fundamentals and History

Introduction and History

The technical developments for telecommunications cable, covering more than a century from the beginnings of the telephone as a revolutionary idea to the sophistication of total communications of the current day, are best recounted in the context of the history of the art. Practitioners work on and solve problems as perceived in their time. Inventions made long before their time frequently languish in dusty archives. Inventions in their time revolutionize the art with a forward leap.

In this article, we wish to capture and preserve some of the exhilaration that attended the inventions and the solution of the then-current problems. To recount the answers and ignore the surrounding flow of history is to do a disservice to the answers and to the readers.

Early Cables and Problems

Telephone cable first served as a means of carrying overhead aerial wire on poles through tunnels, over bridges, or small bodies of water. Single insulated conductors were pulled into iron pipes, or even wrapped with canvas for protection. As early as 1880, four canvas-wrapped cables having seven conductors each carried telephone service across the Brooklyn Bridge.

The ground-return method of transmission and the close proximity of the wires in the cable led to severe interference. Shielding tapes of lead, tinfoil, or a helical wrapping of copper wire were tried as mitigative measures with only limited success.

Insulations available were gutta-percha (see the section on insulation), rubber, and a compound having the trade name of Kerite (a vulcanized compound of oxidized linseed oil and rubber). Such textile materials as cotton, silk, and jute also were employed in helical servings over the copper wires.

The Water Problem

Water leakage into the cable cores was a major problem. Various means for preventing water entry included pulling the cable core into a pipe, filling with oil, and then pressurizing. As another alternative, the pipe containing the wires simply was filled with oil or kerosene. Since the oil or kerosene was lighter than water, the oil or kerosene leaked out; water entered and shorted the cable conductors. The rigid iron pipes used to protect the cables were costly and inconvenient, in addition to being subject to leaks. Cable cores sometimes were pulled into lead pipes. The flexibility of the lead pipes was advantageous during installation.

The First Twisted Pairs

Interference was much reduced when the ground return was replaced with a second overhead aerial wire to make a full metallic pair for each circuit. Twisting the cable pairs reduced interference between pairs. This was an idea that had occurred to Alexander Graham Bell in 1876; he was granted a patent in 1881. Bell's patents were noted for their statements indicating good understanding of fundamental principles, and the patent on twisted pairs was not an exception.

The First Foam Insulation

An engineer employed by the Western Electric Company, W. R. Patterson, pioneered a cable having cotton-insulated wires pulled into a flexible lead pipe. Since telephone plant engineers considered unfilled cable unthinkable, Patterson had to find a filling compound. A saying of the day explained: "It would be as easy to sell a cable filled with air as a sausage filled with air" (1, p. 212).

Patterson discovered that melted paraffin, when charged with gas and allowed to cool under pressure, foamed with tiny bubbles. The paraffin turned white, spongy, and relatively flexible and was used as the cable filling. Cables of this type were called *Patterson cables* and were the first dependable cables available. A description and cross-sections of typical cables are shown in Fig. 1.

Extruded-Lead Sheath

In another thread of the history of cables, John A. Barrett of the Bell System worked with John Robertson, an independent inventor, to develop a means of extruding a lead sheath directly over a cable core. Robertson had invented a lead press to extrude a lead jacket over wires used for electric lights. By 1886, Barrett and Robertson had extended the extrusion principle and built a hydraulic extrusion press for cable sizes up to 2 inches (in) in diameter. Limited production was started. Before enough of the lead presses were available, cable cores were pulled into straight sections of lead pipe, in a backbreaking operation, as shown in Fig. 2.

Unusual Cable Designs

Although the Patterson cables represented an early attempt to solve field problems, many competitive ideas were being offered for telephone plant use. As an

PATTERSON CABLE.

The "Patterson Cable" is a group of conductors, each covered with two or more windings of cotton or jute, or both, saturated with paraffine and protected by a lead pipe. The space between the group of conductors (or core) and the pipe is filled with aerated paraffine.

The core may be made in continuous lengths of 1,500 feet, and the protecting pipe is jointed over it in lengths of 75 to 100 feet. Any flaw which may exist in the pipe and any leaky joint is detected by the process of filling.

The almost invisible globules of gas, scattered uniformly through the mass of the paraffine filling, render it elastic, so that the natural shrinkage of the paraffine in cooling is compensated for by the expansion of these globules, preventing the formation of cracks and longitudinal fissures, through which water would penetrate indefinitely in case of a break in the protecting pipe.

FIG. 1   Description of the Patterson cable, taken from 1893 Western Electric catalog. (Copyright ©, 1975, American Telephone and Telegraph Company. Reprinted by permission, from Ref. 1, p. 213.)

FIG. 2   Pulling telephone cable core into lead pipe before the availability of lead extrusion presses. (Reprinted with permission from American Telephone and Telegraph [AT&T], Ref. 2, p. 324.)

example, a proposed construction consisted of wires threaded through regular shirt buttons spaced at intervals to prevent shorts. Kerosene filling was supposed to keep water out after the button cable was pulled into a protective pipe.

In another example, copper wires were pulled into glass tubes, then into straight lead pipe, then filled with a viscous compound to hold the fractured particles of glass after bending the lead pipe. These constructions were not popular in the field.

The "Conference Cables"

Beginning in 1887, cable designers and users started a series of conferences to discuss field problems. As an outcome of the first one, a specification was issued to standardize cotton yarn insulation, reversed lay cable core stranding, and compound filling to keep water out. A maximum wire-to-ground capacitance of 0.20 microfarads (μF) per mile and a lead sheath alloy of 3% tin also were standardized. These were called the *conference cables* (1, p. 214).

Dry Paper/Lead Cables

Favorable experience with extruded-lead sheath opened the door to consideration of nonfilled cable with paper-ribbon and air insulation. In the 1891 conference, dry paper-ribbon insulation with extruded-lead/tin-alloy sheath was made the new standard, with a wire size of 19 gauge (ga) and a pair capacitance of 0.085 μF per mile.

The dry-core and the extruded-lead sheath concept of the conference cables set the design pattern for more than the next half century. The idea of filling the interstitial spaces among the pairs did not find favor again until the 1960s, well after the introduction of plastic insulation.

Cable Core Types

The clutter and congestion of overhead open-wire lines in cities provided a strong driving force to convert to cable. Higher pair count cables were needed. Finer gauges (22, 24, and 26) were standardized in addition to 19 ga. The maximum number of pairs available in a given wire gauge increased rapidly (see Fig. 3). The largest sheath diameter was standardized at what was called a full-size cable with 2⅝-in outside diameter.

To respond to the public outcry to reduce the "cloud of wires which obscured the sky," cable installation doubled underground pair mileage each year for many years. Also adding pressure on the demand for cable, requests for telephone service grew explosively. The number of installed phones jumped about 500% per decade from 1895 to 1910, tapered off to nearly 100% for years, and to less than 4% per year currently.

After introduction of the conference cables, all cables were reverse-layer stranded. Reverse-layer stranding offered mechanical advantages in that the

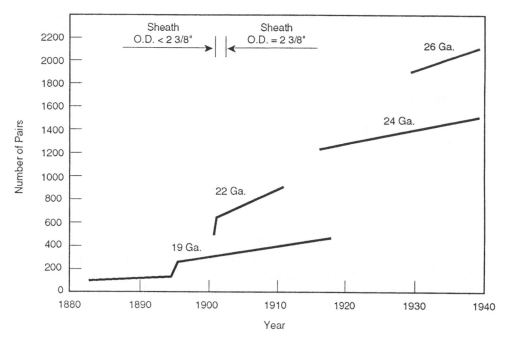

FIG. 3 Increase in telephone cable size with time, maximum number of pairs versus year.

layers tended to remain intact in handling and splicing. More important, the nonparallel relationship between pairs in adjacent layers reduced crosstalk.

As a matter of manufacturing economy, all pairs were made with the same length of twist, that is, 3 in. Such cable was said to have *nonstaggered twists*. Excessive crosstalk results between adjacent pairs of like twist lengths, when the twists fall into "step."

Staggered Twists

Years later, in 1920, the desirability of improved crosstalk performance was deemed to outweigh the cost of the manufacturing inconveniences and staggered twists were introduced. Two different twist lengths permitted use of nonlike adjacent pairs in a layer, with much improved crosstalk. Later, five twists per layer became common. And still later, when plastic cable was introduced, the twist count jumped to 25. This aspect of cable technology is discussed in more detail below.

Pulp Cable

In 1921, development work started on pulp cable, in which wood pulp insulation was applied directly onto copper wires. Eliminated thereby were the papermaking process, the slitting into ribbons, and the helical-wrapping operations. By

1928, 50-wire pulp insulating machines began producing commercial cable. Eventually, 60-wire machines became the standard for the gauges 19 to 26. Such machines continue to serve the telephone industry.

Unit-Type Cable

Pulp insulation has a higher coefficient of friction than ribbon and causes large layer-type fine gauge cable to kink and buckle when bent. This fact and the desire to organize the cores of high pair count cable better stimulated the development of unit-type cable, in which 25, 50, or 100 pair groups (called *units*) are manufactured in one operation. Then, the desired number of units are cabled to form a large unit-type cable. Such cables are suitably flexible. A typical unit-type pulp cable is shown in Fig. 4.

Each of the units in a unit-type pulp cable tends to perform transmissionwise as a separate cable. Unit-to-unit splicing maintains the relative isolation between units. This feature and the general upgrading of cable quality facilitated the use of pulp cable for the T1 digital carrier system, introduced long after the design and manufacture of the cable.

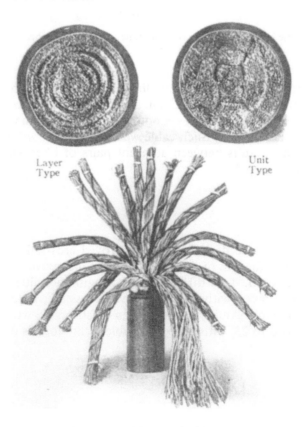

FIG. 4   Unit cable having 1818 pairs. Also, cross-sections of layer-type cable compared with unit-type cable. (Reprinted with permission from American Telephone and Telegraph [AT&T], Ref. 3, p. 443.)

FIG. 5   Example of the ubiquitous open-wire toll line seen along the roadside before its eventual replacement by cable. (Reprinted with permission from American Telephone and Telegraph [AT&T], Ref. 4, p. 526.)

Toll Cable

Except for selected routes like the Boston–New York–Washington, DC, communications corridor, open-wire lines carried essentially all long-distance toll circuits. A typical roadside sight appears in Fig. 5. All eventually were replaced by toll cable, but, as late as the 1970s, overhead intercity lines could be found in many rural areas of the United States.

Toll cable employed a different set of core standards than local or exchange cable. To reduce signal losses, early toll cables were made with 10- and 13-ga wires. Capacitance between wires of a pair was scaled down from the 0.085 conference standard to 0.062 μF per mile. Phantom circuits, in which two pairs can be arranged to provide three voice channels, require quads. A multiple twin quad consists of two twisted pairs twisted together to make up the phantom circuit.

Star Quads

Spiral-four or *star* quads, in which the four wires are twisted around a common axis, could have been used for phantom circuits. The difficulty of meeting balance requirements and the high phantom circuit loss of star quads dictated the choice of multiple twin quads. Star quads have been popular in Europe because they require about 30% less core space for the same pair capacitance.

Before metal–plastic sheaths replaced lead sheaths, core space was considered expensive.

Design and Manufacture

Design and manufacture of multiple twin quads having adequate balance to control crosstalk between the phantom and the side pairs was an art. A high degree of symmetry among the components and special twisting machines provided quads meeting field needs. After about 1911, essentially all toll cable was quadded. Loss reduction through use of inductance loading (see the section on transmission lines) and then gain through vacuum tube amplifiers permitted wire size reduction to 19 ga also. Still later, in the 1930s, cable carrier was introduced. Special quads were developed to meet the stringent crosstalk requirements and are covered in more detail below.

Sheath Types

In the beginning, telephone cable cores simply were pulled into iron pipe, gas pipe, and, to some extent, into lead pipe, as mentioned above. Direct extrusion of lead sheath over cable cores began in 1888. Patterson, in a study of lead, found that an alloy of lead and 3% tin had better strength and corrosion resistance than plain lead. Pure lead is quite soft and is subject to corrosion.

Not long after the beginning of the extrusion of lead directly over paper-ribbon-type cable cores, further studies were undertaken to find a less expensive lead alloy. Work on a number of alloys disclosed that a 1% antimony lead alloy had better physical and corrosion characteristics than the 3% tin lead. The 1% antimony lead was standardized in 1912 and remained the preferred material until the end of the era of lead cable sheath, initiated by the introduction of polyethylene and metal combinations for sheath in 1947.

Use of antimony instead of tin generated a large cost savings and provided an excellent sheathing material. Other lead alloys were explored at different times as possible solutions for several special problems and are discussed in more detail below.

Sheath Protection

Extruded lead provided the desired water and water vapor barriers for keeping paper and pulp insulation dry over their service life. Various protective measures for the lead sheaths were developed to meet a variety of environmental situations. These included stray currents, insects, gophers, squirrels, rocks, gully crossings, boat anchors, and similar hazards. Early in the history of the industry, protective measures were built around use of layers of paper, jute, asphalt, helical steel tapes, and steel armor wires.

In the air, protection against the gnawing attack of squirrels generally was accomplished on location by the addition of guards to discourage gnawing.

Tape armor might be used in severe cases. For buried cable, where rocks and frost heaving could cause sheath damage, a cushioning layer of jute and asphalt commonly were employed. In some situations, buried cable crossed gullies subject to occasional violent water action, and a light wire armor would suffice to control damage.

After polyethylene became available and innovative sheath structures were developed, the older, slower, and relatively messy protective constructions were replaced with high-speed plastic and metal combinations having equal or better performance. Some of the details are covered in this article.

Land, River, and Harbor Cable

Paper–lead-type cables were designed to accommodate a range of field problems. The strategy generally involved application of protective coverings as mentioned above. As an example, cable for conduit and aerial usage permitted plain lead in most cases. In some situations, an asphalt and sissalkraft paper might be employed for enhanced corrosion resistance in an underground conduit plant.

Water Crossings

Crossings of harbors and major rivers required more specific design efforts. For example, deep water (up to 900 or 1000 feet [ft]) employed a lead sheath of extra thickness supported by a paper-ribbon core tooled to give a firm and hard high-density structure. A merit of the high-density design was that in case of a sheath break, water entry caused the paper to swell and restrain the water's flow to a short and repairable distance, for years if necessary.

Such cable always was armored with one or two helical applications of steel wires for added strength and protection from errant boat anchors, water action on rocky bottoms, and the like. The armor technology was carried over from the earlier days of telegraph cable, which was developed and designed to solve the same set of problems.

Ocean Cable

All ocean cable owes a great debt to the pioneers who developed the designs, the special ships, and machinery for laying ocean telegraph cable, generally called *submarine cable* at that time. Names like Lord Kelvin, Wheatstone, and others figured importantly in the solution of many difficult problems.

Following land practices for telegraph, a single wire with ground return served as an adequate transmission path. For river, harbor, bay, gulf, and ocean crossings, the single wire required water-tolerant insulation and increasingly strong and reliable strength enhancement, achieved by one or two layers of helically applied armor wires. Designers built upon past successes and learned from their failures. The mystique of "proven integrity" (PI) evolved.

One of the interesting concepts was that of redundancy. The signal-carrying conductor always was stranded from a number of single wires to guard against breakage and failure of the cable. Copper smelting and refining, and wire drawing had not reached the stage at which inclusions of slag and foreign materials were as well controlled as in today's copper industry. The redundancy in some early telegraph cable was carried as far as the use of stranded individual armor wires.

Insulation

The chief insulating material available for early cables was a natural nonelastic latex compound called *gutta-percha*, from trees of the Malay Peninsula. Unprotected in light and air, gutta-percha oxidized and embrittled quickly, but served as a reliable insulation when immersed in water. A later refinement, from which impurities were removed and rubber and wax were added, was called *paragutta*. Paragutta insulated the conductors of three of the first long, continuously loaded telephone cables placed from Key West to Havana in 1921. Later, ocean telephone cables were insulated with the wonder plastic, polyethylene.

First Coaxial Cable

The first use of a full coaxial return instead of a ground return was in 1931, when a three-voice channel carrier, nonloaded cable spanned the ocean from Key West to Havana. The first use of submerged vacuum tube repeaters occurred in 1950, in still another Key West–Havana telephone cable, for a 24-voice channel system. In this cable, polyethylene replaced the gutta-percha and its derivative, paragutta, as the conductor insulation. The 1950 cable served as a field trial for the design of the first transatlantic telephone cable.

All subsequent ocean telephone cable employed polyethylene insulation. A carefully engineered polyethylene having even lower loss served as the insulant in the more recent high-capacity ocean telephone cables.

First Transatlantic Telephone Cable

The first transatlantic telephone cable, TAT-1, was placed in 1955 for the first of a two-cable, 36-voice channel system; the second cable was placed in 1956. Wire armored for strength and protection, TAT-1 looked much like early telegraph cable, except for the polyethylene insulation and the coaxial outer conductor. One more crossing, TAT-2, spanned the Atlantic with the same armored-cable construction.

Armorless Ocean Cable

The British Post Office pioneered a radical departure from the PI of the past in the form of an armorless cable for the deep water portion of the ocean crossing.

The strength member consisted of a core of special high-strength steel wires in the center of the cable. The coaxial transmission path was on the outside. Figure 6 illustrates the construction of an early trial idea.

The American armorless design employed several departures from past technology. The system included rigid in-line electron tube repeaters. A special cable ship having a linear cable engine instead of the usual drum-type payout capstan was designed and built to accommodate the rigid in-line repeaters.

The armorless family of ocean cables was highly successful. Subsequent crossings included TAT-3 through TAT-7. For a progressive reduction in cable transmission loss, the core diameter was increased in several steps, from 0.62 in for TAT-1 and TAT-2, to 1 in for TAT-3 and TAT-4, to 1.5 in for TAT-5, and finally to 1.7 in for TAT-6 and TAT-7. Repeater spacings were reduced, and transistors replaced the vacuum tubes. Channel capacity reached a top of 4000 voice-frequency circuits on one cable.

The most recent crossing, TAT-8, took place in 1988 with another radical change in technology, the first optical waveguide cable, with 40,000 two-way voice channels on two glass fibers. Details of the engineering and design features of the family of ocean telephone cables are covered in the section, "Ocean Cable."

Maxwell's Equations

Communications signals travel as electromagnetic waves through free space, as by radio, or are guided over a variety of media made up of conductors and dielectric materials. Whether through the air, as for Marconi's famous first

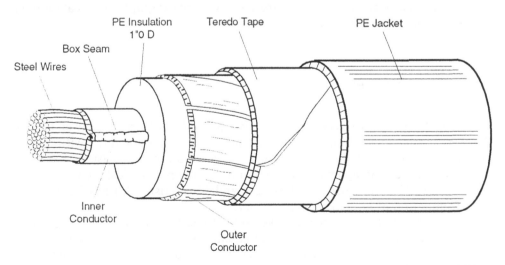

FIG. 6 Early version of design of armorless ocean cable pioneered by the British Post Office. The coaxial path is outside of the strength member.

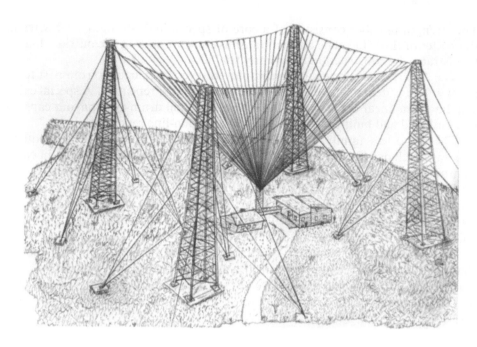

FIG. 7 Photograph of model of Marconi's radio telegraph towers and inverted umbrella antenna, Wellfleet, Massachusetts, that transmitted telegraph signals across the Atlantic.

radio telegraph signals across the Atlantic in 1901, or the telegraph signals over the first successful transatlantic cable in 1864, all obey the laws formally derived by James Clerk Maxwell in 1865. The Atlantic Ocean has long since claimed the towers of Marconi's radio antenna, once high on the bluffs of Cape Cod, Massachusetts, near Wellfleet. Figure 7 shows a photograph of a scale model on exhibition at the site of the radio transmitter.

James Clerk Maxwell

In his *A Dynamical Model of the Electromagnetic Field* (1864), Maxwell (1831–1879) presented field equations for electromagnetic waves credited with traveling at the velocity of light (5). He drew the important conclusion that light consisted of electromagnetic waves, contrary to the mechanical view prevalent at the time. His "Treatise on Electricity and Magnetism" (6) has been called "the foundation of electromagnetic theory."

The laws for electromagnetic waves are defined by what generally are called *Maxwell's equations*. The form of the equations depends upon mathematical concepts and is different for Cartesian, cylindrical, or spherical coordinates. As an example, two of the six equations for the Cartesian coordinates x, y, and z are

$$\frac{\partial E_z}{\partial y} \frac{\partial E_z}{\partial z} = -i\omega\mu H_x \tag{1}$$

$$\frac{\partial H_z}{\partial y} \frac{\partial H_z}{\partial z} = (g + i\omega\epsilon)E_x \tag{2}$$

The second and third sets of the six equations are identical to the first set shown above, except that coordinates x, y, and z and the subscripts of E and H step along to the next letter for the respective equations. That is, x goes to y, y to z, and z to x. E is the electric intensity, H is the magnetic intensity, μ is the permeability, ϵ is dielectric constant, g is the conductivity of the conductors, $i = \sqrt{-1}$, and ω equals $2\pi f$, where f is frequency in Hertz (Hz).

Overview

As an overview, general solutions of Maxwell's equations define how an electromagnetic wave propagates in free space or in a dielectric medium. For the different kinds of telecommunications cables, the equations determine particular solutions that are geometry dependent. In this case, the electrical current flows in the metallic conductors and the electromagnetic wave flows in the dielectric space between the conductors. Almost all of the energy flows in the dielectric space.

For shielded pairs, coaxials, and waveguides, the particular solutions lead to the transmission parameters discussed in the section on transmission lines. These parameters characterize and define cable performance, and facilitate the design of systems for practical application of electromagnetic wave transmission of useful intelligence.

Hertz's Experiments

Maxwell laid the mathematical foundation for the concept of electromagnetic waves, but he did not live to witness experimental proof of their existence. In a series of experiments described as "one of the greatest contributions ever made to experimental physics," (7) Heindrich Hertz (1857–1894) demonstrated the means of generation and detection of electromagnetic waves and substantiated Maxwell's predictions.

Hertz started his renowned experiments in 1884 and completed them in 1888. He employed a spark coil (like those used in Henry Ford's Model T cars) to generate the "Hertzian" waves in a circuit (see Fig. 8). As indicated in the illustration, his detector was a simple loop (about 28-in diameter) of heavy wire with two closely spaced brass balls. He detected reception of waves by observing a feeble spark between the balls.

Early Use of Maxwell's Equations

Many of the early practitioners of Maxwell's equations attacked the transmission line problem for telegraph and telephone use. As mentioned in the history

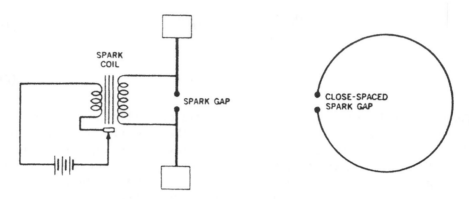

FIG. 8   Schematic of one of Hertz's fundamental experiments demonstrating the generation of short electromagnetic waves (left) and a detector (right). The circular loop of copper wire was about 28-in diameter and the metal plates were about 32-in square. (Copyright ©, 1975, American Telephone and Telegraph Company. Reprinted by permission, from Ref. 1, p. 350.)

section of this article, the idea of loading with inductance to reduce attenuation was one of the important results. Those who worked in this area in the early years of telephone cable include Oliver Heaviside, George A. Campbell, Michael I. Pupin, Lord Kelvin, and others. Practical experience guided much of the early work, Maxwell's theoretical understanding being ahead of the art by many years.

Beginning with the move in the 1930s to higher frequencies for carrier systems and the desire to use high-frequency coaxials, shielded pairs, and other structures of interest, S. A. Schelkunoff, Sallie Pero Mead, S. P. Morgan, and others applied Maxwell's equations to a variety of communications structures. This work provided the theoretical foundations and guidance for much cable development work.

Waveguides

The fundamental studies based on Maxwell's equations led to insights that were not expected intuitively. One interesting example is a waveguide transmission mode that defies the seemingly universal square root of frequency law, for which transmission loss "always" increases as the square root of frequency. The idea of electromagnetic waves inside a pipe was strange enough in the first place because everyone "knew" two separate wires were needed to carry a signal. The early concepts focused on the flow of electrons (current) in the wires and ignored the flow of the electromagnetic energy between the wires. The "water in a pipe idea" held sway for many years.

Work with Maxwell's equations led Schelkunoff to the discovery of a field configuration inside a metallic pipe, for which doughnutlike electromagnetic fields were found to propagate with very low loss. The loss decreased with increasing frequency only in paths approaching near-optical straightness, an exciting idea.

Worldwide interest and research essentially had solved the problems of transmission over nonflat earth with near-optical straightness just in time to be bypassed by another form of waveguide, light waves in lightguides of hair-thin glass fibers.

Shields and Clogston Lines

Out of Schelkunoff's work and that of others came the solution to many engineering problems concerned with the design and use of practical coaxial conductors, shielded balanced pairs, and waveguides. Calculation of the shielding effect of cylindrical shields helped reduce control of crosstalk to a manageable science. Mathematical work by A. M. Clogston showed how, by properly proportioning thin laminations of metal and insulation, the high-frequency currents in a conductor could be enticed to flow in deep layers, contrary to the usual limitations imposed by skin-effect phenomena. Practical problems of achieving many thin laminations of metal and insulation prevented commercial use of Clogston lines, as they were called.

Practical Examples

Transcontinental coaxials and ocean crossings with armorless coaxial cable are practical examples of Maxwell's equations being applied to real problems to provide calculable performance confirmed by measurements on the finished structures. Fascinating examples of recent applications are the use of Maxwell's equations to determine the characteristics of transmission lines and filter elements on microwave integrated circuit chips much too small to be seen without a microscope. Hundreds of references tell of computer and numerical methods of solving Maxwell's equations for a variety of microwave problems.

Perspective

As a closing perspective on Maxwell's equations, early practitioners designed transmission systems without benefit of complete understanding of electromagnetic wave propagation. Conductor resistance and capacitance between the conductors were known to limit transmission distance. Telegraph signals were viewed as limited by the "KR" rule of Kelvin, propounded in 1854, well before Maxwell's famous treatises. The emphasis was on the maximum possible transmission speed for on–off signals, rather than signal loss. Kelvin stated that the maximum rate was inversely proportional to the product of total capacitance K and total resistance R, independent of cable attenuation and applied voltage (1,

p. 241). Telephone signals applied to the input of lines were weak. Transmission loss limited useful distances to those within the sensitivity range of receivers. Even so, the KR idea was assumed to apply and probably delayed deeper insights into the phenomena involved.

Challengers of the KR idea included Oliver Heaviside, whose work was published in *The Electrician* and in the *Philosophical Magazine* beginning in 1873. Heaviside disclosed transmission line theory that led to better understanding of transmission and to the idea of inductive loading as mentioned in the history section of this article.

Conclusion

Bell's invention of the telephone launched an exciting intellectual foment that has engaged the minds and imagination of mathematicians, theorists, engineers, and practitioners, in a golden age of discovery, invention, and the reduction of new ideas to practice. Many famous men, including Maxwell, Hertz, Marconi, and innumerable others have helped fuel the engine of communication that continues to gain speed and power. New problems, new solutions, and new ideas add the spice of excitement to the communication and information age. Much more is sure to come.

Transmission Lines

An interesting property of a very long uniform electrical transmission line is that the line current and line voltage decay at a uniform rate as a signal launched at the input travels along the line. Figure 9 illustrates a signal of voltage E launched on an infinitely long line. The small circle having one cycle of a sine wave within is the symbol for a signal source.

Expressed in mathematical terms, the ratio of the voltage E_2 at the output end of any section of length ℓ to the voltage E_1 at the input of the section is

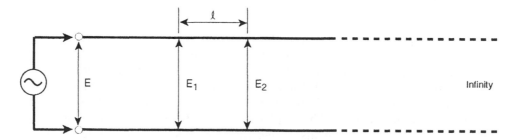

FIG. 9   Signal voltage E launched at the input end of an infinitely long line has the same E_2/E_1 ratio at all points along the line.

equal to the E_2/E_1 ratio for any other section of line of length ℓ. The magnitude of the ratio is

$$\frac{E_2}{E_1} = e^{-\frac{P}{\ell}} \tag{3}$$

where e is the base of the Naperian logarithm ($e = 2.71828\ldots$), and $P = \alpha + j\beta$. P is called the propagation constant and consists of the components α, the line attenuation constant, and β, the phase constant. The attenuation α is in nepers, which become decibels (dB) when multiplied by 8.686. The phase constant β indicates that the phase of the output voltage of section length ℓ lags behind that of the input voltage by $\beta\ell$ radians. (One radian is equal to 57.3°.) A line is said to be one-half wave long when $\ell = \pi$ radians.

The term j is defined as $j^2 = -1$; it is the same as the i employed in the section on Maxwell's equations. Mathematicians use i to indicate the complex operator and engineers use j. Since this usage is well established in the literature, we do not depart from it herein.

Line Elements

Transmission lines may be considered to be made up of a sequence of series and parallel elements (see Fig. 10). R is the series resistance of the conductors, L is the internal inductance of the conductors plus that of the space between the conductors, C is the capacitance between the conductors, and G is the conductance due to the dielectric loss associated with the capacitance.

The parameters R, L, G, and C are called the *primary constants* of the line and are discussed in more detail below in relation to specific physical structures. Maxwell's equations are employed to determine the primary constants of specific structures (e.g., coaxials). Primary constants may be thought of as the interface between cable designers and cable manufacturers.

When the elemental line length in Fig. 10 approaches zero or a differential length $d\ell$, differential equations may be set up and solved to show the relationship between the primary constants and what usually are called the secondary constants. The secondary constants may be thought of as the interface between the cable designers and the communications system designers.

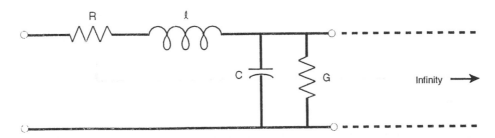

FIG. 10   One section of series and parallel circuit elements that, in sequence, make up an electrical transmission line.

The more important relationships between the primary and secondary constants are summarized below.

Signal Loss

The concept of propagation constant P was mentioned in the introduction to this section on transmission lines. The real component of P, α, is called the line attenuation constant and provides a measure of the signal loss for the line. The solution to the differential equations provides the desired relationships between the primary and secondary constants. The rigorous relationship is

$$P = \alpha + j\beta = \sqrt{(R + j\omega L)(G + j\omega C)}. \tag{4}$$

Practical transmission lines are finite in length. Analyses based on the infinite line apply to finite lengths when the output end is terminated in what is called the characteristic impedance Z_o of the line (defined in a section below); this is illustrated in Fig. 11. Thus terminated, all of the incident wave is absorbed by Z_o. Hence the formula above applies to a line of finite length.

Propagation constant P is called a complex quantity consisting of the real part α and the imaginary part β. Separating real and imaginary components gives the line loss

$$\alpha = \sqrt{RG} \qquad \text{at and near DC (direct current),} \tag{5}$$

$$\alpha = \sqrt{\frac{R\omega C}{2}} \qquad \text{at voice frequencies,} \tag{6}$$

$$\alpha = \frac{R}{2}\sqrt{\frac{C}{L}} + \frac{G}{2}\sqrt{\frac{L}{C}} \qquad \text{at high frequencies.} \tag{7}$$

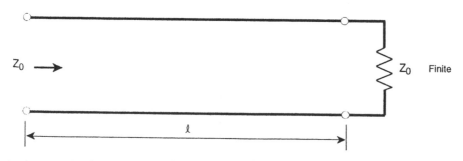

FIG. 11   Transmission line of finite length ℓ terminated in characteristic impedance Z_o has an input impedance of Z_o.

All of the above formulas for α are approximations. Equation (5) depends upon ωL and ωC being much less than R and G, respectively. Equation (6) depends upon ωL being much less than R and is useful at voice frequencies. Equation (7) depends upon $\omega L \gg R$ and $\omega C \gg G$, and applies at high frequencies of roughly above 100 kilohertz (kHz) for the usual multipair cable. For intermediate frequencies, the rigorous formula must be used to avoid errors of about 1% or more.

In these equations, unit length values of R, L, G, and C in ohms, henrys, mhos (reciprocal ohms), and farads (F), respectively, result in nepers per unit length.

Phase Constant

The phase constant β is defined here as the lag in the phase of the voltage at unit distance along a uniform line, referenced to the phase of the input voltage. The value in radians is given by

$$\beta = \sqrt{LC}\left(1 + \frac{R^2}{8\omega^2 L^2}\right) \tag{8}$$

G is assumed to be negligible relative to ωC, which is typical of essentially all insulating materials likely to be employed in multipair cables. Note that line dissipation increases the delay slightly. The $R^2/(8\omega^2 L^2)$ factor amounts to less than 1% for multipair cable pairs in the 100-kHz range.

Polar Representation of α and β

The attenuation and phase constants may be illustrated in a polar plot (see Fig. 12). The length of the vector is proportional to the attenuation, and the angle of rotation is equal to $\beta\ell$. Thus, as the vector rotates through the first quadrant, $\beta\ell$ increases from zero to $\pi/2$ radians. In Fig. 12, the vector is shown in the first quadrant after one full rotation of 2π radians. For each $\pi/2$ of rotation (that is, 1/4 wavelength), the E_2/E_1 ratio as defined here remains constant. Restated, each increment in length equivalent to 1/4 wave increases the total attenuation a fixed number of dB for a given line at a given frequency.

Characteristic Impedance

Characteristic impedance Z_o is defined as the input impedance of a uniform line of infinite length, and is equal to

$$Z_o = \sqrt{\frac{R + j\omega L}{G + j\omega C}} \quad \text{at all frequencies,} \tag{9}$$

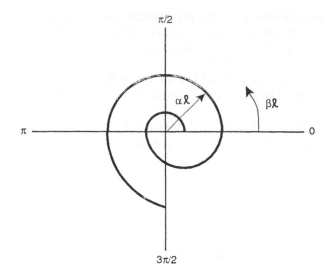

FIG. 12   Polar representation of attenuation α as the radius vector, and of phase constant β as the number of radians or rotation as line length $\beta\ell$ increases.

$$Z_o = \sqrt{\frac{R}{\omega C}} \qquad \text{at voice frequencies,} \tag{10}$$

$$Z_o = \sqrt{\frac{L}{C}} \left(1 - \frac{jR}{2\omega C}\right) \qquad \text{at high frequencies,} \tag{11}$$

$$Z_o = \sqrt{\frac{L}{C}} \left(1 + \frac{R^2}{2\omega^2 L^2}\right) \sqrt{\tan^{-1} \frac{R}{2\omega L}}, \qquad \text{at high frequencies.} \tag{12}$$

Equation (9) for Z_o is derived from the differential equations and is rigorous. Equation (10) is an approximation that applies at low frequencies where $\omega L \ll R$ and $\omega C \gg G$. At high frequencies where $\omega L \gg R$ and $\omega C \gg G$, two approximations are given. Equation (11) shows the value of Z_o in rectangular coordinates and Eq. (12) shows it in polar coordinates.

The term $R^2/(2\omega^2 L^2)$ may amount to several percent for the usual multipair cable in the 100-kHz range. Note that the effect of line dissipation due to R causes a small reactive term to appear in Z_o. The dielectric loss G is much less than that due to R, and thus does not appear in the above approximations.

Velocity

The velocity of an electromagnetic wave in a uniform transmission line is

$$v = \frac{\omega}{\beta} = \frac{1}{\sqrt{LC}} \left(1 - \frac{R^2}{8\omega^2 L^2}\right) \tag{13}$$

For R, L, and C in per-mile units, v is in miles per second. The ω/β term shows the fundamental relationship between velocity and phase. Line dissipation reduces velocity slightly. The term $\sqrt{LC}$ is proportional to the dielectric constant of the insulation, thus justifying the intuitive desire to use insulations having a low dielectric constant.

The reciprocal of the velocity $1/v$ is the transit delay and often is of concern when dealing with long systems. A typical value of $1/v$ for multipair cable is in the range of 7 microseconds (μs) delay per mile versus about 5.4 μs for free space.

A loaded voice frequency cable might have 5 to 10 times more delay because of the addition of inductance coils. As a result, a transcontinental cable might have a delay of about 30 milliseconds (ms) or about 250 ms if loaded. For comparison, the minimum round-trip delay to a geostationary communications satellite is about 240 ms. Delays of this magnitude are noticeable in telephone conversations.

Impedance Matching

When a signal traveling along an otherwise uniform transmission line encounters a change in impedance, some of the incident wave is reflected. The situation is analogous to the reflection of a flashlight beam off a sheet of glass, with the exception that in the case of the transmission line, the reflected wave simply reverses direction and returns toward the input end of the line.

If the terminating impedance is Z_a instead of Z_o, the reflection coefficients are

$$\text{Voltage coefficient} = \frac{Z_a - Z_o}{Z_a + Z_o} \tag{14}$$

$$\text{Current coefficient} = \frac{Z_o - Z_a}{Z_a + Z_o} \tag{15}$$

The magnitude of the reflected voltage E_r is equal to the incident voltage E_i times the voltage coefficient. The net voltage at any point along the line is the vector sum of the incident and reflected voltages. Figure 13 provides an example of incident voltage, reflected voltage, and net voltage for a line terminated in Z_a.

Examination of the relationships between reflected and net voltage reveals interesting properties of wave transmission. For example, if $Z_a = 0$, the coefficient equals -1, that is, the reflected wave equals the incident wave with a reverse in sign. The net voltage is zero at the point where $Z_a = 0$, as would be expected for a short circuit. Correspondingly, if Z_a is infinite (open circuit), the voltage coefficient equals $+1$. The reflected wave then is equal to and has the same sign as the incident wave and the net voltage is double the incident voltage.

A similar analysis for the situation applying to line current leads to the conclusion that the net current at the output end of the line doubles when the line is shorted and goes to zero when the line is open.

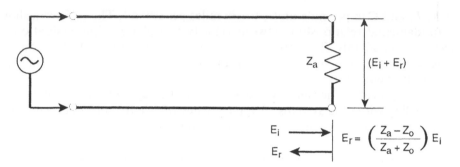

FIG. 13 Transmission line of impedance Z_o showing incident voltage E_i, reflected voltage E_r, and net voltage $(E_i + E_r)$ when terminated in impedance Z_a.

Design of Terminal Apparatus

The design of terminal apparatus necessarily must be concerned with the impedance that the apparatus presents to the line. Details are beyond the scope of this article, but a few words provide the basis for intuitive insights. Splices, as an example, may present impedance irregularities. A wave reflected at the terminal point will travel back to the splice and a portion will be re-reflected and return to the receiving apparatus as a delayed signal.

For pulse-type transmission systems, it is easy to see that the delayed pulses may be confused with the desired incident pulses. The process may repeat, and thus produce several reflected pulses. The idea of multiple reflections of a transmitted pulse is shown in Fig. 14. To avoid performance degradations, impedance matching must be considered in system design.

Although the degradation is illustrated by a digital example because it is easy to visualize conceptually, analog systems are not immune to similar troubles. Multiple reflections distort the shape of attenuation-versus-frequency response, which introduces distortion in the transmitted analog signal. Television

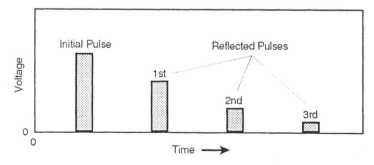

FIG. 14 Successive pulses due to multiple reflection, as from a splice that has introduced an impedance discontinuity.

pictures sometimes show "ghost" images to the right of the main picture due to delayed reflections.

Resonant Lines

Reflected waves as discussed here lead to a special property of transmission line falling under the general category of resonance. Again, detailed treatment of the subject is beyond the scope of this article, but a brief overview provides sufficient insights to dispel some of the mystery.

Almost everyone has had the experience of blowing into the open end of a short length of tubing and enjoying the pleasant whistle that may result. The pitch depends upon the length of the tubing and whether the far end is closed or open. When the length is such that the reflected wave "interferes constructively" with the input wave, a musical note is heard. This is so for all wind instruments, from pipe organs to piccolos, and for the low and mournful moan heard when blowing gently across the mouth of a large jug.

Resonance in transmission lines occurs when the line length is such that the reflected wave adds to the input wave. The terms *constructive interference* and *destructive interference* come from optics and refer to bright bands or dark bands of monochromatic light. The reflected wave combines constructively with the incident wave to produce a bright band or destructively to produce a dark band.

Overview of Resonance

A simple visualization gives an overview of resonance in transmission lines. In Fig. 15, the heavy line at the bottom represents the transmission line and the sinusoid above represents the voltage along the line. Voltage is chosen for the figures because the input impedance of the line is high or low, correspondingly.

In the figures, the far end of the transmission line is on the right; the input end is on the left. Line length is measured in fractional wavelengths, where λ is the symbol for wavelength.

In interpreting the figures, voltage is zero on the right for far end shorted; for $\lambda/4$, voltage is maximum at the input; for $\lambda/2$, voltage is minimum at the input; and for $3\lambda/4$, voltage is maximum at the input. A similar interpretation applies with far end open.

The combined incident and reflected waves as shown in Fig. 15 are called *standing waves*, whether for electromagnetic waves or sound waves. A detector probe moved along the line would indicate voltages as shown and is called a *standing wave detector*. Such detectors provide a sensitive means of determining the degree of mismatch between Z_a and Z_o and serve as means of measuring Z_a.

The velocity of sound in air is about 1.1 ft per ms; that of an electromagnetic wave in air is about 0.98 ft per nanosecond (ns). Hence, resonant electrical lines and pipes for sound are roughly the same length when the electrical frequency is about one million times higher than the sound frequencies.

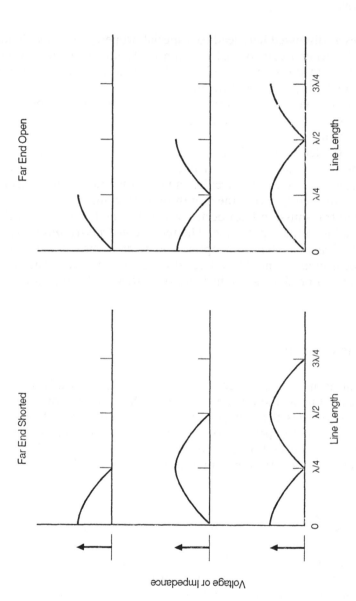

FIG 15 Input voltage (proportional to input impedance) of resonant line, having far end shorted or open, for line lengths of λ/4, λ/2, 3λ/4.

Communications Network and Cable Types

The telephone communications network has evolved for well over a century to meet the needs and wishes of the public. In the beginning, the plant was strictly local, with service from the central office to the subscribers. This plant has been called the local plant, loop plant, subscriber distribution plant, and probably other names.

Individual lines to a subscriber commonly are called *loops*. In the beginning, all subscriber loops were via aerial open wire and pole mounted, following the practice of telegraph technology. The connection from the overhead wire "dropped" to the subscriber's telephone, and hence was called *drop wire*. Party lines were common. Rural lines were single aerial wires and were multiparty; some still are in use.

Subscriber Network

As the transition from open wire to cable occurred and the number of telephones grew, the arrangement of the associated cable plant became more complex. Cables near the central office were the feeders to the subscriber distribution plant. Branch cables branched off the feeders to local areas into block cables. As time and the complexity of connecting millions of subscribers evolved, differing design philosophies were introduced, reduced to practice, or abandoned, as discussed in this section.

In larger cities, the feeder cable plant is placed in underground conduit or ducts. In residential areas, the local distribution cable is installed aerially or may be buried.

Resistance Design

Subscriber distribution cable is usually 26 ga near the central office. Heavier gauges are used to reach more distant customers. Detailed design rules are beyond the scope of this article, but overall principles are reviewed briefly. One limiting factor is the DC resistance of the subscriber's loop and the ability of the central equipment to detect "off hook" or a request for service when the resistance is high.

The off-hook detection sensitivity varies for different types of central-office equipment: manual switchboards in the early days, then step by step, panel, crossbar, and electronic equipment. A typical value for the maximum permissible resistance is 1300 ohms, standard for many years, which is equivalent to about 5 miles of 24-ga wire.

As the loop length increased, increasingly heavy wire was employed to extend the length and to hold the total DC resistance on or below target. Such design strategy was called *resistance design*. As a rough number, the average loop length in the United States has been in the range of two miles for many years and theoretically could be served by 26-ga wire.

Practice Abroad

In general, telephone systems outside the United States have been owned and operated by the government post offices until recent years. The widespread use of private telephones, as in America, has not been common abroad. Canada, Japan, and the Scandinavian countries are exceptions, where the degree of penetration of telephone usage is high; the arrangement of the distribution plant in these countries is similar to American practice. Terms may be different, however; for example, central-office trunk cable is called junction cable in England.

Trunk Cable

Trunk cables connect one central office to another within a city. Inner-city trunk cable generally is installed in an underground conduit plant. The size and gauge commonly used are 900 or 1100 pair, 22 ga. Digital T1 carrier is employed extensively in trunk cable plant. Also, glass fiber lightwave cable is being used significantly for new trunk runs and replacement of existing copper cable in some instances.

Network Planning

The telephone network, including all equipment, represents a huge investment (about $250 billion) and the loop plant is about 25% of the total. In the history of managing the planning for growth in the customer base, the idea that a preferred balance exists between the cost of loops and cost of trunks was recognized at least as early as 1906.

The process of optimizing the balance between the length and cost of subscriber loops versus the length and cost of trunk cable between central offices and the number of central offices has evolved over the years. Computerized methods continuously have improved the methodology of plant planning for minimum costs and for the best service and reliability. Highly sophisticated computer programs provide "what if" alternatives to guide plant management planning.

Multiple Plant Concept

During the years of rapid growth, the subscriber-loop network was designed by what was called the *multiple plant concept* (see Fig. 16). The purpose was to permit rapid addition of new customers. This was done by bridging customers repeatedly along a given pair. If private lines were not available, multiparty service of up to eight parties was offered.

The bridging generally was done during installation of the cable. A given pair was preconnected in sealed terminals having drop-wire binding posts at a

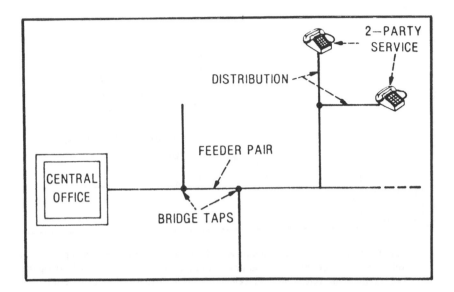

FIG. 16 Plan for multiple plant concept, used mostly for multiparty service in the days of only about 15 telephones per 100 people. (Copyright ©, 1985, American Telephone and Telegraph Company. Reprinted by permission, from Ref. 8, p. 459.)

number of points along the pair. In general, the number of bridging points exceeded the eventual number of customers, and represented a substantial amount of unused prewiring. The multiple concept met with disfavor when customers no longer were satisfied with multiparty service. The time was about 1925 when telephone penetration amounted to about 15 telephones per 100 people.

Dedicated Plant

A new concept, the dedicated plant, was introduced in 1962. The basic idea was to provide one permanently connected pair to a given housing unit or business location. Spare pairs and control and access points were provided to accommodate future demands.

Practice showed this system to be relatively inflexible, especially for meeting unexpected growth. Figure 17 shows the essential features of the dedicated concept. The control and access points were new ideas, but failed to accommodate easily unexpected localized areas of growth in demand.

Serving Area Concept

In 1971, another plan, the permanent plant and serving area concept (SAC) was initiated. The concept is illustrated in Fig. 18. The idea was to separate feeder plant planning from that for subscriber distribution. The subscriber distribution

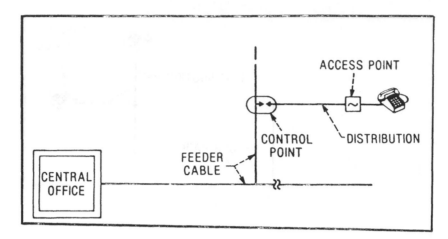

FIG. 17 Plan for dedicated plant concept provided one permanently connected pair per housing unit. Proved to be inflexible and not adaptable to unexpected growth. (Copyright ©, 1985, American Telephone and Telegraph Company. Reprinted by permission, from Ref. 8, p. 460.)

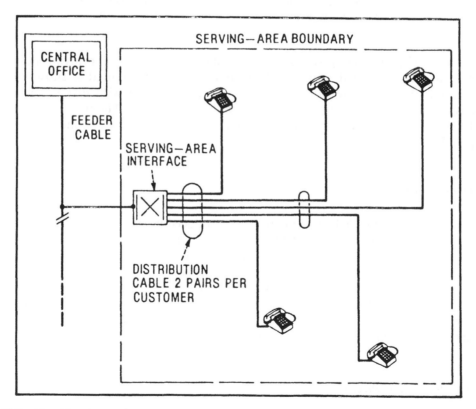

FIG. 18 Plan for serving area concept (SAC) separated the feeder plant from the distribution plant. These met at a feeder distribution interface, greatly facilitating rearrangements, record keeping, and plant stability. (Copyright ©, 1985, American Telephone and Telegraph Company. Reprinted by permission, from Ref. 8, p. 461.)

plant was designed and installed to meet the anticipated ultimate demand. A serving area interface provided the access point for connection to feeder cables, which were to be added at intervals as needed.

The SAC concept was introduced to meet planning needs for suburban areas and has evolved to meet the needs of most of the subscriber-loop network, including the special problems for rural loops. In general, the serving area interface is in an above-ground cabinet and provides good flexibility for connection to feeder cable. Two pairs are dedicated to each housing unit.

Plastic Insulated Cable—Aerial

A revolution in subscriber distribution methods started with the introduction of plastic insulated cable (PIC) in the mid 1950s. In aerial PIC cable, subscriber drops were connected on an as-required basis. Connection of subscriber drops was simple. Removal of the plastic cable sheath at the desired point, installation of a nonsealed ready access terminal, and connection of the customer drop was accomplished in one trip by the installer. The nonsealed concept and associated ready access terminal were invented by P. K. Koliss of AT&T Bell Laboratories in the mid-1950s.

Because sealing was not required, a simple plastic cover provided the desired "roof" overhead at a substantial saving compared with fully hermetically sealed terminals for paper or pulp cable. Millions of ready-access-type terminals including a number of later designs are in use and may be seen in aerial plants around the world.

Trouble Brewing

After a few years of experience with the original ready access terminals, trouble was brewing. Although more than half of the terminals performed well and met the original design goals, the others became a source of increasing upkeep cost. The problem was that some terminals were being used as access points for rearrangements, testing, and other plant operations. A good idea of the resultant "rat's nest" of wires is seen in Fig. 19.

The solution was twofold. First came a partial retreat to the older techniques. "Fixed-count" terminals presented the craftsperson with a panel of binding posts for customer drop wires already connected to a limited number of cable pairs. The rest of the cable passed behind the panel out of harm's way. The photograph in Fig. 20 shows a typical fixed count design, quickly named the *lunch box*. A second component of the solution accrued from the stabilizing influence of the SAC principles. The SAC concept greatly reduced the need to attempt plant rearrangements in terminals.

In a variation of the same fixed-count idea, the terminal may include a short stub cable spliced into pairs of the main cable in a splice case far enough from the pole to discourage impromptu rearrangements. Trends include better covers to keep out dust, dirt, spiders, and water. Insulation displacement connection systems to speed drop-wire installation are offered as an option.

FIG. 19   Ready access terminals sometimes were incorrectly used for plant rearrangements and troubleshooting. A "rat's nest" of wires was the result. (From Ref. 9, p. 250. Reprinted with permission. Copyright © 1976.)

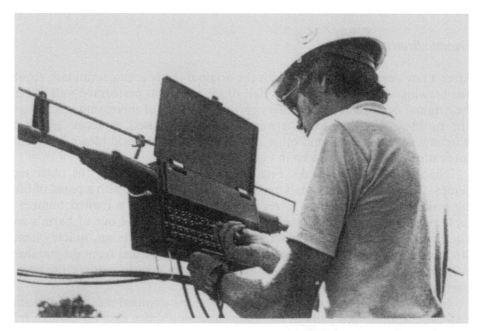

FIG. 20   "Fixed-count" terminals provide prewired binding posts for convenient subscriber drop-wire connections and reduce the temptation to try rearrangements at the terminal. (From Ref. 9, p. 250. Reprinted with permission. Copyright © 1976.)

Plastic Insulated Cable — Buried

Shortly after the introduction of aerial PIC cable, the idea of an all-buried or underground subscriber distribution system was promoted based on the presumed waterproofness of plastic insulation.

Cable utilization practices for all-buried plant required new installation methods and machinery, new hardware, and new equipment and careful coordination with other utilities, in particular, power distribution that joined the parade to the buried plant. Environmental pressures, beautification, and freedom from storm interruptions were among the driving forces for change. Aerial drops disappeared and became buried service wire instead, for example.

Air-core PIC cable buried directly in the earth found a very harsh environment due to exposure to all kinds of digging (human and animal), installation cuts and bruises, and lightning pinholing. A disastrous flood of water inside the cables and much transmission trouble and maintenance expense resulted.

Waterproof Plastic Insulated Cable

The eventual solution was waterproof cable, in which the PIC core was filled with a special compound of petroleum jelly blended with polyethylene. The polyethylene reduced the greasiness and the tendency of the compound to run out of the cables at elevated temperatures. Compound-filled cable promptly was dubbed "icky PIC," suggested by D. J. Thomson of AT&T Bell Labs. The name quickly swept the communications cable industry. Waterproof cable eventually captured close to 90% of the buried cable market.

Intercity Cable

Intercity cable ties cities and towns together and into the telephone network. Much of the intercity cable is aerial and many miles of such plant can be seen on pole lines along highways today. Intercity cable frequently is loaded at 6000-ft intervals. Pole-mounted loading coil cases can be seen along many country highways (see Fig. 21).

N Carrier — First Short-Haul Carrier

A popular intercity carrier cable was Type N, the first carrier system (12 channels) designed for short-haul usage. A high-performance 19-ga PIC cable having 66 nanofarads (nF) per mile was introduced in the early 1950s to serve this need. A characteristic repeater housing can be seen along N-carrier routes in one of the early applications of vacuum tubes in an unattended outdoor environment. Figure 22 shows an early repeater housing for vacuum-tube-type N-carrier lines. The cupola on top of the housing provides ventilation to avoid overheating in the summer and keeps the snow out in the winter.

FIG. 21   Many pole-mounted loading coil cases may be seen along roadsides. The coils increase the "reach" of cables in longer subscriber loops and in intercity trunks by adding inductance in series with each side of the pair.

Digital Carrier

The first digital carrier was introduced as the 24-channel T1 system in the early 1960s. Many intercity routes are provided by T1 carrier in both aerial and buried plant. A typical T1 repeater housing of this era is shown in Fig. 23. Painted white and pole mounted at waist height, the T1 housings are easily recognizable. A special feature of T1 carrier is its applicability for use on existing cable, generally 22 ga. T1 carrier initiated one of the early steps in "mining" increased channel capacity from existing voice frequency cable via electronics, and set a trend in motion that has accelerated with time.

FIG. 22 N-carrier repeater housing for the first use of vacuum tubes in an outside environment. The cupola on top lets air circulate to cool the housing and keeps the snow out in winter. (From Ref. 10, p. 399. Reprinted with permission. Copyright © 1953.)

Screened Cable

Single cable operation of T1 is feasible when the cable core is divided into two groups. T-screen cable includes a diametrical aluminum tape or a variety of arrangements that separate the core into two compartments, one for eastbound and the other for westbound transmission. One example of the diametral corrugated aluminum shield is shown in Fig. 24. Note that the shield provides two compartments for pairs.

Single cable operation, on the other hand, is common in larger unitized pulp trunk cable. Opposite directions of T1 transmission are placed in units on

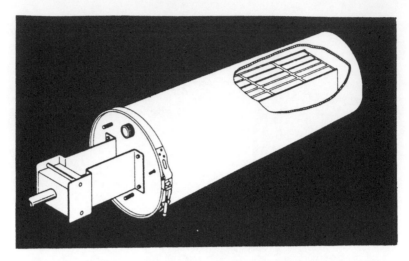

FIG. 23   T1 carrier housing, an early design of characteristic appearance. Painted white and pole mounted at waist level, they are often seen along roadsides. T1 is a digital system and continues to be the most popular and widely used of all carrier systems. (From Ref. 11, p. 113. Reprinted with permission. Copyright © 1971.)

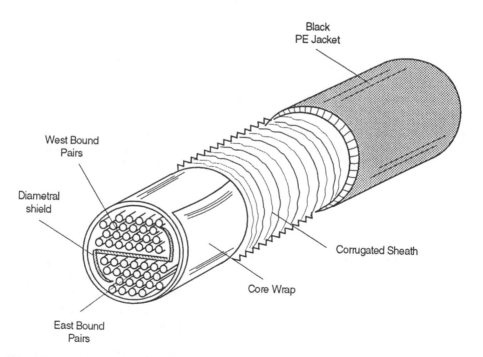

FIG. 24   T-screen cable having a diametrical shield. The two cable compartments permit single cable operation of T1 carrier; otherwise, separate cables would be required for West–East and East–West transmission.

opposite sides of a multiunit cable. The intervening units provide enough shielding to control crosstalk.

Metropolitan-Area Trunk Cable

A special low-loss T1 cable was developed for metropolitan-area trunk (MAT) applications. Features include 25-ga conductors insulated with foamed plastic covered with thin skin (foam-skin) and having a capacitance of 60-nF per mile. The cable cores are unitized and include a special moisture sensor as a part of the sheath maintenance system.

Special Cable Types

Other short-haul digital carrier systems were developed, including special cable. One of these was a 96-channel T2 carrier with a special LoCap (low capacitance) cable (22 ga, 38 nF per mile). Another was a T1C system (48 channels) with ICOT (InterCity and Out-of-state T carrier) cable (60 nF per mile for filled cable, 24 ga). Although both were successful as development projects, field usage was not extensive. Intercity needs also have been provided by short-haul digital microwave radio and more recently with glass fiber lightwave cable systems.

Long-Distance Network

As discussed in the history section of this article, the long-distance cable network evolved from 19-ga, 62-nF-per-mile, long-pair-twist, quadded voice-frequency toll cable, the first with lumped loading on pairs and phantoms. The quad construction was unusual in that each pair first was paralleled as nontwisted pairs and bound with a light cotton thread to attain equality of wire lengths. To assure minimum difference in wire diameters, two adjacent lengths of bare wire from a supply reel of wire made up each pair. Also, the wires of a pair were insulated with adjacent ribbons from a given roll of paper, to reduce differences in paper thickness. Two such pairs then were given pair and phantom twists in a special single-operation machine. The long-pair twists facilitated higher throughput during twisting. The heyday of these cables was in the 1920s.

K-Carrier Cable

K carrier (12 channels) on 62 nF per mile, short-pair-twist, quadded toll cable began in the 1930s. The designation was simply *K cable*. Special high-speed machines were developed for twisting four wires into K quads in one operation. K carrier systems provided about one-third of the total Bell System voice circuit mileage by 1947. By 1950, the mileage amounted to about 7 million voice circuit miles, equivalent to roughly 10,000 miles of K cable.

K cable was the first and only paired (or quadded) cable to go fully transcontinental in the United States, on a route not far from the old Lincoln Highway (U.S. 30). The earlier voice-frequency, long-pair-twist quads made it only as far as Chicago. K cable was the first paired or quadded cable to be used for multichannel carrier, and was a highly successful development. Phase out finally took place in the 1970s after a cycle life of more than 40 years.

Coaxial Cable

Coaxial cable was tested in field trials before World War II. After the war, many miles of multicoaxial toll cable were installed at roughly 1000 cable miles per year. Four transcontinental coaxial cable runs and many interconnecting coaxials provided the long-distance links coast to coast and state to state. Coaxial cable provided the first television link from New York to Chicago and was the scapegoat for any studio trouble. "It's trouble on one of the pipes!" was the usual explanation.

Microwave radio relay also joined forces with coaxial cable to meet the growing demand for more long-distance telephone capacity. The march of technology brought satellite telephone links. And the newest arrival on the scene has been glass fiber lightwave cable systems, with an attendant explosion in channel capacity.

Ocean Cable

Demand for intercontinental telephone connections led to development of ocean telephone cable to supplement a very limited and unreliable channel capacity via transoceanic radio. The first transatlantic coaxial cable, designated TAT-1, spanned the North Atlantic in a West–East and East–West crossing in 1955 and 1956 (Fig. 25). Many ocean coaxial cables followed to meet the exploding demand for intercontinental connections. Growth in demand has continued unabated at about 27% per year, more than doubling every 3 years.

The first optical ocean telephone cable, designated TAT-8, brought a 1000-to-1 increase in channel capacity, compared with the highest capacity ocean coaxial cable, TAT-7.

Local Cable and Wiring

Special cables and wires have been developed to meet the particular requirements for local plant near or at customer and subscriber premises, including central offices, apartment buildings, office distribution computer cables, apparatus case cables, and many more. Interesting and important designs and applications are covered briefly below to complete the overview of the communication network.

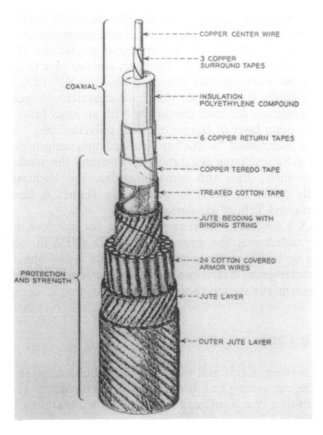

COAXIAL

COPPER CENTER WIRE

3 COPPER SURROUND TAPES

INSULATION POLYETHYLENE COMPOUND

6 COPPER RETURN TAPES

COPPER TEREDO TAPE

TREATED COTTON TAPE

JUTE BEDDING WITH BINDING STRING

PROTECTION AND STRENGTH

24 COTTON COVERED ARMOR WIRES

JUTE LAYER

OUTER JUTE LAYER

FIG. 25 Sketch of the cable used for TAT-1, the first telephone cable to cross the Atlantic Ocean. All earlier cables were for telegraph signals only and were incapable of voice transmission. (Reprinted with permission from American Telephone and Telegraph [AT&T], Ref. 12, p. 181.)

Inside-Wiring Cable

The problem of providing the wiring for telephones inside multistory office buildings and multifamily apartment houses has been met by a type of cable called simply, inside-wiring cable. Chief among the cable characteristics is the flame-retardant, color-coded insulation and flame-retardant plastic jackets. The plastic of choice in early cables has been polyvinyl chloride (PVC). Coded as Type D cable, the acronym DIW (D inside wire) has been used. When burned, PVC releases chlorine, forms hydrochloric acid, and may corrode metals (e.g., central-office switches) severely. More recently, fluoropolymer insulations have found inside-wire applications because of less toxic by-products in case of fire.

Local-Area Network Cable

Linking of desktop PCs (personal computers) and computer-driven worksta-tions in local-area networks (LANs) has evolved as a new challenge for cable and wire. Early systems employed coaxial cable. Shielded multipairs came next. These cables are costly and difficult to install and interconnect.

Nonshielded multipairs having 100-megabits-per-second (Mb/s) capability have been developed, are much less expensive, and are easy to install. An example is an AT&T Network Systems' design, SYSTIMAX™ PDS LAN Cable (see Fig. 26). PDS stands for premises distribution system. A general-use design employs dual insulation for each conductor with an inner layer of polyethylene covered with an outer layer of flame-retardant polyethylene.

An additive nicknamed "rocks" (pulverized magnesium or aluminum hydroxide rock) achieves flame resistance by stopping the tendency of burning polyethylene to melt, drip, and spread the fire. The hydroxide decomposes endothermically (takes up heat) to quench the flame. A design for plenum applications employs Teflon® insulation.

SYSTIMAX LAN cables meet requirements for low loss, low capacitance, and resistance unbalance, low near-end crosstalk (NEXT), and support 100-Mb/s links for up to 100-meter lengths. Impedance is 100 ohms and EMI (electromagnetic interference) effects are reduced. The superior-grade dielectric materials assure essentially square-root frequency loss, that is, copper loss only.

Interface or Tip Cable

The interface between inside and outside plant cables occurs in what is called a *tip cable* for the connecting link between an outside cable in the entrance vault in the basement of a central-office building and the main distributing frame on

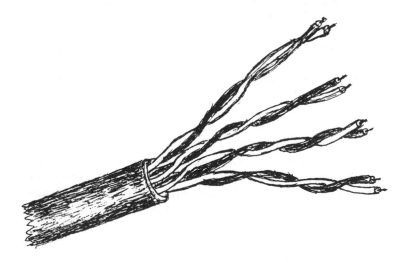

FIG. 26   A high-performance cable, SYSTIMAX™ PDS LAN Cable provides a 100-Mb/s connection to desktop computers and computer-driven workstations in business and engineering offices. (Adapted from Ref. 13.)

an upper floor. Early tip cables employed lead-covered 22- or 24-ga tinned copper wire cores, insulated with silk or acetate and cotton servings. Lacquer coatings reduced the sensitivity of insulation resistance to high humidity.

In the late 1960s, the textile insulation on tip cables was replaced with a dual insulation of polyethylene plus a 4-mil skin of colored PVC for fire retardance. The capacitance of the dual-insulated cables was the standard 0.083, reduced from 0.1 μF per mile with the textile insulation. Also, crosstalk was improved to permit runs as long as 500 ft. A PVC jacket bonded to an inner corrugated 9-mil aluminum tape was developed to replace the older lead sheath. The new sheath was given the acronym of ALVYN (aluminum-vinyl sheath).

Apparatus or Stub Cables

Stub cables serve as interfaces between regular outside cables and apparatus. Examples include stub cables for loading coil cases, repeater housings, and cross-connect terminals. In general, stub cables accomplish the insertion of some device, say a loading coil, in series with a cable pair and hence require "in" and "out" pairs, sometimes associated with each other as quads. These cables are specific and are matched to the problem at hand.

Central-Office Cable

Before the advent of electronic switching, the equipment in central offices was interconnected with color-coded and unitized, PVC-insulated, tinned copper wire cables. In general, these cable conductors were soldered to terminal blocks, hence the use of tinned wires. Early cables employed helically applied silk or acetate and cotton yarn servings, with lacquer over all. Color coding was achieved by use of yarn groups of different colors, giving the characteristic "barber pole" stripe.

Extruded PVC insulation eventually replaced the helical textile servings. The relatively complex color coding standardized with the textile insulation was simplified by adopting a code based on one or two spaced ink bands to identify tip and ring of a pair. Five colors (blue, orange, green, brown, and slate) were standardized for the PVC conductor insulation. Four band colors (white, red, black, and yellow) distinguished the four color groups employed in the color code for the standardized 20-pair units.

Most of the mechanical robustness and cut-through resistance of the old textile-insulated wires is attained by cross-linking the PVC insulation with high-energy electron bombardment. Such wire achieves enhanced high-temperature cut-through resistance during soldering and may be employed for use in the more physically stressful situations.

Station Wires

Station wire provides the connection between the terminal and protector block at the entry point of a customer's premises and the outlet or terminal for plug in of the line cord to the telephone. Typically, station wire is a spiral-four or star quad of 22-ga copper, with PVC insulation and jacket for flame retardance. Conductor insulation is color coded red, green, black, and yellow for the four wires.

Drop Wire

Self-supporting drop wires for aerial applications evolved as a pair of insulated copper-plated steel wires. Crude insulation, rubber and painted cotton servings, were common in early days, and provided a continuing maintenance problem as the wires aged in the outdoor environment.

A neoprene-jacketed parallel pair insulated with synthetic rubber came next. A helical serving of cotton cord over the rubber provided added tensile strength. The neoprene-jacketed design, the so-called C drop wire, reduced maintenance costs to the vanishing point and was a very successful development. The neoprene design eventually was replaced by a drop wire consisting of a single extrusion of a special PVC plastic over two copper–steel wires, lamp-cord style.

Buried Service Wire

For customer drops from buried cable, a jacketed star quad plus aluminum tape shield and an outer jacket may be employed. Where rock crushing or rodent damage is a problem, the aluminum shielding is replaced with bronze or stainless steel tape. Compound-filled versions of four conductor service wires are available for use with waterproof cable installations.

Another "wire" structure sometimes used for relatively long (500 ft or more) service drops consists of a twin extrusion (like lamp cord) of polyethylene over two 19-ga wires protected with a helical bronze tape and an outer PVC jacket.

Special Wires

A large number of highly specialized single-insulated wires or pairs are found in use for a variety of special applications in the telephone network. Ground wires are a simple example — different types for specific problems.

Another example is a special pair that serves as a jumper between the main distributing frame and the input side of a central-office switch. When a customer moves, a new jumper pair is installed between the customer's new position on the main distributing frame to the original input side of the central-office switch to avoid a number change.

The list of wires that stitch the telephone system together includes all of the cable and wire described above and many more for miscellaneous applications beyond the scope of this section.

Cable Development Problems and Solutions

The secondary transmission line constants (attenuation phase constant β, phase velocity v, and characteristic impedance Z_o) are the tools of the transmission system designer and are discussed in the transmission line section. The performance of a given transmission system depends on the secondary line constants, their variation with cable length and among pairs or units in a cable, and the crosstalk between the pairs or units and the interference from external sources of noise.

The primary electrical constants determine the values of the secondary constants. Each of the primary constants R, L, G, and C relate to the physical, dimensional, and electrical properties of the line, that is, the installed cable. The crosstalk and interference also relate physical, dimensional, and electrical properties of the pairs or coaxials in a cable. The primary constants are the tools cable designers employ to define and characterize cable performance.

In this section, we show the relationship between the primary constants and cable dimensions and material properties. Insights into the deeper concepts are explored to uncover the unity among the means of guiding the propagation of electromagnetic waves for communication over metallic media.

Coaxial Resistance

The high-frequency resistance of a single conductor having a cylindrical coaxial return conductor rises above the DC resistance value for the conductor when a phenomenon called *skin effect* comes into action. At DC or zero frequency, the electrical current is distributed uniformly across the cross-section of the conductor and the resistance has its lowest value. As the signal frequency is increased, the portion of the electromagnetic wave traveling in the metal of the conductor is reduced and the wave flows increasingly on the surface of the conductor.

Correspondingly, the return current flows increasingly on the inside of the return conductor. For the theoretically perfect conductor, the electromagnetic wave flows exclusively in the dielectric space between the outer surface of the inner and the inner surface of the outer conductors for coaxial conductors. The cutaway of the inner and outer conductors shown in Fig. 27 illustrates the decay of current at depths below the respective inner and outer conductor surfaces.

The resistance R of the coaxial conductors versus frequency is calculated from a solution of Maxwell's equations. Since these solutions are complex and not well adapted to use in everyday work, approximations generally are used in design work. An approximation applicable at high frequencies illustrates the insights we wish to discuss:

$$R\ (\text{inner}) = \frac{0.00401}{d}\ \sqrt{\rho\mu f} + \frac{0.00794\rho}{d^2}, \tag{16}$$

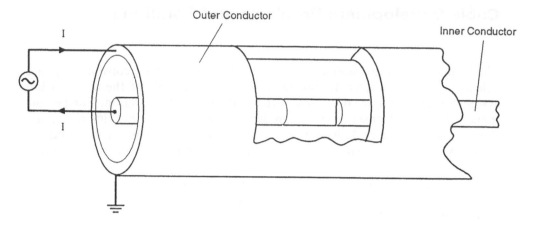

FIG. 27 Cut-away of coaxial showing decay of current in conductors due to skin effect, maximum values on inner surfaces.

$$R \text{ (outer)} = \frac{0.00401}{D} \sqrt{\rho\mu f} - \frac{0.00794\rho}{D} \left(\frac{3}{D_1} + \frac{1}{D} \right), \qquad (17)$$

$$R = R \text{ (inner)} + R \text{ (outer)} \qquad \text{ohms per mile.}$$

where
$\quad d$ = Diameter of central conductor, in
$\quad D$ = Inside diameter (ID) of outer conductor, in
$\quad D_1$ = Outside diameter (OD) of outer conductor, in
$\quad \mu$ = Permeability (equals 1 for a vacuum)
$\quad \rho$ = Resistivity of conductors, microohm centimeters (cm)
$\quad f$ = frequency in Hertz.

The communications coaxial mentioned in the section on cable types has typical dimensions of $d = 0.1004$ in, $D = 0.371$ in, $D_1 = 0.395$ in, with polyethylene disk insulation 0.085-in thick and 1-in spacing center to center. For copper having 1.7315 microohm cm at 70°F, a typical value for R (inner plus outer) is 53.9 + 14.1 = 58.0 ohms per mile at a frequency of 1 megahertz (MHz). The DC resistance is 5.94 ohms per mile, showing that the skin effect at 1 MHz has increased R by almost 10 to 1.

Curvature Effects

The term $0.00794/d^2$ for the inner conductor is equal to one-quarter of the central conductor DC resistance. This term is a curvature correction to take account of the fact that current flowing in the layers below the outer surface encounters less space, which causes an increase in resistance. Correspondingly, the similar term for the outer conductor is one-quarter of the DC resistance of the outer conductor. Here, the current below the inner surface of the outer conductor encounters more space, and the curvature correction is subtracted to reduce the resistance.

The x *Tables*

An exact solution for the resistance of a single cylindrical conductor having a coaxial return has been tabulated as a function of parameter x where

$$x = 0.3569d \sqrt{\mu f/r} \tag{18}$$

Tables of values of the ratio $R/(R\,DC)$ are given in handbooks (e.g., those for chemistry and physics, also Ref. 13). For a coaxial having the dimensions given in the preceding section, $x = 27.29$ at 1 MHz and the ratio $R/(R\,DC) = 9.90$. For the 0.1004-in inner conductor, the DC resistance is 5.456, which gives R (inner) $= 9.90 \times 5.456 = 54.01$, versus 53.88 by the approximation above.

Pair Resistance

The resistance of a pair of conductors inside a cylindrical metal shield is calculable by formulas derived from Maxwell's equations. The formulas are too complex to be included here. For a pair as shown in Fig. 28 driven by a balanced source of voltage at a given frequency, the approximate resistance is the sum of that of the conductors plus the resistance of the outer shield as though it had coaxial return.

At communications frequencies, the pair induces currents in the shield that result in losses that must be supplied by the signal source. Stated another way, the electromagnetic signal is guided by the two conductors and the outer shield. Most of the energy flows in the dielectric space. The portion of the wave that flows in the two conductors and the shield causes the resistance.

For a balanced pair as in Fig. 28, the skin effect discussed for coaxial conductors also applies to the shielded balanced pair. Another effect, called *proximity effect*, increases the resistance above that calculated for skin effect

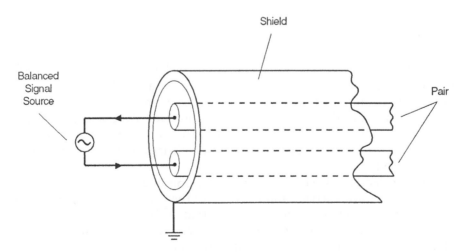

FIG. 28 Shielded balanced pair having a balanced signal source.

only. The current in the conductors follows the path of minimum impedance. As a result, the current tends to flow in the nearest faces of the go-and-return conductors.

The same proximity effect causes the induced currents to concentrate on the inside of the cylindrical shield in the regions nearest the two conductors. For a typical shielded pair at frequencies in the low MHz range, proximity increases the expected resistance of the conductors (based on skin effect only) by about 1% or less. The effective resistance of the outer shield is roughly doubled. The net result is that the high-frequency resistance of a shielded balanced pair is about 6% above the skin-effect value for electrically thick shields.

Balanced pairs sometimes are constructed by folding an aluminum-coated sheet of Mylar® directly over a twisted pair of insulated wires. For typical fractional mil thickness of the aluminum layer, the surface resistance may be 10 times or more than an electrically thick layer and thus add substantially to the total resistance (and attenuation) of the pair. In the 1- to 10- or 20-MHz range, the total pair resistance might be double that of a pair having optimum dimensions and an electrically thick shield of copper.

Multipair Resistance

For multipair cable, each pair is surrounded by similar pairs that may be visualized as forming a "cage shield" around a given pair. Comparing such a pair with the shielded balanced pair discussed above, skin effects, proximity effects, and shield losses would be expected.

A mathematical description of the cage shield has defied analysis. Based on measurements, the high-frequency resistance of pairs in a multipair cable is between 25% and 40% higher than the skin-effect value for two conductors. The implication is that the cage shield has an effective resistivity substantially higher than a shield of solid copper, for which the excess resistance is only about 6%.

Skin Depth

Currents flowing in coaxial conductors have their highest values at the surface and decrease exponentially in a radial direction inward. The decay obeys the transmission line rule, $P = \alpha + j\beta$, for propagation in the radial direction, but with a difference. For the diffusion of the current into a conductor, the attenuation α in nepers and the phase constant β in radians are equal numerically. At a depth where $\beta\ell = \pi/2$, the current at that depth lags 90° behind the surface current (Fig. 29).

At greater depths, the current flows in the opposite direction and subtracts from the net forward current, and thus causes the resistance to increase. For a tubular conductor having a thickness such that $\beta\ell$ is equal to $\pi/2$ radians, a minimum resistance is observed, about 9% lower resistance than a thinner or much thicker wall. This point is called the optimum thickness. When a tube has a thickness such that $\beta\ell$ is equal to 1 radian, the high-frequency resistance of

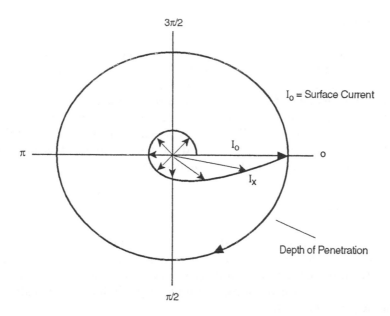

FIG. 29   Polar plot showing exponential decay and phase shift as current diffuses into a conductor. Current concentration on the surface is called skin effect.

the tube is approximately equal to its DC resistance and the thickness is called 1 skin depth. Skin depth is calculated as

$$\text{Skin depth} = 1.98 \sqrt{\rho/\mu f} \qquad \text{in,} \qquad (19)$$

where ρ is resistivity in microohm cm and f is frequency in cycles. For copper and a frequency of 1 MHz, skin depth is equal to 0.0026 in.

Ohms per Square Concept

A useful concept is that of the ohms per square. For conductive sheets of metal or resistance material, the resistance between opposite sides of a square is independent of the size of the square. Thus, for a tubular conductor of copper, a square is formed by unfolding a length of tubing equal to the circumference πD. Calculating the resistance of a length πD of copper tubing of diameter D from the formula for the resistance of a coaxial outer conductor gives $R = 0.26$ milliohms per square at 1 MHz.

The ohms per square concept permits estimating the high-frequency resistance of any conductor by simply converting the D length to one mile by the ratio $12 \times 5280/\pi D$. Doing this for the 0.1004-in coaxial central conductor gives R (inner) $= 52.2$ versus 53.9 by the formula. Other frequencies are calculated by applying the square root of the frequency ratio.

Design of Resistors

The ohms per square concept provides insight into how to make resistors having resistance independent of frequency. Layers having a thickness of 1 skin depth or less have an alternating-current-to-direct-current (AC/DC) resistance ratio equal to 1, independent of frequency, for lower frequencies including DC. For a resistivity 100 times that of copper, skin depths would be 0.00026 in or 0.26 mil and a resistance of 2.6 milliohms per square. In integrated-circuit chips, resistivity, thickness, and line width lead to a wide range of adjustability of resistance for resistors in microdimensional space.

Coaxial Inductance

The inductance of a pair of coaxial conductors is made up of the space inductance for a wave traveling in the dielectric space and the internal inductances for the portion traveling in the conductors. For dimensions as defined for the example of resistance calculations, the unit length inductance is calculated as (note that log is to base 10):

$$L \text{ (space)} = 0.7411 \log (D/d), \tag{20}$$

$$L \text{ (inner)} = \frac{0.638}{d} \sqrt{\frac{\mu f}{\rho}}, \tag{21}$$

$$L \text{ (outer)} = \frac{0.638}{D} \sqrt{\frac{\mu f}{\rho}}, \tag{22}$$

$$L = L \text{ (space)} + L \text{ (inner)} + L \text{ (outer) millihenrys per mile.}$$

The formulas for L (inner) and L (outer) are approximations suitable for frequencies above 100 kHz. For the communications coaxial having the dimensions used for the resistance calculations above, calculated values of space inductance and internal inductances of inner and outer conductors at 1 MHz are given by

$$L = 0.4207 + 0.00836 + 0.00226$$

$$= 0.4313 \text{ millihenrys per mile.}$$

At 1 MHz, the total internal inductance is 0.0106 out of a total of 0.4313 millihenrys per mile, showing that $1 - (0.0106/0.4313) = 97.5\%$ of the wave travels in the dielectric space.

Shielded Pair Inductance

The inductance of a shielded pair such as that discussed above for resistance is roughly equal to the inductance of two coaxials driven in balanced mode. For this arrangement, the space inductance would be twice that of one coaxial and correspondingly for the internal inductance. The space inductance for a shielded pair is

$$L \text{ (space)} = 1.482 \log \left(\frac{2S}{d} \times \frac{D^2 - S^2}{D^2 + S^2} \right) \quad \text{millihenrys per mile,}$$

where

D = Inside diameter of shield, in
S = Interaxial spacing of conductors, in
d = Conductor diameter, in.

The formulas for internal inductance of the conductors and the shield are too complex to be included here. The internal inductance of the conductors is roughly two times that of one conductor with a coaxial return and that of the shield is roughly equal to that of the two conductors. Those approximate values apply when the pair is close to optimum, discussed below. For a typical balanced pair, about 97% of the energy flows in the dielectric space, versus 98% for the coaxial.

Multipair Cable Inductance

For multipair cable, the formulas for the inductance of a shielded balanced pair given above may be applied, the shield in this case being the "cage" of surrounding conductors. Study of capacitance measurements discussed below support the conclusion that the effective inside diameter of the cage is roughly 2.5 times S, the interaxial conductor spacing.

Interaxial separation is subject to precise electrical determination via low-frequency inductance measurements in which the shield is electrically "transparent." At frequencies of about 1 kHz and below, inductance is constant and is given as

$$L = 1.482 \log (2S/d) + 2L \text{ (inner) millihenrys/mile.} \tag{23}$$

The value of S can be determined quite precisely from the measured low-frequency L and the known conductor size. L (inner) is constant at low frequencies and is equal to 0.0805 millihenrys per mile.

Based on high-frequency measurements of total pair inductance and subtracting the space inductance calculated with the balanced-pair formula, the cage is found to have an internal inductance roughly equal to that of the two conductors of the pair. The internal inductance of the cage shield thus is much higher than the inductance of a solid copper cylinder, possibly due to the coil effect of the helical stranding of cable components. Alternatively, the effective diameter may be a function of frequency.

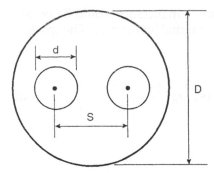

FIG. 30   Cross-section of balanced pair showing dimensional notations.

Coaxial Capacitance

Following the dimensions noted above, the capacitance of coaxial conductors is

$$C = \frac{38.89\epsilon}{\log \dfrac{D}{d}} \quad \text{nF per mile.} \tag{24}$$

The dielectric constant ϵ is calculable for disk insulation, employed in communications coaxial to reduce the net dielectric constant. For a disk thickness a, a disk center-to-center spacing b, and a dielectric constant k for the disk material,

$$\epsilon = 1 + a(k - 1)/b.$$

This expression applies to flat disks that fit the inner and outer conductors. If slots are used, or if the disks are smaller than the inside diameter of the outer conductor, adjustments are required. The dielectric constant also is calculable from measured capacitance C and space inductance L (space):

$$\epsilon = 0.03470 \, C \times L \text{ (space)},$$

where C is in nF per mile and L (space) is in millihenrys per mile. Since the internal inductance of the conductors is calculable, L (space) is readily determined by a measurement of the total inductance and subtracting out the internal inductances.

Shielded Pair Capacitance

For shielded balanced pairs having the dimension S for interaxial spacing of the two conductors, d for conductor diameter, and D for inside diameter of the shield (see Fig. 30), a good approximate capacitance is

$$C = 19.44\epsilon/\log \left(\frac{2S}{d} \frac{D^2 - S^2}{D^2 + S^2}\right) \quad \text{nF per mile.} \quad (25)$$

Multipair Cable Capacitance

For multipair cable, capacitance may be estimated by use of the formula given above for a shielded balanced pair. The interaxial conductor spacing is determined approximately by the physical geometry. For example, with solid insulation over each conductor, S is equal to the diameter over the dielectric dimension (DOD) of the insulation, plus an allowance for any gap between the insulated conductors. The gap may be as much as 5 mils for 19-ga conductors and essentially zero for 26 ga. The gap also varies with twist length, being larger for the longer twists. A precise determination of S may be accomplished by measurement of low-frequency inductance as described above.

Cage Diameter

An estimate of the cage diameter follows from Fig. 31, showing an insulated pair of conductors surrounded by the conductors of adjacent pairs, all having plastic insulation, so that dimensions are known. Examination of the cable cross-section supports the arrangement shown in the sketch. If much shorter pair twists were employed, a space around each pair would be expected. The average diameter of the cage is estimated to be 2.5 DOD.

The calculation of C for a pair in a multipair cable requires knowledge of the value of the dielectric constant. Since cables generally are not filled uniformly with a material having a known dielectric constant, ϵ must be measured.

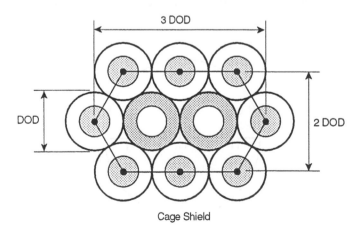

FIG. 31 Pair in multipair cable showing the equivalent "cage" shield made up of the conductors of surrounding pairs.

Direct Capacitance

Here, we introduce a useful concept, direct capacitance (see Fig. 32). The conductors 1 and 2, inside a shield 3, have three direct capacitances, C_{12}, C_{13}, and C_{23}. Mutual capacitance C is that seen by a balanced source driving the balanced pair, and is defined as

$$C = C_{12} + \frac{C_{13} \times C_{23}}{C_{13} + C_{23}} \qquad \text{nF per mile.} \qquad (26)$$

A clever method determines ϵ with good accuracy, based on the following relationships:

$$C = f(S, D, d, \epsilon), \qquad (27)$$

$$C_{13} = C_{23} = F(S, D, d, \epsilon), \qquad (28)$$

where f and F are two known functions. The function f is that given above for capacitance C, (Eq. [25]). Separation S is measured by the low-frequency inductance method. Conductor diameter d is known. C and C_{13} are measured for the cable in question. Thus, we have two equations from which we solve for the two unknowns, D and ϵ. The value of ϵ is in the range of 1.85 for low-density, polyethylene-insulated, air-core cable. Details of the method are described in Ref. 15.

Conductance

The conductance of coaxials and pairs is determined from the definition of the power factor of insulating materials:

$$\text{Power factor (PF)} = G/(\omega C) \qquad \text{(Mho, F),}$$

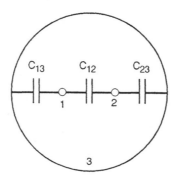

FIG. 32   Cross-section of shielded pair showing the three direct capacitances. Mutual capacitance is defined in the text.

$$G = 0.006283\,(\text{PF})\,fC \qquad \text{micromho per mile,} \qquad (30)$$

where f is in hertz and C is nF per mile. This formula applies to coaxials, shielded pairs, and multipairs.

Power Factor

In general, the power factor of the basic materials is known. If a composite of several insulating materials is involved, direct measurement of conductance may be required if the power factor cannot be rationalized from the properties of the insulating materials and the dimensional proportions. The power factor of insulating material like polyethylene is very low, and may approach 0.00002, as for the best polyethylene for ocean cable. Pigments such as those for color and stabilizers to protect the polyethylene from oxidative degradation increase losses.

The power factor for commercial insulating compounds like polyethylene and polypropylene is likely to be in the 0.0002 range. The conductance contributes about 0.9% to the total attenuation of the polyethylene-disk insulated 0.375 coaxial cable at 100 MHz. With solid insulation, the loss might be 9% or 10%.

Optimum Dimensions

The attenuation of a coaxial having a given size of outer conductor depends on the diameter of the inner conductor. An optimum conductor size or ratio of D/d leads to a minimum attenuation. The preferred D/d ratio is 3.61 for inner and outer conductors having the same resistivity. If the outer conductor has a higher resistivity, the preferred D/d ratio is increased. For transmission of maximum power, the preferred D/d ratio is 1.65.

The same idea applies to shielded balanced-pair structures. Minimum attenuation is attained when $S/D = 0.44$ and $D/d = 5.6$. Other optima exist. For example, the capacitance of a shielded balanced pair has a minimum value when $S/D = 0.485$. For the estimated cage diameter and $S/D = 0.4$, the resulting multipair capacitance is about 4% higher than the minimum.

Minimum-Cost Cable

Another optimum of interest for multipair cable is the minimum cost for a given attenuation. Holding the number of pairs constant at 50, the 3.15-MHz attenuation at 22 dB per mile, and using foamed polyethylene insulation, the curve in Fig. 33 shows the relative cable cost versus ga. Cable cost shows a distinct minimum at 22 ga and 38 nF per mile. Relative cable cost shows about a 6-to-1 increase between 38 and 83 nF per mile.

The minimum-cost idea is simply a tradeoff between copper costs and polyethylene costs, sheath included. The 83 capacitance with 15.3-ga cable would be about 2 times larger than its 30-nF, 22-ga equivalent and uses 4.5 times more

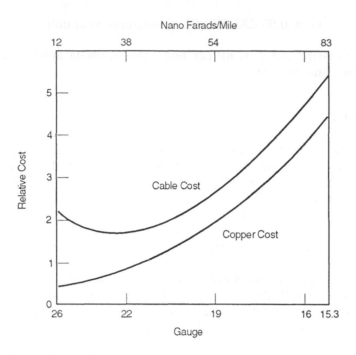

FIG. 33   Plot showing relative cable cost versus gauge and capacitance. Attenuation is held constant at 22 dB/mile at 3.15 MHz. Minimum cost is seen to occur at a gauge size of 22, 38 nF per mile.

copper. The optimum cost with Teflon insulation would occur with a capacitance of about 83 nF per mile since Teflon is more costly than polyethylene. The idea of an optimum cost for PIC-type cable was proposed by A. S. Windeler of AT&T Bell Laboratories.

Early practitioners were faced with a different set of conditions in regard to optimum cost versus gauge and capacitance. Paper insulation always has been relatively low cost, much less than polyethylene on a unit volume basis, costing roughly 5 cents per pound versus a range of around 20 cents per pound for polyethylene; however, lead sheath costs would quickly stop any large move toward low capacitance.

Multipair Crosstalk

Metallic communications media generally are subject to some degree of pickup of signals from outside sources. The challenge is to determine the tolerable levels of extraneous signal or noise and to design the medium to meet the objectives.

In telephone technology, pickup from adjacent systems in the same cable is called *crosstalk* and pickup from outside sources is called *noise* or *interference*; both noise and crosstalk also are called *electromagnetic interference* (EMI). The phenomena responsible for crosstalk are electromagnetic in nature.

In multipair cable, control of crosstalk in the cable design and manufacturing stage is facilitated by considering the components, electrostatic (ES) crosstalk and electromagnetic (EM) crosstalk. ES crosstalk is that caused by capacitive coupling between two adjacent pairs and EM by magnetic coupling or simply "transformer" coupling.

Components of Crosstalk

The ES and EM components of crosstalk are illustrated in Fig. 34 for two adjacent, parallel, nontwisted or nontransposed lines in air. A disturbing source is shown as applying a signal voltage *E* to Disturbing Pair 1 and generating a crosstalk response in Disturbed Pair 2. Typical of the practical situation, the ends of both pairs are assumed terminated in the characteristic impedance Z_o.

For simplicity of illustration, the ES coupling is shown as a single capacitor *CU* concentrated at the midpoint for capacitance unbalance; this is defined below. The *CU* coupling supplies a disturbing current *I*(ES) that divides to produce current *I*(ES)/2 to the left and to the right, shown as the solid arrows. Two parallel rectangular sets of conductors couple by simple mutual inductance or transformer action to generate the *I*(EM) current that flows around the disturbing pair. By Lens' law, the direction of flow is opposite to that of the disturbing current *I*.

Total crosstalk is the sum of the two components. We see that at the near end, the *I*(ES) and *I*(EM) components add and, at the far end, the components subtract. In normal multipair cable, addition at the near end occurs about 80% of the time; within a star quad, the near-end addition occurs 100% of the time.

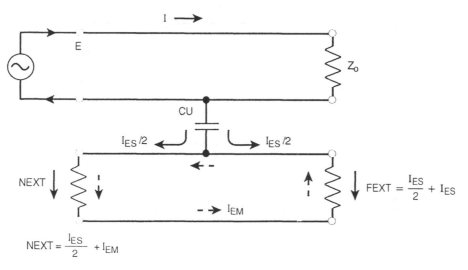

FIG. 34 Two parallel pairs illustrating that NEXT is the sum and FEXT is the difference of the two components of crosstalk (electrostatic and magnetic).

If the near end of a star quad is not terminated in Z_o, the reflected NEXT may masquerade as high FEXT (far-end crosstalk).

Measurement Methods

The simplified view of crosstalk components provides an insight for a means of measuring the two components on an electrically short length of cable ($\ell \ll \lambda/4$). The disturbing pair is energized as shown in Fig. 35. One end of the disturbing pair is left open circuited and the detector measures the total I(ES) at the opposite end. For I(EM), the disturbed pair is shorted at one end and the

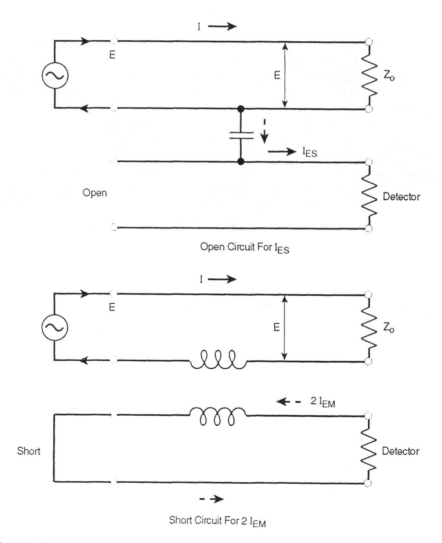

FIG. 35 Measurement of crosstalk components, open circuit for electrostatic and shorted for magnetic components.

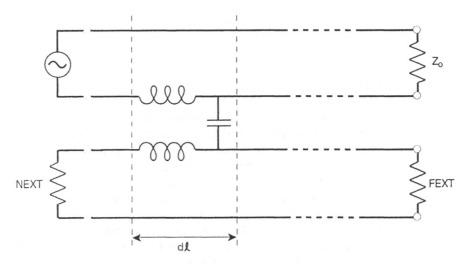

FIG. 36 Schematic of two pairs showing how the differential components of crosstalk are summed to obtain NEXT and FEXT.

detector is placed at the opposite end to measure $2 \times I$(EM), assuming the detector impedance matches the line. Since the test cable is electrically short, the detector may be placed at either end. Also, the open end reduces the I(EM) component to zero for the ES measurement. The short reduces the ES current to zero for the I(EM) measurement. Since capacitive currents lead by 90° and inductive currents lag by 90°, the two components either add directly or subtract.

Summation of Components

For a long-line calculation, the I(ES) and I(EM) components are assumed to apply over a differential length $d\ell$ (see Fig. 36). The capacitance unbalance CU becomes $d(CU)$. Inductive coupling is indicated as the mutual inductance coupling dm for the differential length $d\ell$. The total crosstalk is summed or integrated over the cable length.

The total current seen at the far end, generally called FEXT, when compared with the received current on the disturbing pair is called output–output FEXT. If the comparison is made between the input signal and the output crosstalk, the result is called input–output FEXT and would be a larger number of dB by the amount of the total dB attenuation for the cable length involved.

For NEXT, the summation along the length, encounters constructive and destructive interference, already discussed for transmission lines. Assuming uniform differential couplings, those one-quarter waves down the line will have 180° of phase shift compared with the near end and thus will cancel. As a result, NEXT shows deep dips (high dB values) versus frequency for long lengths at multiples of quarter wave. In normal multipair cable, the crosstalk couplings along length include some degree of randomness, with the result that the deep dips in NEXT usually are not distinct.

Capacitance Unbalance

Capacitance unbalance between two adjacent pairs is measured on multipairs in the factory to indicate design and manufacturing quality. The single coupling capacitor shown in the sketches for discussions of I(ES) and I(EM) is derived from the direct capacitances among the four conductors shown in a bridge form in Fig. 37 as follows. Note that the disturbing pair is Conductor 1 and 2 in the bridge and the disturbed pair is 3 and 4. Six additional direct capacitances are encountered, that is, one each between the conductors of the two pairs and four to ground for the four conductors. They are not shown because they do not cause pair-to-pair capacitance unbalance. Solving for the detector current of the bridge network shows that capacitance CU is

$$CU = (C_{13} + C_{24}) - (C_{14} + C_{23}) \tag{31}$$

The crosstalk component I(ES) is related to capacitance unbalance CU and I(EM) is related to mutual inductance M as:

$$I(\text{ES})/I = 0.125 \ Z_o\omega(CU) \tag{32}$$

$$I(\text{EM})/I = 0.5\omega M/Z_o \tag{33}$$

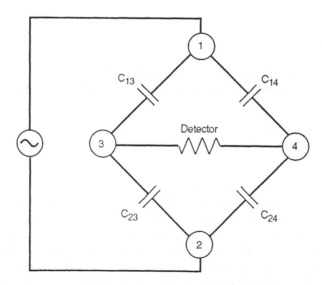

FIG. 37   Bridge arrangement of two pairs showing the four direct capacitances that cause capacitance unbalance when all are not equal. The definition of capacitance unbalance appears in the text.

where I is the disturbing current, CU is the capacitance unbalance in F, M is the mutual inductance in henrys, Z_o the characteristic impedance and ω is 2π (frequency).

Crosstalk — Factory Measurements

Crosstalk is measured on factory cable lengths on a sampling inspection basis. Test results on the individual values of crosstalk between the $n(n - 1)/2$ pair-to-pair combinations in a cable having n pairs generally are reduced to a single number for a given frequency and to a standard length of 1000 ft. The root-mean-square (rms) value of all possible cable pair-to-pair combinations was calculated in the days of analog carrier.

The average power sum (PS) crosstalk has come into use since the introduction of digital carrier systems. A popular cable type is the waterproof design with plastic insulation consisting of foamed polyethylene plus a skin of solid polyethylene, called foam/skin or DEPIC™. The interstitial air spaces among the conductors are filled with a suitable compound to exclude water. Typical values of output–output NEXT and FEXT measured on commercial cable are given in the table below.

Frequency (MHz)	Average PS FEXT (dB/1000 ft)		Average PS NEXT (dB/1000 ft)	
	22 Ga	24 Ga	22 Ga	24 Ga
0.150	65.6	65.9	60.7	61.3
0.772	52.2	52.4	51.1	51.7
1.60	45.9	46.2	46.7	47.0
3.15	40.0	40.4	43.1	43.3
6.30	34.0	34.1	39.0	39.1

Tertiary Crosstalk

Crosstalk in long cable systems is more complex than the views given above would indicate. One reason is that any given pair has ES and EM couplings to other circuits in the cable (sometimes called *tertiary circuits*). We wish to provide some insight into the nature of the problem.

Consider layer-type cable construction for simplicity. Certain manufacturing processes distort the geometry of the helix of a twisted pair in a fashion such that the pair may have a high degree of unbalance to other paths in the cable core. One such path is a circuit made up of the preceding pair P and the following pair F in a layer.

When the P and F unbalances are large and opposite in sign, their difference $(P - F)$ will be large (see Fig. 38). The capacitance unbalance from the pair to the P and F circuit is seen to be equal to the difference $(P - F)$. The unbalances P and F change sign when the helix of the pair is asymmetrical. The high unbalance causes a large crosstalk current to flow in the PF circuit. Because the attenuation of the PF path is low, a large disturbing current may be carried to the end of the system, where the normal received current is low and thus vulnerable to a strong disturber.

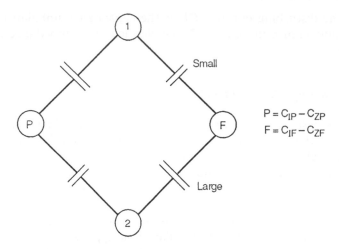

$$P = C_{IP} - C_{ZP}$$
$$F = C_{IF} - C_{ZF}$$

FIG. 38   Schematic showing the definition of P and F capacitance unbalance, and how inequalities lead to crosstalk to the P and F tertiary circuit.

The coupling is illustrated in Fig. 39, in which each pair and the tertiary circuit, as the PF path would be called, are shown as single lines. Another tertiary path consists of the inner layers as a group, with return on the outer layers as a group, also a low attenuation path.

Cause of Pair Distortion

The $P - F$ type of helix distortion may be caused by a longitudinal sliding or rolling force along a pair. An example would be an eyelet or guiding roller with a high tension on the pair. Figure 40 illustrates an exaggeration of the effect in which one conductor of a pair is assumed straight and the other is in a uniform helix around the first. When the helix is uniform, P and F have the same sign, so that $P - F = 0$.

The cause of the helix distortion is illustrated in Fig. 40. Before the sliding or rolling object under force f passes over the pair, the loops of the helix are

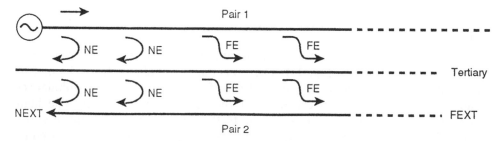

FIG. 39   Schematic showing how crosstalk to a third (tertiary) circuit may worsen crosstalk between two pairs.

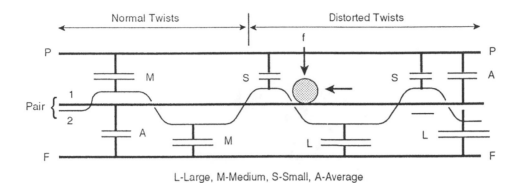

L-Large, M-Medium, S-Small, A-Average

FIG. 40   Generation of helix distortion by a guiding eyelet or roller, thus causing $P - F$ capacitance unbalance.

uniform, as shown on the left; afterward, they are short–long, and so on. On the left of Fig. 40, the $P - F$ difference is zero; on the right, the $P - F$ difference is large, showing that the distortion has caused the pair to be unbalanced to the tertiary circuit. Force f distorts the helix independent of whether the two conductors have unequal lengths as in Fig. 40.

Surface Transfer Impedance

Multicoaxials as employed in the toll telephone network are stranded in one or two layers as shown in Fig. 41, a photograph of the cross-section of a standard design. Crosstalk between any two adjacent coaxials depends upon the concept of surface transfer impedance.

Consider a single coaxial having a tubular outer conductor of solid metal. A current I of unit amplitude flowing into the inner and returning on the outer conductor generates a voltage E per unit length on the outer surface of the outer conductor. As described in the section on skin depth, the current is highest at the inner surface and decays in the outward radial direction. The ratio of voltage E to current I has the dimensions of impedance Z_{12}, the subscripts indicating a transfer from Inner surface 1 to Outer surface 2. The surface transfer impedance is calculated

$$Z_{12} = 2Z_{11}e^{-h} \tag{34}$$

$$Z_{11} = 0.00401 \sqrt{\rho\mu f}/D + j(0.638 \sqrt{\rho\mu f}/D) \tag{35}$$

$$h = 0.1252 \sqrt{\mu f/\rho}(D_1 - D) \tag{36}$$

The impedance Z_{11} is the surface self-impedance of the inner surface of the outer conductor. At high frequencies, h is large for copper outer conductors and the e^{-h} multiplier reduces Z_{12} to a very low value, so that coaxials are said

FIG. 41   Cross-section of 22 coaxial cable, COAX-22. (Reprinted with permission from Ref. 16, p. 207.)

to be self-shielding at high frequencies. These formulas apply when h is three or more skin depths.

At low frequencies, the outer conductor is relatively transparent. In the voice range, Z_{12} is equal to the DC value of Z_{11}, that is, the DC resistance of the outer conductor. Signals in the coaxial path readily couple with other coaxials. Impractically thick outer conductors would be required to prevent significant voice-frequency crosstalk.

Crosstalk between Coaxials

The mechanism of energy transfer between coaxials is illustrated in Fig. 42 for two parallel coaxials. The disturbing current I on the inside of the outer conductor generates a voltage on the outer surface of Coaxial 1 via the surface transfer impedance. The voltage on the outer surface of Coaxial 1 drives a current around the path made up of the two outer conductors. The impedance of the pair of outer conductors is

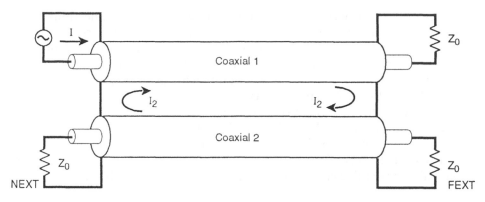

FIG. 42   Crosstalk between two coaxials. Current I into Coaxial 1 generates Current I_2 around the outer conductor path and then NEXT and FEXT in Coaxial 2.

$$Z = 2R \text{ (outer)} + j\omega(L \text{ (space)} + 2L \text{ (outer))} \tag{37}$$

and applies even when the two outers are in continuous contact. The process is reciprocal: the current I on the outer surface of Coaxial 2 generates a voltage on the inner surface of Coaxial 3 and causes NEXT and FEXT crosstalk. Since the coupling is uniform, NEXT has frequency-dependent, deep-destructive interferences as discussed for near-end multipair crosstalk. At high frequencies, coaxial-to-coaxial crosstalk becomes vanishingly low.

Coaxial Electromagnetic Interference

Pickup of signals from outside sources generally is called EMI. The mechanism for the entry of external disturbing signals into coaxials is straightforward and is discussed first. All multicoaxial communications cables include a metallic element in the sheath that is intended to reduce disturbances in the signal paths of the coaxials. We visualize a simplified situation with a single coaxial inside the cylindrical metal member of the sheath (Fig. 43) as follows.

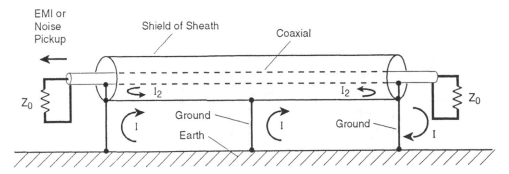

FIG. 43   Mechanism of electromagnetic interference entry via surface transfer impedance into a coaxial inside a cable having a metallic sheath.

The sheath is shown connected to earth at intervals. Such connections would be a matter of safety in an aerial line. For the more usual buried case, the sheath-to-earth path at any point tends to be self-terminating by the continuing line to the right and left. Interfering current flows around the sheath-to-earth path, induced by radio stations, power equipment, lightning, and many other sources whether the cable is placed aerially or buried.

The world is electrically noisy. The noise currents in the sheath produce a voltage via Z_{12} on the inside of the sheath that drives a current I_2 around the sheath-to-coaxial outer conductor path, and then into the signal path of the coaxial. Unfortunately, Z_{12} provides the least protection at the lowest frequencies at which power line effects and lightning are common problems.

Multipair Electromagnetic Interference

The electromagnetic interference situation is quite different for pairs, whether or not they are shielded individually (see Fig. 44). Assume an unshielded pair inside the shield of the cable sheath, which is typical of multipair cable. To aid in the visualization, the pair is not twisted. Also, Conductor 1 is shown to have a large capacitance C_{13} to Sheath 3, Conductor 2 has a smaller capacitance C_{23} to sheath. Thus, the pair is assumed to have what is called a *capacitance unbalance to ground*.

The net result is that a current I_a will flow in Conductor 1 and I_b in Conductor 2 as a result of an external disturbing current I on the outside of the sheath. The difference $I_a - I_b$ drives a current through the terminations and thus causes an interfering signal. The obvious answer is to balance C_{13} and C_{23}.

Bell had a clear perception of the merits of a twisted pair when he wrote his patent application in 1878 (1, p. 204). Manufacturing technology included efforts to reduce capacitance unbalance to ground not long after the introduction of paper-ribbon insulation in the conference cables in the 1890s. Wire for current-day plastic-insulated cables is manufactured under automatic capacitance control. Such cables have values of unbalance to ground approaching the vanishing point in well-managed cable plants.

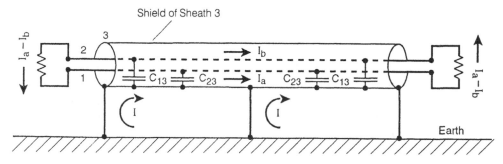

FIG. 44   Mechanism of electromagnetic interference entry via capacitance unbalance to ground into a pair inside a cable having a metallic sheath.

Common-Mode Interference

In electronics technology, current flowing in the same direction on both conductors of a pair generally is called *common-mode interference* and the ability of a pair to attenuate the effect on the desired signal is called *common-mode rejection*. A rejection level of 60 dB or better is typical of cable pairs. Coaxials have but little common-rejection capability at low frequencies unless the outer conductors are isolated from each other.

A most important point is that the rejection of noise signals by twisted pairs improves as the frequency is reduced, in contrast to the situation for coaxials. The improved rejection of noise in pairs at low frequencies more than compensates for the reduced effectiveness of shields at low frequencies. The attempts of early practitioners to shield each conductor in cables, as described in the history section of this article, met with limited success for reasons related to the concepts discussed herein.

Pairs have a large advantage in EMI rejection compared with coaxials. To retain that advantage, terminal apparatus must be balanced also. The advantages of pairs are lost when they are driven with unbalanced (one side grounded) sources into unbalanced receivers.

Resistance Unbalance

Following a rationale as above for capacitance unbalance to ground, a simple demonstration shows that a difference in the resistance of the conductors of a pair (resistance unbalance) also contributes to pickup of extraneous signals. This was perceived in the early days of long-distance cable development.

In general, the shorter runs in a subscriber plant permit a less-stringent requirement for both resistance and capacitance unbalance. The upgrading in quality made possible through the use of plastic insulation has greatly improved the performance of the longer loops.

Circuit Density

The number of analog transmission paths at voice frequency that can be placed in a cable of a given size depends upon the desired unit-length transmission loss and the physical structure. Choices in physical structure include pairs or quads, shielded pairs, and coaxials. Different considerations apply when high-frequency systems are being designed.

The comparisons in this section are intended to illustrate fundamental principles and are limited to voice-frequency transmission only for two structures designed to match the attenuation of 19-ga, 83-nF-per-mile, polyethylene-insulated multipair air-core cable at 1 kHz. The two alternative designs are multicoaxial cables and shielded pairs for multipair cables.

Space Sharing — Nonshielded Pairs

Nonshielded pairs have the advantage that all the space among the pairs is employed for signal transmission. Core-space requirements are reduced because the pairs share the physical and electrical space of other pairs.

Physically, the twisted pairs nest with each other such that essentially all insulated conductors in the cable cross-section appear to be surrounded uniformly by other conductors. This idea is illustrated in Fig. 31 for estimating the capacitance of a nonshielded pair. The nesting would be reduced for pair twist lengths much shorter than the nominal range of 2 to 6 in that is in current use.

For multipair cables, the cross-sectional sheath space assigned to one non-shielded pair is illustrated in Fig. 45. The proportions are roughly to scale for plastic-insulated air-core cable. Cross-sectional areas generally are measured in circular mils equal to the square of the diameter, expressed in mils, divided by 1000 to give kilo circular mils (kCM).

A typical space for a 19-ga, 83-nF, polyethylene-insulated pair (60-mil DOD) in an air-core cable is about 12.3 kCM, for example, equivalent to a diameter equal to 111 mils. Without nesting, a pair would require a sheath space of 2(DOD) squared, equal to 14.4 kCM for this example.

Unused Interstitial Space

The interstitial areas among such stranded multielement cylindrical structures as coaxials and shielded pairs does not contribute to the signal transmission space. The wasted space amounts to about 32% of the cross-sectional area, equivalent to about 15% of the diameter. This is shown in the following table.

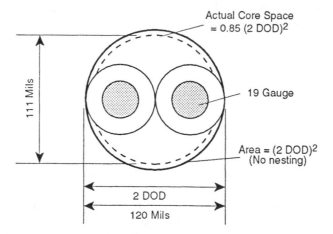

FIG. 45   Space per pair for a nonshielded pair in a multipair cable. The actual space required is less than the space within a circle having a diameter equal to (2 × DOD) due to the sharing of interstitial space.

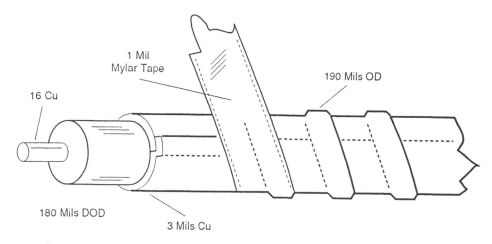

FIG. 46 Design details of a coaxial having a capacitance of 83 nF per mile.

No. of Layers	n	Rel. CD	Rel. CA	Rel. CA/n	$\sqrt{\text{Rel. CA/n}}$
1	1	1	1	1.00	1.00
2	7	3	9	1.29	1.13
3	19	5	25	1.32	1.15
4	37	7	49	1.32	1.15

n = Number of coaxials or shielded pairs
Rel. CD = Relative core diameter
Rel. CA = Relative core area
Rel. CA/n = Relative core area per coaxial or pair
$\sqrt{\text{Rel. CA}/n}$ = Relative diameter of core area per coaxial or pair

For the comparison, a coaxial and a shielded twisted pair are designed to match the 1.30-dB-per-mile attenuation of a 19-ga nonshielded pair. The thinnest practical copper tape (3 mils) is proposed for the outer conductor and for the shield. The overlapped longitudinal seam is bound with a 1-mil MYLAR tape. The two structures are shown in Figs. 46 and 47.

Core space for each of the three candidate voice-frequency structures is summarized in the table below.

	Core Space, kCM
Nonshielded multipair	14
Multicoaxials, 1.32 × OD	48
Shielded multipair, 1.32 × OD	41

The table shows that the nonshielded multipair requires substantially less care space than either the coaxial or the shielded pair. The sharing of space with other pairs accounts for the space per pair being only 14 kCM, versus 49 and 41, respectively, for multicoaxials and shielded pairs. These space savings translate into about a 2-to-1 smaller core diameter for multipair than for shielded structures because shielded structures do not use the interstitial areas to reduce

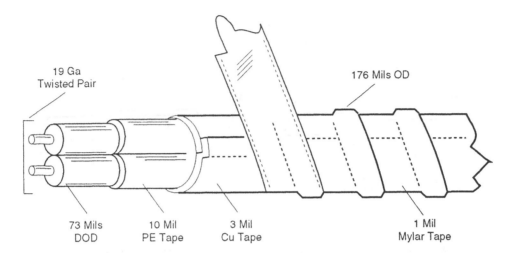

FIG. 47   Design details of a shielded pair having 93 nF per mile.

transmission loss. The coaxial appears slightly less efficient in space utilization than the shielded pair because its dielectric space includes no air whereas the shielded pair includes some air and thus has a slightly lower dielectric constant.

Nonshielded star quads are not computed for the comparisons shown above. Shielded star quads would suffer the same 32% space penalty due to inefficient use of core space, of course. Shielding would not reduce the high side-to-side capacitance unbalance within the quad (a problem with quads), or reduce the phantom capacitance that is always high because of the adjacency of the wires. Multiple twin quads are slightly more efficient in core-space utilization than pairs, but not enough to overcome the added manufacturing complications.

Comparison of Crosstalk

Coaxials have high crosstalk at frequencies from DC to between 10 and 100 kHz because the surface transfer impedance of the outer conductor is unfavorable. Crosstalk for multipairs improves as frequency is lowered. Estimated coaxial crosstalk would be about 30 dB when the outer conductors are connected together at the ends, compared with 80 to 100 dB for the crosstalk between adjacent nonshielded pairs in the voice range.

Fine Gauges

Finer gauges increase maximum pair counts, of course. Use of 26 ga permits more than 4000 pairs in a cable sized for a standard 4-in duct. Reduction to 28 ga would push the nonshielded pair count to more than 5000 pairs. Sizes finer than 26 ga have been tried but have not been popular in America. Gauge 28 has found some use abroad.

Two-Wire versus Four-Wire

Early telephone lines employed a single aerial wire with ground return patterned after telegraph technology, as described in the history section. The need for a full "metallic" line to reduce and control noise became apparent rather quickly. In 1881, J. J. Carty demonstrated the superior noise performance of an aerial two-wire metallic telephone line between Boston and Providence.

The two-wire view prevailed, nevertheless. The first transcontinental aerial line was constructed with about five million pounds of 8-ga copper in a full two-wire metallic circuit in 1915. This spectacular feat for the first voice communication across the country would have failed had not two other inventions — the vacuum tube and the hybrid coil or hybrid transformer — come along to lend critical support.

Next, we wish to explore how the hybrid coil contributed in a major way to the success of the two-wire transcontinental telephone service and, in addition, paved the way for the doubly extravagant use of four-wire circuits.

Two-Wire Lines and Amplifiers

A problem arises when one pair of wires is used for communication in both directions at the same time. Amplifiers inherently are one-way devices, with an input port for a weak signal and an output port for a strong amplified copy. Inserting such an amplifier into a two-wire line would improve, say, East–West transmission, but would kill the West–East connection.

This dilemma attracted practitioners in the days of relatively unsuccessful efforts to develop mechanical voice-frequency amplifiers. Thomas A. Edison, for example, filed a patent in 1884 on a method of converting one amplifier into a repeater having bilateral gain, long before the vacuum tube amplifier was invented (U.S. Patent No. 340,707; filed Dec. 15, 1884, issued Apr. 27, 1886 [Ref. 1, p. 264]).

The hybrid coil principle is unique, powerful, and worthy of a brief overview. We begin with a simple center-tapped transformer winding coupled to a second winding (see Fig. 48). To illustrate the finesse of this simple arrangement, we start the discussion with a signal source connected to the center tap. Currents I_1 and I_2 flow to the left and right, respectively. When the load impedances Z_1 and Z_2 are equal, the detector coil that responds to the difference $(I_1 - I_2)$ has no output because $I_1 - I_2 = 0$.

The signal source could be a telephone transmitter, the detector would be a telephone receiver. The performance is the same when the transmitter and receiver are interchanged. The idea is equally applicable when two center-tapped coils are used for a balanced two-wire line instead of the unbalanced one-wire and ground example shown in Fig. 48.

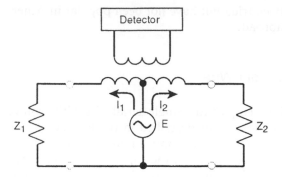

FIG. 48   Schematic of hybrid coil showing how a signal source E drives currents I_1 and I_2 into Z_1 and Z_2, respectively. When $Z_1 = Z_2$, the detector current is zero.

The Hybrid Junction

The principles outlined embody what is sometimes called a *hybrid junction* or *magic tee* in microwave technology. We summarize the properties of the hybrid junction here. In Fig. 49, the black box labeled H has the properties shown by the arrows inside the box. That is, signals at the input port go only to Z_1 and Z_2 and not to the output port. Conversely, signals from Z_1 or Z_2 or both appear only at the output port. These properties follow from the coil arrangement shown in Fig. 48. Note also that we simplify by showing a single line to represent a pair.

Bilateral Repeaters

The hybrid junction can be combined with one amplifier to make a bilateral repeater (see Fig. 50). Signals from West are sent to the amplifier input, and

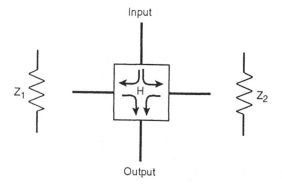

FIG. 49   "Black box" defining properties of a hybrid junction. Input to the box goes only to Z_1 and Z_2. Signals from Z_1 or Z_2 go only to the output, just as in a hybrid coil.

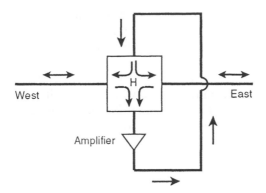

FIG. 50 Bilateral repeater in which one amplifier and a hybrid junction amplify two-way signals on one line. If the East and West lines are not well matched, the amplifier may "howl."

the amplifier output is sent to the upper port of H to exit into West and East equally. Thus, East hears an amplified signal from West, and vice versa for East-to-West transmission.

If the line impedance to West does not equal (match) the East line, some of the amplified signal will be reflected from West and will return to the amplifier input. Because the amplifier will oscillate or "howl" if the two lines are not well balanced, this scheme is useful only for low-gain amplifiers.

W. L. Richards in 1895 invented a letter system, again illustrated with the hybrid junction idea and single lines to represent one pair as in Fig. 51. Richards introduced a second hybrid coil and a second amplifier. The networks N_1 and N_2 balance the West and East lines, respectively. West could be cable and East open wire. As long as N_1 and N_2 balance their respective lines, the amplifiers do not howl.

Four-Wire Lines

In preparing for the first transcontinental line, George A. Campbell in 1912 undertook a study of the problem of bilateral gain and the Richards double-

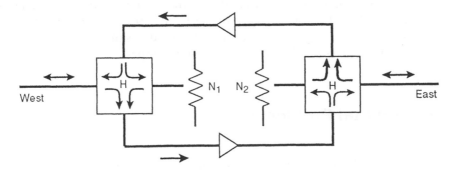

FIG. 51 Richards repeater, in which two hybrid junctions and two amplifiers solve the limitations of the bilateral repeater of Fig. 50.

hybrid coil idea. The outcome was described as an "unbelievably radical" idea. Campbell noted that the bilateral repeater could be turned into a four-wire system to great advantage. His plan is redrawn in Fig. 52.

Campbell further demonstrated that the system would not sing as long as at least one of the N networks matched its line reasonably well. This radical proposal was put into commercial service in 1915 in a 450-mile Boston-to-Washington cable circuit. The high amplifier gains made possible by the four-wire idea facilitated reducing the copper more than enough to pay for the second pairs of wires.

Electronics versus Copper

The tradeoff of electronics for copper thereby was launched. By 1925, four-wire circuits were in operation between New York and Chicago. The technical problems for going transcontinental on cable had been solved. However, the great distances over cable, having relatively high loss, uncovered three new problems:

1. The need to equalize the frequency response for better voice fidelity
2. The necessity of regulating gain to compensate for temperature effects on cable loss
3. The need to suppress echoes from mismatches at the far ends of the lines

The story of solutions to these problems is a fascinating one, but beyond the scope of this article.

We conclude this section by noting that Campbell's four-wire idea initiated the trend to the substitution of electronics for copper that has continued unabated to this day.

Equivalent Four-Wire

One further transmission type should be mentioned—equivalent four-wire. In carrier systems, East–West transmissions may be sent on a low-frequency band, and West–East on a high-frequency band over the same two-wire circuit. Thus, copper is cut by 2 to 1 in the tradeoff of electronics for copper. The ultimate tradeoff is the use of highly refined silicon dioxide (sand!) in the form of lightguide optical fibers for the full elimination of copper.

Environmental Problems

Environmental considerations, as viewed broadly, include the impact of the telephone plant as well as the protection of the plant. Concern for the visual environment began early in the history of transmission media for the growing

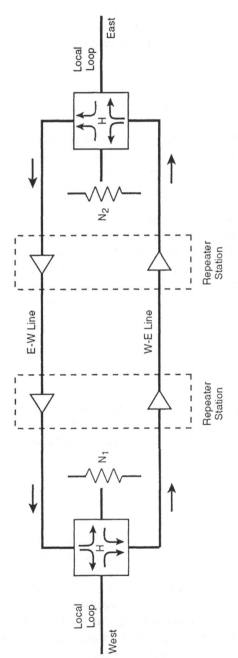

FIG. 52   Four-wire line proposed by Campbell in 1912 and used around the world since then. This invention initiated the first major step in the substitution of electronics for copper.

telephone system. Overhead wires on tall poles in the larger cities were said to "obscure the sky" as soon as five years after Bell's invention of the telephone.

The extreme sensitivity of aerial plant to adverse weather—sleet, ice, and wind—provided a strong incentive to find a better solution to protect the continuity of service to customers. Aerial cable and then cable in buried conduits provided steps along the way to protect service continuity, as well as improving the visual environment.

Building for "Out of Sight" Plants

The cost of rebuilding the plant, the loss in revenue, and the inconvenience to customers were major considerations for storm, fire, and other damage. No matter how desirable, change could not come until the supporting technology brought solutions within economic reach. Achievement of "out of sight" plants was accomplished in two steps.

The first step, conduit cable for metropolitan areas, came early and by the 1890s conduit plant, essentially as it is known today, was common. The second step, buried subscriber distribution cable, did not fall within economic reach until much later, after the development of PIC cable. Each of the solutions, aerial cable, conduit cable, and buried cable brought its own set of protection problems. We touch upon problems and solutions below.

Lightning

Definition

American terminology defines a *stroke* as the event of lightning between cloud and earth, and a stroke may include only 1 discharge or as many as 40 in unusual cases. The average is about three. The terminology varies in different parts of the world.

Frequency of Lightning

The U.S. Weather Bureau publishes what is called an *isokeraunic* map periodically. In an isokeraunic map, geographic points having the same number of thunderstorm days per year are connected by lines, "iso" meaning equal. The idea is the same as an isobaric map showing lines of equal barometric pressure (referenced to sea level) at a given time. The contours shown in Fig. 53 give the mean number of thunderstorm days per year for the United States and provide a rough measure of the incidence of strikes to ground. A *thunderstorm day* is defined as any day during which thunder is heard. The number of thunderstorm days per year ranges from about 10 on the West Coast to about 100 in Central Florida. The maps are said to vary year to year. Most of the continental United States is subject to a large number of lightning storms each year. Storms in the South tend to have strokes carrying very high current to ground.

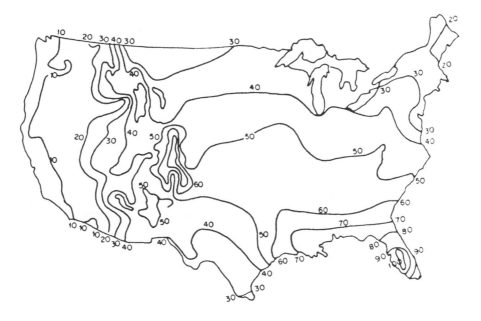

FIG. 53 Isokeraunic map of United States showing lines of equal number of thunderstorm days per year. Highest levels occur in the Southeast. (Reprinted with permission from Ref. 17, p. 27-2, Fig. 27-1.)

Protection

Lightning is the single natural environmental exposure that commands more careful attention than any other in the protection of the telephone plant and telephone customers. For example, every telephone drop has a lightning protector at the point of entry into the customer's premises. Obviously, the name *lightning protector* is a bit of reverse English.

In the beginning, protectors tended to follow telegraph practice. A simple form consisted of two silk-insulated wires twisted together and connected between the conductor and ground. A surge of voltage caused by lightning would break through the silk insulation and provide a path to ground.

Next came a short air gap between the sawtooth edges of two small metal plates. With these, the metal tended to melt, bridge the gap, and permanently short the line. It was found that small carbon blocks flashed over with a wider air gap for the same voltage as for the metal and were self-cleaning because carbon particles tended to burn instead of bridging. The design of carbon-type protectors progressed through a series of steps to improve performance and reduce cost, and stabilized on a simple screw-in cartridge by the 1950s.

Vacuum arresters were tried as long ago as 1914, without success. A gas-tube protector consisting of a gap between two appropriately shaped metal electrodes inside a glass cylinder filled with a special gas found limited use beginning in 1964. The gas tube eventually replaced the carbon type completely. Carbon gaps sometimes generate static noise after firing and thus were a maintenance problem, whereas gas tubes clear or "reset."

The introduction of solid-state repeaters and other apparatus in the outside plant brought a need for more stringent measures to protect the transistors and other low-voltage components from harm. These problems were solved without too many blown transistors along the way. Solid-state protectors are likely to replace gas tubes in the march of technology.

Effects on Pulp Cable

Subscriber drop wires connect to pulp-insulated aerial telephone cable via sealed, airtight terminals with binding posts for connecting the drop wires. Lightning cartridge-type protectors at this point of entry into the cable are designed to keep lightning out of cable cores. In spite of the voltage-limiting capability of protectors, lightning sometimes causes arcing and shorting between conductors.

Plastic–Metal Sheaths

The introduction of plastic–metal cable sheaths soon demonstrated the pervasive nature of lightning. The outer polyethylene jackets of cable sheaths were punctured with many small pinholes in a phenomenon not encountered with lead sheaths. Lightning discharges to the cable-support strand of aerial cables generated sufficient voltage between the strand and the aluminum shield (e.g., in an alpeth sheath) to cause many punctures of the outer jacket. Alpeth (aluminum–polyethylene) is an acronym for a sheath consisting of a corrugated-aluminum-tape shield covered with an extruded-polyethylene jacket. The aluminum is folded over the cable core with an overlapped longitudinal seam.

Attempts to maintain intercity aerial runs of alpeth-sheathed cable under air pressure as a maintenance strategy proved impractical because new pinholes were generated as fast as the old ones were repaired. One discharge to the strand could produce a number of pinholes. In aerial cables, the greater part of a stroke enters the earth by flashing from the strand to earth, and thus does less damage to the cable than otherwise.

Effects on Buried Cable

Lightning discharges to ground at points near a given buried cable produce a large elevation of the potential of the earth due to the local IR drops, and generate numerous pinhole punctures in the outer jacket of a nearby cable. The potential in the earth is equal to the lightning current I in amperes times the earth resistance R in ohms. A rough idea of the mechanism of damage is shown in the overhead views given in Fig. 54. The discharge is assumed to have hit at a point on the earth as indicated. In the absence of cable, injecting a high current results in a radial current flow, indicated by the radial arrows. Assuming a uniform earth resistivity, circles of equal potential above remote ground are

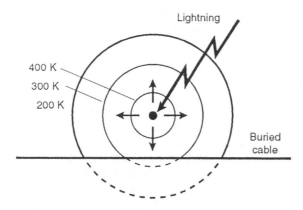

FIG. 54 A buried cable is shown in the path of the radially moving pulse of lightning current. Pinholes may be punctured in outer plastic jackets.

generated as the wave of discharge current moves radially outward. For purposes of illustration, the circles are labeled in arbitrary values of kilovolts above remote earth.

When a cable is in the path of the radially moving current pulse shown in Fig. 54, the outer jacket of the cable at the nearest point will be punctured as soon as the local voltage exceeds the dielectric strength of the polyethylene, in the range of 25,000 to 50,000 volts. Current will flow into the aluminum to the right and to the left. We could assume that the first pinhole captures all the arriving current, much less than the lightning discharge current, of course.

As the radial pulse of current continues outward, the potential will rise to the dielectric strength of the polyethylene jacket and cause two more pinholes, one on either side of the first. This continues until the IR drop caused by the current wave is too small to generate further pinholes. Depending on the magnitude of the lightning discharge current, only a few or a number of pinholes may be generated. Pinholes have been observed as close as about 10 ft apart, and spaced with surprising uniformity. Sometimes a section of jacket is peppered with pinholes, probably because the stroke was nearby. Direct hits to cables "drill" holes in the polyethylene jacket up to 1/16-in in diameter or more and may do considerable core damage.

Buried Lead-Covered Cable

Buried lead-covered cable, being in contact with the earth, does not permit build-up of potential and does not exhibit an effect similar to the pinholing in polyethylene jackets. Direct strokes to a cable may melt a hole in the lead but, in general, lead sheaths tolerate strokes approaching 100,000 amperes based on laboratory studies. About 1% of lightning strokes to cable are 100,000 amperes.

Polyethylene-jacketed cables having corrugated steel perform well in lightning exposures, provided that the pinholing of the polyethylene is tolerable.

The pinholes may permit entry of lots of damaging water into air-core PIC cables or wet the cores of paper-pulp-insulated cores.

Crushing Effects

A curious phenomenon caused by lightning was not observed in buried cable until multicoaxial cable was introduced. Under certain conditions of soil type and moisture content, a direct strike to a buried coaxial cable may cause cable crushing, sometimes shorting several coaxials, due to the explosive action of vaporized moisture at the point of contact. Extensive laboratory studies characterized the performance of different sheath and armor constructions versus a range of current levels for simulated lightning strokes. Current levels as high as 100,000 amperes were studied and it was shown that lead sheath alone (no armor) was vulnerable to crushing.

The best protection for coaxial turned out to be one or two 3/8-in bare copper conductors placed above the cable at the time of plowing or trenching to intercept the lightning. The function of the grounded copper conductor is similar to the ground wires used in power transmission line construction. The ground wire or wires placed in the path of the lightning stroke intercept the stroke just as lightning rods protect a house or other structure.

Buried Multipair Cable

With buried paper cable, similar crushing effects probably took place but were not found because cable failure did not occur. Paper cores are tolerant of considerable deformation since paper is very tough and crush resistant. Air-core PIC cable, on the other hand, would suffer crushed and split polyethylene insulation. The entry of water at the point of jacket puncture would lead to service outage and need for repair. Because of bounce-back of the plastic sheath, crushing as such would not be noted during repair.

Estimating Number of Hits

We complete the discussion of lightning-caused problems in cable by presenting a view of the method of estimating the number of direct strokes to a given buried cable per year. The isokeraunic map shows how many thunderstorm days to expect. A given statistic is that the estimated number of strokes to ground is about 2.5 per square mile per 10 thunderstorm days.

A cable "attracts" lightning when the earth is hit within a given distance, depending upon the soil resistivity. Soil resistivity usually is measured in meter-ohms and ranges from 100 or more for soils like that found in Iowa, to 1000 or more for the well-washed sandy soils of the Southeastern states. The "attraction" band of width d depends on soil resistivity. The likelihood of a cable being hit is simply the ratio of the area of the band of width d to one square mile. Reciprocals of the ratios are tabulated below.

Resistivity (Meter-Ohms)	Band Width d (Ft)	Reciprocal Ratio, 5280/d
100	25	240
1000	45	117
10,000	140	38

Thus, for a locality having 30 thunderstorm days per year and a soil resistivity of 100 meter-ohms, the strikes in 100 miles of cable would be $100 \times 2.5 \times 30/(10 \times 240) = 3.1$ hits per 100 miles per year.

Number of Troubles

Not all hits cause trouble. The number of troubles would be the hits reduced by a percentage of strokes causing a cable failure. Suppose that the cable construction was such that 99% of the strokes could be tolerated; then, the troubles become $3.1 \times (100-99)/100 = 0.031$ per 100 miles per year, or about 1 trouble per 3000 miles per year in a 30-thunderstorm-day region.

Actual calculations for a specific cable design and field conditions are more complex than the example above. It generally is not economical to design a system so robust that all lightning strokes are tolerated.

By the rationale outlined above, hits per acre are $2.5 \times 3/640 = 0.012$ per 30 thunderstorm days per year or once every 85 years; however, put up an attracter like the Empire State Building and the hits may run about 40 times a year or even a dozen times in one storm.

Corrosion

Corrosion effects range from mild deterioration to complete destruction. The financial loss for replacement and protection of structures is estimated at about $15 billion annually in the United States. Corrosion is a complex reaction involving chemical or electrochemical effects and may include physical and mechanical factors. Much research effort has been expended to understand, mitigate, and control corrosion effects.

Our concern herein is chiefly with the protection of telephone cable sheath. As mentioned in the history section, Patterson found that alloying lead with 3% tin improved the corrosion resistance. Lead has been used since ancient times.

Stray Currents

A common cause of corrosion in lead sheaths of telephone cables is stray direct current flowing from the lead to moist soil or water. According to Faraday's law, one ampere of current removes about 75 pounds of lead per year. The lead in one foot of full-size cable weighs about 4.8 pounds. Many years ago, DC was used for electric streetcars, trains, and elevators. Cathodic protection of systems for other structures also provided ample stray current sources in the earth. Some of the stray current sought a return path on the buried or conduit telephone cable plant and corroded lead sheaths via Faraday's law.

Another source of corrosion includes the presence of galvanic cells consisting of two dissimilar metals (e.g., manhole hardware) in contact with an electrolyte. Soil water contains enough ions to provide the electrolytic action. The dissimilarities may be caused by subtle effects such as local differences in the amount of oxygen dissolved in the water along the cable sheath. Abraded areas in the lead may cause the formation of local corrosion cells.

Control of Corrosion

Corrosion of lead sheath is controlled by three mitigative measures.

1. Return the stray current over a metal drain wire.
2. Maintain the cable sheath at negative potential relative to the surroundings by applying a source of voltage.
3. Cover the sheath with an insulating layer.

Application of a source of voltage as from a rectifier to maintain the negative potential is called *cathodic protection*. Sacrificial anodes connected to the lead may be used to provide the desired potential. Polyethylene jackets isolate the metals of the cable and prevent electrolytic corrosion. Before the plastics were available, paper and asphalt coverings sometimes were employed.

Plastic–Metal Sheaths

The question of corrosion of the metals in the composite plastic–metal cable sheaths has been the subject of many laboratory and field studies. The aluminum of buried alpeth air-core PIC cable was expected by many to be vulnerable to corrosion after lightning pinholing of the jacket and the entry of water. Field experience showed the aluminum to perform very well, with one exception. In instances where massive conductor insulation damage to the core had taken place, the exposed copper conductors with central-office battery (49 volts) were affected by Faraday's law, which led to relatively severe corrosion of the corrugated aluminum over a length of several feet.

Many samples of corrugated steel and aluminum composite sheaths such as stalpeth and PASP (polyethylene-aluminum-steel-polyethylene, in order of application over cable core) were buried in several laboratory test plots and followed over a period of years. In general, results of these studies showed acceptable corrosion performance and supported the earlier design decisions of the practitioners of that era. For more recent designs, plastic-coated aluminum and steel have been developed and permit bonding to polyethylene jackets and reduce corrosion effects.

Conductor Corrosion

Another problem area involving electrolytic corrosion was encountered in the pedestal terminal system employed with buried PIC cable for providing access to the cable core for connection of subscriber drops. All connections to the core conductors were covered with insulation. However, insulation damage due to handling the exposed core conductors or use of insulation-piercing tools during troubleshooting in search of water-induced troubles in the cable core resulted in relatively frequent conductor "cutoff."

The mechanism of cutoff is another example of Faraday's law. By its nature, polyethylene, in spite of its exceptional waterproofness as an insulating material, attracts a thin layer of water onto its surface in the presence of high relative humidity. The humidity inside the pedestals tends to be high and at night condensation is assured. Surface leakage current flows over the surface of the insulation from the plus conductor to the minus conductor.

Laboratory tests showed that Faraday corrosion cutoff occurred at pinholes in the insulation after about 5-microampere years for the finer conductors. The current-time product is small because the pinhole restricts the Faraday action to a very small volume of metal. Conductor openings were observed in the field after several years.

Similar openings occurred along the cores of wet, buried PIC cables, especially when both conductors of a pair had been damaged in the factory or during field installation. In this case, the openings occurred in a relatively short time. The flow of leakage current between insulation pinholes also generated a popping and spitting noise that could be heard on the line.

Special Problems

Many special protection problems demanded attention over the span of time since Bell's invention of the telephone. Not all were exotic or glamorous, but nevertheless many have required good insight and diligence to devise acceptable solutions. A few interesting cases are recalled.

Toredos

Initially, telegraph ocean and harbor cables were incapacitated by a sea worm called a *toredo* that bored through the conductor insulation. Copper toredo tape became an important component of oceanic telephone cables. Appropriate work demonstrated that polyethylene tape could be substituted for the copper. The polyethylene performed well and eliminated a more costly item that often corroded away.

Gophers

West of the Mississippi, a burrowing rodent called a gopher gnaws on buried cable and wire to keep its teeth from growing too long. It turns out that gophers can open their mouths only so far. Cables larger than a certain size are saved, but smaller ones require "gopher" protection, consisting of various metal tapes.

Wood Preservation

Aerial telephone plants depend upon wooden poles. Much development work on protective treatments led to methods of treating the wood to preserve the poles and also to treated wood for household decks and the like, which are now taken for granted.

Other Problems

Telephone cables are subjected to still more natural hazards and environmental stresses that we can only mention. Asian and Australian termites are said to bore into polyethylene jackets. Squirrels like to chew the lead off aerial lead-covered cables. An unusual type of worm bores holes in lead sheath.

The wind sometimes sets aerial cable into a "dance" so violent that poles may be shattered. Fungi rot poles and other wooden structures. Hurricanes and sleet rip aerial cables and wire from their moorings. The bright sunlight may embrittle and crack polyethylene jackets and wire insulation unless carefully compounded. Industrial pollutants may induce stress cracking in polyethylene and corrosion in metals. Fire and flood take their toll. Seemingly, the list is endless.

Conclusion

We close the discussion of environmental factors with a look from the "other side of the fence." Electrical noise, generally called EMI, has an impact on cable and the associated telephone system. Through cooperative efforts, the providers of our power and of our communications services have worked together to achieve mutual compatibility.

Second thoughts on the possible environmental impact of EM fields are renewing questions long thought settled. These questions concern the possible health impacts of the electromagnetic fields that surround us, emanating from power distribution systems and many household appliances—all the everyday things of the modern good life.

Compared with other sources, communications cables are not thought to present a health hazard, but some of the facilities tapped into the communications network may be open to review. It is too early to say where the reexamination will lead or what will be uncovered. Understanding is the beginning of steps toward mitigation if there really is a problem.

Loading Concepts and Loading Coils

Starting with the knowledge of transmission-line equations for the transmission loss suffered by voice-frequency signals on a cable pair or aerial line pair, we easily conclude that such pairs "need" additional series inductance to improve transmission efficiency, that is, to reduce loss.

The Challenge

As discussed here, line loss is approximated by Eq. (7):

$$\alpha = \frac{R}{2}\sqrt{\frac{C}{L}} + \frac{G}{2}\sqrt{\frac{L}{C}}, \qquad \text{when } \omega L \gg R \text{ and } \omega C \gg G.$$

The characteristic impedance Z_o of a line is $Z_o = \sqrt{L/C}$. Hence, we may write

$$\alpha = \frac{R}{2Z_o} + \frac{GZ_o}{2}$$

When the resistance loss (first term on the right of the equation) exceeds the dielectric loss (second term), which is the usual situation, line loss may be reduced by increasing Z_o. Z_o is increased by increasing the line inductance L. Loading may be thought of as reducing current flow loss by transmitting the same power but at lower current and higher voltage, just as overhead power transmission lines operate at very high voltages to reduce current losses. The challenge is to find a way to increase L and thence Z_o in a practical and economic manner.

Loaded Lines

The first to accept the loading idea was John Stone Stone of AT&T, who proposed continuous loading in 1894 by wrapping iron wire over the cable conductors to increase the inductance. G. A. Campbell, assigned to carry on the work on loading, soon concluded that a better strategy was to insert series inductance in the form of "loading" coils at intervals. In 1900, experimental loading coils were tried successfully on a 24-mile circuit in Boston and on a 670-mile open-wire line.

Coil Design

Campbell's experiments with loading coils on cable and on open wire resolved the issue of coils versus continuous loading for general use on land lines. He provided the design rules for field use, coil spacing, and coil inductance. The first coils were of the air-core solenoid type. Their large size and large external field caused interference among loaded circuits. A toroidal or doughnut-shaped core soon was adopted.

A neat idea, that of dividing the toroidal coil into two equal parts, one for each side of a cable pair, added the mutual inductance to the self-inductance of the two halves. This permitted reducing the number of turns and thus coil resistance. Coils also have the advantage that inductance increases as the square of the number of turns.

Continuous versus Coil Loading

A point easy to pass over lightly is that continuous wrapping of fine iron wire over each wire of a pair is no inconsequential factory task. At roughly 2500 turns per foot of pair, the added production time easily makes wire wrapping a bottleneck operation. Also, the increase in core diameter would be a significant cost disadvantage. The iron wrapping also increases the pair resistance, offsetting some of the attenuation reduction obtained through loading.

Continuous loading increases the pair inductance by only 5 or 6 to 1, whereas the 88-millihenry coils increase the average value of pair inductance by nearly 60 to 1. The nominal 10 times higher inductance translates to about 3-to-1 lower loss for lumped loading versus continuous loading.

Benefit of Loading

A set of calculations prepared in 1926 demonstrates the benefit of loading a 19-ga toll cable (see Fig. 55) (17). The calculations were made for a loading of 174 millihenrys (6000 ft) instead of for 88 millihenrys standardized in later

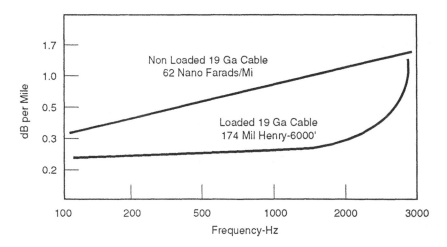

FIG. 55 Comparison of line loss for nonloaded and loaded 19-ga pair. Note that loading levels the loss over the voice-frequency band. (Copyright ©, 1975, American Telephone and Telegraph Company. Reprinted by permission, from Ref. 1, p. 246.)

years. The heavy loading explains the cutoff shown at 2750 instead of 3000 or more for 88 millihenrys.

In the midrange of frequency at which the energy of the voice is highest, 88-millihenry loading reduces line loss by almost 2.5 to 1 and thus permits use of smaller copper conductors to achieve the same line loss. The offset seen in the loss curve above 1400 Hz occurs at the transition between the low- and high-frequency attenuation formulas, when inductive reactance becomes larger than the resistance.

The copper saving is impressive but the most significant effect of loading is the leveling of the line loss over the voice-frequency band. The normal square root of frequency slope of line loss without loading causes appreciable distortion of the voice. For a log–log plot of loss versus frequency, the $\sqrt{f}$ relationship between α and f shows as a straight line. A convenient aspect of log–log plots is that the tangent of the slope angle is equal to the exponent of the abscissa, in this case frequency. That is, having a slope of 26.56°, $\tan 26.56 = 0.5$, gives 0.5 for the exponent of f, thus equivalent to the square root of f.

Limitations and Strategies

Loading initially was employed on long open-wire toll routes as the only means of delivering useful signal levels. The first coils tried were simply air-core solenoids. High resistance, large size, and strong external fields created loss and pickup problems. The toroidal core was in use by 1901.

Dielectric loss due to the line conductance G became a problem with heavy conductors, say in the 165-mil range. As noted in the equation above, increasing Z_o reduces resistance losses but worsens dielectric losses. Aerial lines generally were loaded at 8-mile intervals.

Porcelain versus Glass

A curious finding led to the use of glass insulators instead of the usual porcelain. The darkness inside the petticoats of porcelain encouraged insects to build nests that degraded dielectric performance. Lightning created a double problem— breakdown of coil insulation and change in magnetic properties of the cores due to the DC surges. Solutions for these problems and the complexity in loading the phantom circuits evolved over the next 10 years.

Impact of Cable Carrier

When cable carrier came into use beginning in the 1930s, loading was restricted to shorter voice-frequency cable runs. Because signal cutoff occurs when the coil spacing is one-half wavelength, the typical 6000-ft spacing used for voice lines is much too large for use at carrier frequencies. Tolerable spacing would be about 300 ft for K carrier and only 70 ft for N carrier.

The longer voice-frequency loops typically are loaded as a strategy for meeting signal level goals with the finer gauge cables. Loading of trunk cable has served as a step in the strategy for management of growth. Newly installed trunks would be loaded at the usual 6000-ft intervals. As circuit demand grew, loading would be removed and T1 repeaters installed at the loading points to provide pair gain or increased circuit capacity without installing more cable. The idea could be called graceful growth.

Estimated Copper Savings

A point of interest is the total number of loading coils put into use in the U.S. telephone system from the beginning. A very rough guess is about 100 million coils. For 6000-ft spacing, the implication is that over 100 million miles of cable pair were loaded during the roughly 80 years of relatively large-scale use.

To achieve the same attenuation without loading would have required roughly 2 to 1 more copper. Assuming a copper weight saving of 10 pounds per mile (22-ga copper weighs 10 pounds per mile), the total would amount to 1 billion pounds of copper.

Continuous Loading

Interest in deep-water telegraph cable was intense long before transatlantic telegraph cable was placed in 1866. By the time that telephone practitioners were conducting experiments to prove that loading coils reduced transmission losses in telephone cable, continuous loading was being tried for deep-water telegraph cables, with a goal of increasing transmission speed.

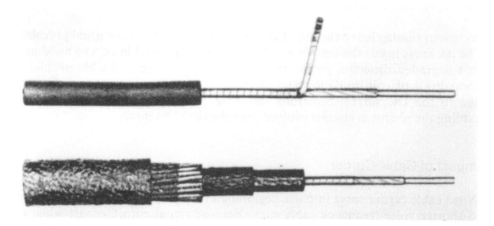

FIG. 56 Continuous loading of permalloy tape on telegraph cable for service between New York and the Azores, placed in 1924. The total inductance was more than 10 times higher than for a cable having a wrapping of iron wire. (Copyright ©, 1975, American Telephone and Telegraph Company. Reprinted by permission, from Ref. 1, p. 813.)

The first continuously loaded telegraph cable was proposed by Karup and successfully placed between Denmark and Sweden in 1902. The loading on Karup's cable consisted of soft iron wire wrapped around the copper conductor.

A Western Union telegraph cable, designed by the Western Electric Company and manufactured in England under Western Electric supervision, employed permalloy tape supplied by Western Electric. A photograph is shown in Fig. 56. Two permalloy tape applications are seen over the copper central conductor in the upper figure. A jute rove serving protects the core insulation and functions as a bedding for the armor wires. This cable was placed between New York and the Azores in 1924. The added inductance permitted a transmission speed of 1900 letters per minute, an almost 8-to-1 improvement over nonloaded cable. Numerous other deep-sea cables were placed during this period.

Conclusion

In this article on the fundamentals and history of electrical communication cables, we present an overview designed to give the reader a feeling for the challenges to and the accomplishments of many dedicated and creative people. From Bell's invention of the telephone to the creation of a new and exciting industry, we recount the significant events, problems, and solutions spawning the communication age.

The dream of voice communication at a distance surely is hundreds of years old. Bell's first successful telephone dates from 1876. The next century saw an explosion in the desire and the demand for voice and picture service to meet our innate wish to communicate.

Electrical telecommunications cables have played a crucial and pivotal role in the stitching and weaving of our country and the world into a unified community. We have selected highlights among the fundamental problems to provide useful insights for practitioners and for those who seek an overview. We touch upon the concurrent history to paint the diorama of the important mileposts along the way and the interactions among the technical contributions of the practitioners who have helped make it all happen.

Acknowledgments: SYSTIMAX PDS LAN Cable and DEPIC are trademarks of AT&T. Teflon and Mylar are registered trademarks of DuPont, Inc.

Bibliography

Introduction and History

Charlesworth, H. P., General Engineering Problems of the Bell System, *Bell Sys. Tech. J.,* 515–586 (October 1925).

Clark, A. B., Telephone Transmission over Long Cable Circuits, *Bell Sys. Tech. J.*, 67–94 (January 1923).

Fagen, M. D. (ed.), *A History of Engineering and Science in the Bell System, the Early Years (1875–1925)*, Bell Telephone Laboratories, Murray Hill, NJ, 1975.

Gerardi, B., and King, R. W., Joseph Henry. The American Pioneer in Electrical Communication, *Bell Sys. Tech. J.*, 1–10 (January 1926).

Hart, C. D., Recent Developments in the Process of Manufacturing Lead-Covered Cable, *Bell Sys. Tech. J.*, 321–342 (April 1928).

Jewett, F. B., The Modern Telephone Cable, *AIEE Trans. 82*, Part 2 (June–December 1909).

Padowicz, H. N., Cable Sheath—A Review, *Western Electric Engineer* (October 1950).

Pilliod, J. J., Philadelphia–Pittsburgh Section of the New York–Chicago Cable, *Bell Sys. Tech. J.*, 60–87 (July 1922).

Rhodes, F. L., *Beginnings of Telephony*, Harper and Brothers, New York, 1929.

Shea, J. R., Developments in the Manufacture of Lead-Covered Paper-Insulated Telephone Cable, *Bell Sys. Tech. J.*, 432–471 (July 1931).

Shea, J. R., and McMullan, S., Developments in the Manufacture of Copper Wire, *Bell Sys. Tech. J.*, 187–216 (April 1927).

Wade, L. G., Electrical Drying of Telephone Cable, *Bell Sys. Tech. J.*, 209–220 (April 1940).

Walker, H. G., and Ford, L. S., Pulp Insulation for Telephone Cables, *Bell Sys. Tech. J.*, 1–21 (January 1933).

Maxwell's Equations

Brainerd, J. G., *International Encyclopedia of Communications*, Vol. 2, 1989.

Clogston, A. M., Reduction of Skin Effect Losses of Laminated Conductors, *Bell Sys. Tech. J.*, 30:491–529 (July 1951).

Fagen, M. D. (ed.), *A History of Engineering and Science in the Bell System, the Early Years (1875–1925)*, Bell Telephone Laboratories, Murray Hill, NJ, 1975.

Morgan, S. P., Mathematical Theory of Laminated Transmission Lines, Parts I and II, *Bell Sys. Tech. J.*, 31:883–949 (September 1952), 31:1121–1206 (November 1952).

Schelkunoff, S. A., The Electromagnetic Theory of Coaxial Transmission Lines and Cylindrical Shields, *Bell Sys. Tech. J.*, 13:532–579 (October 1934).

Schelkunoff, S. A., *Electromagnetic Waves*, D. Van Nostrand, New York, 1943.

Snyder, R. V., All the World's a Filter, *Microwave Theory and Techniques Newsletter* (Summer 1990).

Woodruff, A. E., *Encyclopedia of Philosophy*, Macmillan and Free Press, New York; Crowell Collier & Macmillan, New York, 1967.

Transmission Lines

Dworsky, L. N., *Modern Transmission Line Theory and Applications*, 1979.

Fink, D. G., and Beaty, H. W. (eds.), *Standard Handbook for Electrical Engineers*, 11th ed., McGraw-Hill, New York, 1978.

O'Neill, E. F. (ed.), *Transmission Systems for Communications*, rev. 4th ed., Western Electric Company, Technical Publications, Winston-Salem, NC, 1971.

Ramo, S., Whinnery, J. R., and von Duzer, T., *Fields and Waves in Communication in Electronics*, John Wiley and Sons, New York, 1984.

Terman, F. E., *Radio Engineers' Handbook*, McGraw-Hill, New York, 1943.

Schelkunoff, S. A., *Electromagnetic Waves*, D. Van Nostrand, New York, 1943.

Communications Networks and Cable Types

Abraham, L. G., Progress in Coaxial Telephone and Television Systems, *Trans. Amer. Inst. Elec. Eng.*, 67:1520–1927 (1948).

Caruthers, R. S., The Type N1 Carrier Telephone System, Objectives and Transmission Features, *Bell Sys. Tech. J.*, 30:1–32 (January 1951).

Dobbin, D. P., Distribution Terminals Cut Costs in the Loop Plant, *Bell Lab. Record*, 54:549–552 (October 1976).

Green, C. W., and Green, E. I., A Carrier Telephone System for Toll Cables, *Bell Sys. Tech. J.*, 17:80–105 (January 1938).

Green, E. I., The Coaxial Cable System, *Bell Lab. Record*, 15:274–280 (May 1937).

Gresh, P. A., Dedicated Outside Plant, *Bell Lab. Record*, 39:26–29 (January 1965).

Kelly, M. J., Rasley, Sir G., Gilman, G. W., and Halsey, R. J., Transatlantic Telephone Cable, *Proc. Inst. Elec. Eng. London*, 102B:117–130 (March 1955).

Koliss, P. K., A New 'Ready-Access' Distribution Terminal, *Bell Lab Record*, 36:411–413 (November 1958).

Lally, W. J., The Changing Pattern of Exchange Outside Plant, *Bell Lab. Record*, 39:422–430 (December 1961).

Leberl, A. W., Fischer, H. B., and Biskeborn, M. C., Cable Design and Manufacture for the Transatlantic Submarine Cable System, *Bell Sys. Tech. J.*, 189–216 (January 1957).

O'Neill, E. F. (ed.), *A History of Engineering and Science in the Bell System, Transmission Technology (1925–1975)*, AT&T Technologies, Indianapolis, IN, 1985.

Cable Development Problems and Solutions

Appleyard, R., Origin and Development of the Transmission Equation, *Electrical Communication*, 183 (January 2, 1924).

Burrows, C. R., Exponential Transmission Line, *Bell Sys. Tech. J.*, 7:555 (October 1955). First published in *Bell Lab. Record*, 18:174–176 (February 1940).

Carson, J. R., Mead, S. P., and Schelkunoff, S. A., Hyper-frequency Wave Guides— Mathematical Theory, *Bell Sys. Tech. J.*, 310–333 (April 1936).

Duncan, R. D., Jr., Characteristic Impedance of Grounded Open-Wire Transmission Lines, *Commun.*, 19:10 (June 1938).

Morrison, J. F., Transmission Lines, *Western Electric Pickups* (December 1930).

Norris, F. W., and Bingham, L. A., *Electrical Characteristics of Power and Telephone Transmission Lines*, International Textbook, Scranton, PA.

Ramos, S., and Whinnery, J. R., *Fields and Waves in Modern Radio*, 2d ed., John Wiley and Sons, New York, 1953.

Schelkunoff, S. A., The Electromagnetic Theory of Coaxial Lines and Shields, *Bell Sys. Tech. J.*, 13:532–579 (October 1934).

Schelkunoff, S. A., *Electromagnetic Waves*, D. Van Nostrand, New York, 1943.

Shea, T. E., *Transmission Networks and Wave Filters*, Van Nostrand, New York, 1929.

Sterba, E. J., and Feldman, C. B., Transmission Lines for Short Wave Radio Systems, *Proc. IRE*, 20:1163 (July 1932).

Terman, F. E., *Radio Engineers' Handbook*, 1st ed., McGraw-Hill, New York, 1943.

Environmental Problems

Dobbin, D. P., Distribution Terminals Cut Costs in the Loop Plant, *Bell Lab. Record*, 54:249–252 (October 1976).

Encyclopedia of Science and Technology, 6th ed., s.v. "Communications Systems Protection," Vol. 4, McGraw-Hill, New York, pp. 217–219.

Fink, D. G., and Beaty, W. W. (eds.), *Standard Handbook for Electrical Engineers*, 11th ed., McGraw-Hill, New York, 1978.

Jansky, K. G., Electrical Disturbances Apparently of Extraterrestrial Origin, *Proc. IRE*, 21:1387–1398 (October 1933).

Parente, M., The Energy No One Wants on a Telephone Line, *Exchange*, 2 (January–February 1986).

Schweitzer, P. A. (ed.), Corrosion and Corrosion Protection Handbook, American Society for Metals, Metals Park, OH, 1983.

Smith, A. R., Designing for the Environment, *Bell Lab. Record*, 12–16 (February 1984).

Loading Concepts and Loading Coils

Arnold, H. D., and Elmen, G. W., Permalloy, An Alloy of Remarkable Magnetic Properties, *J. Franklin Inst.*, 195:621 (May 1923).

Buckley, O. E., The Loaded Submarine Telegraph Cable, *Bell Sys. Tech. J.*, 4:355 (July 1925).

Campbell, G. A., On Loaded Lines in Telephone Transmission, *Philosophical Magazine*, 5:313 (March 1903).

Gherardi, B., The Commercial Loading of Telephone Circuits in the Bell System, *AIEE Trans.*, 30(3):1748 (June–December 1911).

Gilbert, J. J., Determination of Electrical Characteristics of Loaded Telegraph Cable, *Bell Sys. Tech. J.*, 6:387 (July 1927).

Green, E. I., The Transmission Characteristics of Open-Wire Telephone Lines, *Bell Sys. Tech. J.*, 9:730 (October 1930).

Hill, R. B., Birth of the Loading Coil, *Bell Lab. Record*, 30:263 (June 1952).

Legg, V. E., and Givem, J. J., Compressed Powdered Molybdenum Permalloy for High Quality Inductance Coils, *Bell Sys. Tech J.*, 19:385 (July 1940).

Shaw, T., The Conquest of Distance by Wire Telephony, *Bell Sys. Tech. J.*, 23:337 (October 1944).

Shaw, T., The Evolution of Inductance Loading for Bell System Telephone Facilities, *Bell Sys. Tech. J.*, 30:149 (1951).

References

1. Fagen, M. D. (ed.), *A History of Engineering and Science in the Bell System: The Early Years (1875–1925)*, Bell Telephone Laboratories, Murray Hill, NJ, 1975.
2. *Bell Sys. Tech. J.*, 324 (April 1928).
3. *Bell Sys. Tech. J.*, 443 (July 1931).

4. *Bell Sys. Tech. J.*, 526 (October 1925).
5. Maxwell, J. C., A Dynamical Model of the Electromagnetic Field, 1864. In: *Scientific Papers*, Vol. 1 (W. D. Niven, ed.), Cambridge, 1890.
6. Maxwell, J. C., *Treatise on Electricity and Magnetism*, Vol. 25, Oxford, 1873.
7. Thomson, J. J., *Encyclopaedia Britannica*, 4th ed., (1912), p. 203.
8. O'Neill, E. F. (ed.), *A History of Engineering and Science in the Bell System: Transmission Technology (1925–1975)*, AT&T Technologies, Indianapolis, IN, 1985.
9. *Bell Lab. Record*, 250 (October 1976).
10. *Bell Lab. Record*, 399 (October 1953).
11. *Bell Lab. Record*, 113 (April 1971).
12. *Bell Sys. Tech. J.*, 181 (January 1957).
13. *AT&T Network Systems, Types 1061 and 2061A Cable*, Product brochure 4137 FS, Technical Support Group, Norcross, GA.
14. National Bureau of Standards, Circular 74.
15. Maupin, J. T., Interaxial Spacing and Dielectric Constant of Pairs in Multipair Cables, *Bell Sys. Tech. J.*, 652–667 (July 1951).
16. *Encyclopedia of Science and Technology,* 6th ed., Vol. 4, McGraw-Hill, New York, 1987, p. 207.
17. Fink, D. G., and Beaty, W. W. (eds.), *Standard Handbook for Electrical Engineers*, 11th ed., Vol. 4, McGraw-Hill, New York, 1978, p. 27–2.

MERLE C. BISKEBORN

Electromagnetic Signal Transmission

Electromagnetic Waves

Basic Concepts

The practice of telecommunications, namely, communication at a distance, relies on the use of electromagnetic (e-m) waves, and e-m waves are characterized by time-varying electric and magnetic fields. An electric field exerts a force on a charged particle in accordance with

$$\mathbf{f} = q\mathbf{E} \quad \text{N,} \tag{1}$$

where $\mathbf{f}$ is force in Newtons (N), q is charge in coulombs, and $\mathbf{E}$ is the electric field acting on the particle in volts/meter. E has been commonly referred to as electric field intensity but that usage is now deprecated. Bold letters are used here to stand for vector quantities, and Eq. (1) indicates that force and electric field intensity are in the same direction, as q is a scalar quantity. Vector quantities have both directions and magnitudes, whereas scalar quantities have magnitudes only. Another example of a scalar quantity is temperature. If a force described by Eq. (1) exists in a region, an electric field of magnitude E exists in that region. That is, Eq. (1) can be considered to define an "electric field." Another electric field quantity $\mathbf{D}$, electric flux density, is related to $\mathbf{E}$ in isotropic media by $\mathbf{D} = \epsilon\mathbf{E}$, where $\epsilon = \epsilon_o K$ with $\epsilon_o = 8.854 \times 10^{-12}$ farad/meter and K is the relative dielectric constant. K is a nondimensional quantity having the values of 1 for a vacuum (free space), close to 1 in the lower atmosphere, from 2 to 10 for common dielectric materials, and near 81 for water for a certain range of radio frequencies.

A charged particle moving with a velocity $\mathbf{v}$ in a magnetic field experiences a force given by

$$\mathbf{f} = q(\mathbf{v} \times \mathbf{B}) \quad \text{N,} \tag{2}$$

where $\mathbf{v}$ is the velocity in meters per second (m/s) of a charged particle having a charge of q coulombs and $\mathbf{B}$ is magnetic flux density in teslas or webers/square meter (m^2). The force $\mathbf{f}$ is perpendicular to both $\mathbf{v}$ and $\mathbf{B}$ and imparts circular motion to the particle. If a force as defined by Eq. (2) exists in a region, a magnetic field of flux density $\mathbf{B}$ exists in the region, and Eq. (2) can be considered to define a magnetic field. Another magnetic field quantity $\mathbf{H}$, commonly called magnetic field intensity, is measured in amperes/m. It is related to $\mathbf{B}$ by $\mathbf{B} = \mu\mathbf{H}$, with $\mu = \mu_o$ for nonmagnetic materials and $\mu_o = 4\pi \times 10^{-7}$ henry/m. In general, $\mu = \mu_r\mu_o$, where μ_r is relative permeability. In plane electromagnetic waves, the $\mathbf{E}$ and $\mathbf{H}$ vectors are perpendicular to each other and the transmitted power density is given by the product EH for root-mean-square (rms) values of E and H.

A thorough, rigorous treatment of e-m wave propagation requires reference to Maxwell's famous equations; space does not permit such a treatment here.

147

Some examples of useful textbooks dealing with electromagnetic fields and waves are those by Johnk (1), Jordan and Balmain (2), and Ramo, Whinnery, and Van Duzer (3).

A basic relation among wavelength, frequency, and velocity for any kind of a wave can be derived directly on the basis of the definitions of the quantities involved, but we simply state the relation here. It is

$$\lambda f = v_p, \tag{3}$$

where λ is wavelength in m, f is frequency in hertz (Hz), and v_p is phase velocity in m/s. For an e-m wave in a lossless medium propagating in the $+z$ direction and having an electric field of E,

$$E = E_o e^{-j\beta z}, \tag{4}$$

where β is the phase constant and E_o is the value of E at $z = 0$. The equation shows that the phase of the wave lags with increasing distance z. The parameter β can be expressed in several ways as shown by Eq. (5).

$$\beta = 2\pi/\lambda = \omega/v_p = \beta_o n \tag{5}$$

The quantity ω is angular frequency ($2\pi f$, where f is frequency in Hz) and β_o is the phase constant of empty space. The relation introduces the important quantity n, the index of refraction. Index of refraction is defined by

$$n = c/v_p, \tag{6}$$

where c is the "velocity of light" and has the approximate value of 3×10^8 m/s and v_p is phase velocity of the e-m wave in question in the medium being considered. A more accurate value of c is 2.9979×10^8 m/s, and, since 1983, c has been taken to have the exact value of 299,792,458 m/s. In discussions of propagation in the following sections, the index of refraction often plays an important role. Assuming $\mu = \mu_o$, the index of refraction n and the relative dielectric constant K are related by $n = K^{1/2}$. That is, n is the square root of K. Discussions of propagation can be carried out in terms of n or K, but we most commonly will prefer to use n. Equation (6) can be rearranged to show that $v_p = c/n$ and, by reference to Eq. (3), it can be seen, taking f to have a fixed value, that $\lambda = \lambda_o/n$, where λ_o is free-space wavelength and λ is wavelength in the medium of interest, such as the earth's troposphere or ionosphere.

In the troposphere, n is only slightly greater than unity, but variations in the value of n due to temperature and humidity variations can cause important effects on radio wave propagation, especially on nearly horizontal paths. In the ionosphere, n is commonly less than unity, and reference to Eq. (6) indicates that in this case the phase velocity v_p is greater than c. The phase velocity, however, is not necessarily the velocity with which information or energy is transmitted. When the phase velocity is greater than c, it develops, for the cases of interest here, that the velocity of transmission of information or energy is the group velocity v_g and that $v_g v_p = c^2$. This relation shows v_g to be less than c.

The Spectrum of Electromagnetic Waves

Electromagnetic waves exist over a tremendously wide range of frequencies extending from nearly zero to as high as 10^{23} or more. Included in this range are radio, infrared (IR), visible, and ultraviolet (UV) waves (see Fig. 1). Radio waves are used very commonly for telecommunications and remote sensing, and interest in the use of IR, visible, and even UV waves for such purposes is increasing. The rapidly increasing use of optical (visible or near visible) frequencies in optical fibers for telecommunications is indeed a major technological development of our times. Within what is considered to be the radiofrequency band itself, interest in higher frequencies also is increasing. Such microwave bands as the 6- and 4-gigahertz (GHz) bands for example, have been used widely for satellite communications, and interest now is developing in the use of millimeter waves having frequencies as high as 100 and 200 GHz or higher for communications and/or radar.

Satisfactory operation of a telecommunications system requires a sufficient signal-to-noise ratio, and low signal attenuation and low noise contribute to this end. For communication through the earth's atmosphere, frequencies in the 1-to-10 GHz band tend to be ideal because they are characterized by both low attenuation and low noise of natural origin. It is of interest to note that attenuation and noise do not necessarily occur independently of each other, as dissipative attenuation is accompanied by the emission of noise. Propagation through rain, for example, is treated in a section below, but we mention here that it is characterized by both attenuation and noise. For a low-noise system, the degradation in signal-to-noise ratio due to noise emitted by rain may be greater than that caused by the attenuation of the signal due to the rain.

The electromagnetic spectrum is an extremely valuable natural resource, and there is much competition for its use. In the United States, the National Telecommunications and Information Administration (NTIA) of the Department of Commerce plays a leading role with respect to use of the spectrum by the federal government; within NTIA, there is the Office of Spectrum Management. The Institute of Telecommunication Sciences, a research organization located in Boulder, Colorado, is also a part of NTIA. Another important agency of the federal government for dealing with telecommunications is the International Radio Advisory Committee (IRAC). IRAC consists of representatives of major U.S. governmental agencies and departments. For nongovernmental allocations and uses of the electromagnetic spectrum, the Federal Communications Commission (FCC) plays a leading role. Congress is involved considerably in oversight of the FCC. Liaison exists between the FCC and IRAC.

At the international level, a mechanism is needed to try to ensure that the electromagnetic spectrum will be used for the maximum benefit of all nations, and the International Telecommunication Union (ITU) fills that role. Of major interest here are two consultative committees of the ITU, namely, the International Radio Consultative Committee (CCIR, corresponding to the initials in the

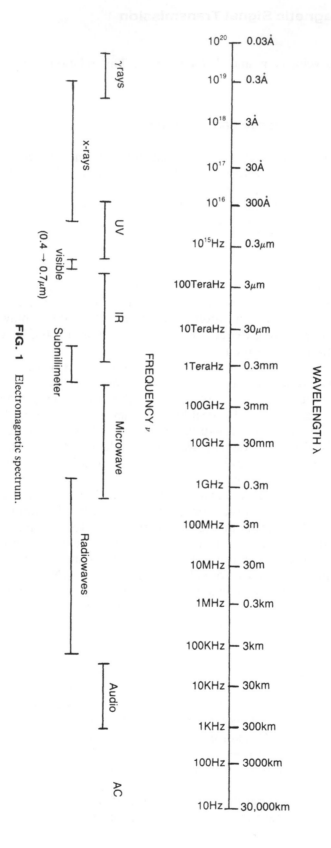

FIG. 1 Electromagnetic spectrum.

French version of the title as in the case of several other such organizations) and the International Telegraph and Telephone Consultative Committee (CCITT). Both the CCIR and CCITT have active study groups in the various technical areas of the ITU. The efforts of these study groups result in volumes of reports and recommendations, which are published every four years. These universally are held in high respect. The recommendations and reports of the CCIR, published in English as *Green Books*, are an excellent source of information on radio wave propagation for professionals and advanced students. A number of references to CCIR reports, particularly those of Study Group 5, appear in the following sections. Another important part of the ITU is the International Frequency Registration Board (IFRB). The International Scientific Radio Union (URSI) and the Antennas and Propagation Society (APS) of the IEEE (Institute of Electrical and Electronics Engineers) also make important contributions to the subject of electromagnetic wave propagation. The U.S. National Committee of URSI normally holds winter meetings in Boulder, Colorado, every year except for every third year following the International URSI General Assembly. Joint meetings of URSI and APS commonly are held every year in the spring or summer at various locations in the United States or Canada. A paper by Smith and Kirby (4) describes the CCIR, and the ITU is the subject of a text by Codding and Rutkowski (5). The *Telecommunication Journal* is a monthly publication of the ITU.

Transmission Lines, Waveguides, and Optical Fibers

Transmission Lines

Electromagnetic waves utilized for telecommunications may travel from one location to another by propagation through such an essentially unbounded medium as the earth's atmosphere or they may be guided by transmission lines, waveguides, or optical fibers. Even when most of the path from one location to another is through the atmosphere, transmission lines or waveguides almost invariably form the connecting links between the transmitter and the transmitting antenna and the receiver and the receiving antenna. We take transmission lines to be two or more conductor lines, such as coaxial lines, parallel-wire lines, or microstrip lines. The term *waveguide* refers to hollow metallic guides. Optical fibers are actually dielectric waveguides, but we generally refer to them as fibers.

Coaxial lines are an important form of transmission line; they have the advantage of the electric and magnetic fields of the line being confined entirely to the space between the inner and outer conductors of the line. The usual spatial orientations of the electric and magnetic fields are shown in Fig. 2. The electric and magnetic fields are everywhere perpendicular to each other and to the axis or length of the line in the lossless case. Such a wave is known as a TEM wave, with T standing for transverse, E for electric field, and M for magnetic field. Coaxial lines can operate from DC (direct current) up to frequencies in the GHz range. Attenuation increases with frequency and eventually

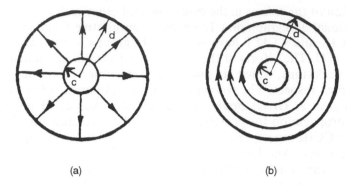

(a)   (b)

FIG. 2   Configuration of fields in coaxial transmission line of inner radius a and outer radius b:
a, lines representing electric field **E** in the transverse plane, perpendicular to the length
of the line; b, lines representing the magnetic field **H** in the transverse plane.

becomes a limiting factor. An expression for the electric field E in a coaxial
line, in terms of inner and outer radii and voltage V between the two conductors,
is given by

$$E = V/[r \ln (c/d)], \tag{7}$$

where r is the radial coordinate, c is the radius of the inner conductor, d is the
radius of the outer conductor, and ln stands for natural logarithm. Also, with I
the current flowing in the line and H the magnetic field value at the radius r
from the center of the inner conductor,

$$H = I/2\pi r. \tag{8}$$

The ratio of V^+ to I^+, where the + symbols refer to values for a positively
traveling wave, is known as the characteristic impedance of the line and is
indicated here by Z_o. A practical significance of characteristic impedance is
that, if the line is terminated by a load impedance Z_L equal to the characteristic
impedance, all of the energy incident upon the load impedance is absorbed and
no energy is reflected. If, however, the load impedance does not equal the
characteristic impedance, some energy is reflected, a wave traveling in the negative direction results, and the combination of positive and negative traveling
waves generates a standing wave on the line. Analysis of this situation discloses
that the standing wave ratio (SWR), namely, the ratio of maximum to minimum
voltage, is given by

$$SWR = [1 + |\Gamma_L|]/[1 - |\Gamma_L|], \tag{9}$$

where Γ_L, the reflection coefficient, is given by

$$\Gamma_L = [Z_L - Z_o]/[Z_L + Z_o]. \tag{10}$$

Commercially available coaxial lines have such standard values of Z_o as 50 ohms, 75 ohms, and so on. For transmission lines and waveguides other than coaxial lines, the field configurations and the expressions for values of characteristic impedance are different, in general, but the relations between SWRs, reflection coefficients, and impedances are the same.

Waveguides

Hollow metallic waveguides are used extensively for microwaves and millimeter waves, for example, in radar systems. Metallic waveguides may be rectangular, circular, or elliptical in cross-section, but we give attention here only to the commonly used TE_{10} mode in rectangular guides. Considering the waveguide to extend in the z direction, a TE mode has the electric field confined to the transverse plane, namely, the x-y plane. Let the guide have a width of a in the x direction and a width of b in the y direction (see Fig. 3). The electric field of the TE_{10} mode is only in the y direction and goes to zero at $x = 0$ and $x = a$, as shown in Fig. 3-b. The subscript "10" indicates that the electric field has one maximum between $x = 0$ and $x = a$ and does not vary in magnitude in the y direction. The electric field goes to essentially zero at $x = 0$ and $x = a$ because the electric field is tangential to the metallic wall at these positions, the metallic wall is very nearly a perfect conductor, and the tangential component of electric field at the surface of a perfect conductor must be zero.

Waveguides have the interesting property of having a cutoff wavelength and cutoff frequency. Only waves having a shorter wavelength than the cutoff wavelength (and therefore a higher frequency than the cutoff frequency) can propagate in a waveguide of the type considered here. A valid depiction of wave propagation in the TE guide considered here is shown in Fig. 4. The diagram suggests that the TE_{10} mode propagates in a guide by being reflected back and forth from one side of the guide to the other. At this point, the concept of λ_g, the wavelength in the guide, as distinguished from λ_o, the wavelength in the

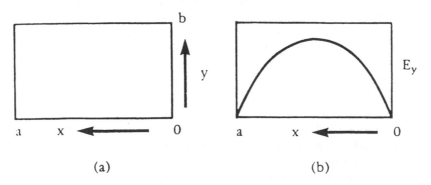

(a)   (b)

FIG. 3   Rectangular waveguide: a, geometry; b, plot of E_y versus x for TE_{10} mode.

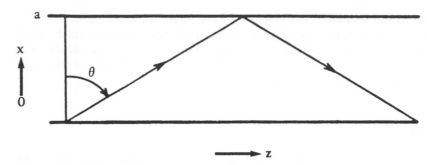

FIG. 4   Illustration of a wave that is reflected back and forth from one side of a metallic waveguide to the other.

direction a wave actually is propagating, will be introduced. The wavelength λ_o is what one ordinarily thinks of when wavelength is mentioned, and it is the wavelength used in considering wave propagation in an unbounded medium. It is also the wavelength in the direction of the diagonal lines of Fig. 4. By definition, wavelength in any direction is merely the distance between two locations that differ in phase by 2π radians or 360°, and wavelength is shortest in the direction a wave actually is propagating. It develops that $\lambda_g = \lambda_z$ and that λ_z is given by

$$\lambda_z = \frac{\lambda_o}{\sin \theta}. \tag{11}$$

The wavelength in the x direction λ_x, however, is given by

$$\lambda_x = \frac{\lambda_o}{\cos \theta}. \tag{12}$$

Setting the waveguide width a equal to $\lambda_x/2$, as is the case for the TE_{10} mode, it can be determined that $\cos \theta = \lambda_o/2a$. If $\lambda_o < 2a$, $\cos \theta < 1$, and $\theta > 0$, reference to Fig. 4 shows that a wave can propagate along the guide. If, however, $\theta = 0$, a wave just bounces back and forth from one side of the guide to the other without making any progress in moving along the length of the guide. The condition $\lambda_o = 2a$ for which $\theta = 0$ is of importance, and the wavelength that satisfies this condition is known as the cutoff wavelength λ_c. Waves with wavelengths that satisfy this condition or are greater than λ_c will not propagate along a waveguide, but waves with wavelengths less than λ_c (or frequencies greater than $f_c = c/\lambda_c$) will propagate.

Waveguides also are available that have circular and elliptical cross-sections. The mathematics needed to analyze these guides is more complicated than for rectangular guides, but they have the same characteristic of having cutoff wavelengths and frequencies. The elliptical guide has the flexibility advantage such that it can be bent and twisted more readily than the rectangular guide. It does not have to be cut precisely to accurate lengths in order to join sections of guide together.

Optical Fibers

In the case of propagation in a hollow metallic waveguide, the wave that bounces back and forth from one side of the guide to the other experiences total reflection at the metallic walls. An optical fiber is a variety of dielectric waveguide, and total reflection can take place at the wall of a dielectric guide also. Reference to a text on basic e-m theory (3) or to a treatment of optoelectronics (6) shows that total reflection will take place in propagation from a medium having an index of refraction n_1 (Medium 1) to a medium having a lower index of refraction n_2 (Medium 2) if the angle of incidence θ_i, measured from the perpendicular, is greater than θ_c, where

$$\theta_c = \sin^{-1}(n_2/n_1).\qquad(13)$$

This situation is suggested in Fig. 5. Actually, the e-m fields do not drop to zero immediately in Medium 2. Instead, the fields decay exponentially. No energy is transmitted away from Medium 1 if the second medium extends to infinity. The type of wave that exists in Medium 2 in such a case is called an *evanescent wave*, the term *evanescent* meaning fleeting or vanishing. The fact that e-m fields can exist in Medium 2 even though they are evanescent can disturb the transmission characteristics of a dielectric guide if other objects are close to the boundary. In practice, however, such a problem can be eliminated

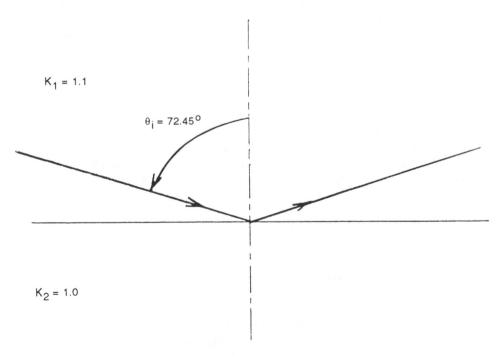

FIG. 5 Total reflection takes place in this example when the angle of incidence, measured from the perpendicular, is 72.45° or greater. This value of θ_i is the critical value θ_c of Eq. (13), in which, for this case, $n_1 = \sqrt{K_1} = 1.049$ and $n_2 = \sqrt{K_2} = 1$.

by introducing a cladding layer outside the original dielectric guide. The characteristics of the cladding are such that e-m fields have dropped to insignificant values at the outer edge of the cladding. The discussion to this point is appropriate for planar dielectric guides and for meridional rays in cylindrical optical fibers. Meridional rays progress along the length of the guide by reflection from one side to the other without experiencing displacement or rotation in azimuth. Skew rays constitute another category of the rays in optical fibers. These rays, which describe approximations to helices as they progress along the fiber, are more difficult to analyze than meridional rays.

The concept of total reflection at a dielectric boundary has been known for a long time, but, even after the development of cladding, the idea of communication by optical fibers did not become promising immediately because of the high values of attenuation encountered in such fibers. Attenuation values of 1000 decibels/kilometer (dB/km) were typical at one time, but values below 1 dB/km now can be achieved readily (6), and the application of optical fibers to telecommunications is proceeding at a rapid rate.

Optical fibers can be step-index or graded-index fibers. In a step-index fiber, the index of refraction changes abruptly from one value to the other at the boundary between the core and the cladding. This type of fiber waveguide is the most similar to the planar waveguide. In a graded-index fiber, the index of refraction changes gradually from the center of the core to the cladding. In terms of the modes, fibers can be multimode or single mode. The different modes in multimode fibers travel with slightly different velocities, with a single sharp impulse, for example, experiencing dispersion (broadening) in time. A principal reason for using graded-index fibers is that they have considerably less dispersion than step-index fibers. The number of modes N that propagate in a fiber is given approximately (7) by

$$N = \frac{v^2}{2} = \frac{2\pi a}{\lambda} [n_1^2 - n_2^2]^{1/2},$$ (14)

where a is the radius of the core of the fiber, n_1 is the index of refraction of the core, and n_2 is the index of refraction of the cladding.

Use of a single-mode fiber avoids the problem of intermodal dispersion and allows the use of large bandwidths. Equation (14) shows that the number of modes is a function of a, the radius of the core, and it develops that to obtain single-mode operation a must be very small. A value of a in the order of a few μm, or even one micrometer (μm), ordinarily is utilized. Higher accuracy is required in the alignment of splices when using single-mode fibers than when using larger multimode fibers, for which a may be around 60 μm. Also, relatively low amounts of power can be carried by single-mode fibers, and it is essential that transmission losses be kept to a minimum when using single-mode fibers. Although single-mode fibers are not subject to intermodal dispersion, material dispersion and waveguide dispersion still occur in single-mode fibers. Material dispersion is due to the variation of the index of refraction with wavelength, and waveguide dispersion refers to the variation of the phase constant of a waveguide with wavelength. All signals have finite bandwidths and therefore involve a range of wavelengths, and the variation of fiber parameters with wavelength contributes to dispersion. Profile dispersion also takes place in graded-index fibers as the variation of index of refraction with radius that is optimum for a certain wavelength may not be optimum for another wavelength.

Attenuation in optical fibers is due to bending losses, Rayleigh scattering, the presence of impurities, and absorption that is inherent to the molecules and lattice structure of fibers. Rayleigh scattering arises from irregularities of structure or composition and varies as λ^{-4}. Attenuation due to Rayleigh scattering therefore decreases quite rapidly with increasing wavelength. Metallic or hydroxyl ions are serious types of impurities, and much of the decrease in attenuation of optical fibers that has taken place has been due to better control of impurities. Some of the lowest attenuations, around 0.2 dB/km at a wavelength of 1.55 μm, have been achieved by use of germanium dioxide (GeO_2) doping of silicon dioxide (SiO_2) fibers. This value of 0.2 dB/km is not much larger than the minimum value of attenuation due to Rayleigh scattering at this wavelength (6).

Reference 6 has good chapters on fiber optical waveguides and optical communications systems. Other recent references on optical fibers and optoelectronics are those by Cheo (8), Yariv (9), and Yeh (10). A comparison of e-m wave propagation by use of fibers and satellites is discussed in a separate section of this article.

Basic Concepts of Ionospheric Propagation

Before the commercial advent of satellites in the 1960s, high-frequency (HF) (3–30 megahertz [MHz]) transmissions, propagated by reflection from the ionosphere, satisfied a significant fraction of very-long-distance communication needs. By very long distance, we refer especially to transmission from one hemisphere or continent to another but also to transmission across a considerable extent of a continent or to distant islands. For about two decades after the introduction of the first commercial geostationary satellite, INTELSAT I, in 1965, satellites took over ever-larger shares of long-distance communication. More recently, optical fiber systems have supplied increasing competition to satellites and other systems of telecommunications. Before satellites and optical fibers, transcontinental service commonly was supplied in the United States and elsewhere by coaxial lines and terrestrial microwave links. Such systems continue to be used but are no longer preferred for most new applications. Interest in ionospheric propagation lessened when satellite service became widespread, but it later experienced somewhat of a revival due to the need for maintaining communications in military situations in which satellites and certain other systems might be vulnerable to attack. Also, ionospheric effects may be important in satellite and deep-space telecommunications and in using such systems as GPS (Global Positioning System) for precise determination of distance or location.

The earth's ionosphere extends from about 50 km to about 1000 km or more above the earth's surface. A small fraction of the atoms and molecules in this region become ionized, mostly by ultraviolet and x-ray radiation from the sun, to form free electrons and positive ions. We are concerned here mostly with the free electrons, as they have the largest effect on the electromagnetic waves that are used for telecommunications. Different portions of the solar spectrum interact with different constituent atoms and molecules of the ionosphere to form different layers, designated as the D, E, and F layers. The F layer, which has the highest electron densities of any of the layers, often is separated in the daytime into the F_1 and F_2 regions. The D layer, extending from about 50 to 90 km above the earth's surface, is the lowest layer and has the lowest electron densities. The maximum electron density of the D layer, about $10^9/m^3$, occurs between about 75 to 80 km. Despite the low electron density, the D layer has a high collision frequency and causes the familiar effect of inhibiting distant reception of amplitude modulation (AM) broadcast stations in the daytime but allowing distant reception at night. (Attenuation in the ionosphere is proportional to both electron density N and collision frequency v [11].) The E layer typically has peak electron densities of about 1 to $1.5 \times 10^{11}/m^3$ in the 90- to 140-km height range. The F_2 layer has the highest electron densities of the regular layers in the ionosphere (up to $2 \times 10^{12}/m^3$) and retains quite high densities even at night when solar radiation is lacking. Long-range HF propagation commonly involves reflection from the F_2 layer but may involve reflection from the E and F_1 layers as well.

A very recently completed text by Davies (12), written to replace two earlier texts by the same author, provides a comprehensive treatment of ionospheric propagation. Other treatments of the ionosphere or ionospheric propagation include those provided by Budden (13), Kelso (14), Ratcliffe (15), and Rishbeth and Garriott (16). A brief introduction to ionospheric propagation by Flock (11) is based upon the approach of Allis, Buchsbaum, and Bers (17) to propagation in a plasma (an ionized medium).

Linear and Circularly Polarized Waves

The ionosphere is permeated by the earth's magnetic field, and this magnetic field affects the propagation of electromagnetic waves. For example, wave propagation parallel to the magnetic field is different from propagation perpendicular to the field. Also, linear and circularly polarized waves are affected differently. The term *polarization* has several meanings. Here, a linearly polarized wave has its electric field **E** oriented in a fixed direction in space, whether vertical, horizontal, or in between. The magnitude of **E** varies sinusoidaly in the case of a linearly polarized wave. A circularly polarized wave has an **E** vector of fixed length that rotates with a uniform angular velocity of ω, with $\omega = 2\pi f$ and f the frequency of the wave in Hz.

The very important parameter n, the index of refraction, and the closely related parameter K, the relative dielectric constant that equals n^2, were introduced in the discussion of basic concepts of e-m waves. We now consider the values of K for the waves that propagate in such an ionized medium as the ionosphere. This type of medium commonly is referred to as a plasma, obviously an entirely different kind of plasma than that found elsewhere. Although we are unable to provide a general treatment of Maxwell's equations here, we now introduce one of Maxwell's equations in the following form.

$$\nabla \times \mathbf{H} = \mathbf{J} \tag{15}$$

Here the symbol ∇ is a vector operator, $\mathbf{H}$ is the magnetic field, and $\mathbf{J}$ is total electric current density. $\mathbf{J}$ includes any and all current densities that are present, including the vacuum displacement current density. Assuming sinusoidal time dependence J_o, the current density in a vacuum, has the form of $j\omega\epsilon_0 E$, with E the electric field. In a plasma, free electrons are present, and these are subject to a force due to the electric field and achieve a velocity parallel to the electric field. Electrons in motion constitute an electric current. To account for the effect of the plasma, it is necessary to determine an expression for this current density. We do this first for the case of no magnetic field. For this purpose, recall that the force on an electron, $\mathbf{f} = q\mathbf{E} = m\mathbf{a}$, where q is the charge of the electron, m is the electron mass, and $\mathbf{a}$ is acceleration. Bold type indicates a vector quantity. Acceleration is the derivative of velocity, and for sinusoidal time dependence $\mathbf{a} = j\omega\mathbf{v}$. Making use of this relation and the fact that $\mathbf{J} = Nq\mathbf{v}$ allows solving for $\mathbf{J}_1$, the convection current density due to the electrons. Then J can be set equal to $J_o + J_1$, and a person with experience with Maxwell's equations will recognize that J can be set equal to $j\omega\epsilon_o KE$, where $K = n^2$ is the relative dielectric constant, in this case for wave propagation in a plasma in the absence of a magnetic field. Carrying out this determination shows that

$$K = 1 - \omega_p^2/\omega^2, \tag{16}$$

with $\omega_p^2 = Nq^2/m\epsilon_o$ and ω_p known as the (angular) plasma frequency. The expression actually applies in the presence of a magnetic field also, if the electric field E is parallel to the magnetic field, because the force due to the magnetic field is zero if the velocity v due to E is parallel to the magnetic field [Eq. (2)]. Consideration of Eq. (16) shows that K is less than unity and, assuming that K also is greater than zero, and recalling that n, the index of refraction, is the square root of K, it is evident that n is less than unity also. Therefore, the phase velocity v_p is greater than c [Eq. (6)]. This condition might at first seem impossible, as it commonly is assumed that no velocity can be greater than c. But the correct statement is that information and energy cannot be transmitted with a velocity greater than c. This point is discussed at the end of the basic e-m waves concepts section.

The analysis for determining the value of K for circularly polarized waves is not given here, but the results are stated below for K_l and K_r, the relative dielectric constants for the left and right circularly polarized waves, respectively.

$$K_l = 1 - \frac{\omega_p^2}{\omega(\omega + \omega_B)} \tag{17}$$

$$K_r = 1 - \frac{\omega_p^2}{\omega(\omega - \omega_B)} \tag{18}$$

Here ω_B is the angular gyrofrequency of electrons (the angular frequency of their rotation about magnetic field lines) and is given by $|qB/m|$, with q the charge of the electron, m the mass of the electron, and B the value of the earth's magnetic field in teslas or webers/m^2 (Wb/m^2) at the location in question. Because of the difference in the values of K_l and K_r, left and right circularly polarized waves propagate with at least slightly different velocities and have at least slightly different wavelengths.

Circularly polarized waves frequently are used for satellite communications, for example, but one reason for introducing them at this point is to treat the phenomenon of Faraday rotation of linearly polarized waves. To a first approximation, a linearly polarized wave propagating through the earth's ionosphere will remain linearly polarized, but the linear E vector of the wave will experience rotation. This rotation can be analyzed by recognizing that a linearly polarized wave can be represented as the sum of two circularly polarized waves of equal amplitude. Consider a linearly polarized wave propagating in the z direction with the electric field **E** directed along the x axis at $z = 0$. Such an electric field can be generated by left and right circularly polarized waves oriented symmetrically but rotating in opposite directions with respect to the x axis and each having an amplitude of $E/2$. As the linearly polarized wave progresses in the z direction, the same general picture applies except that the phases of rotation of the circularly polarized components differ with the result that after a certain distance the vectors may be oriented as in Fig. 6. The linearly polarized wave is shown to have rotated through an angle of ϕ. E_l and E_r, the electric fields of the left and right circularly polarized waves, respectively, are shown as vectors having half the amplitude of E and are identified by auxiliary arrows showing their directions of rotation. The electric field of the linearly polarized wave is always halfway between the instantaneous positions of the two circularly polarized components. The angle of Faraday rotation of **E** is given by

$$\phi = (\beta_l - \beta_r)/2 \qquad \text{rad,} \tag{19}$$

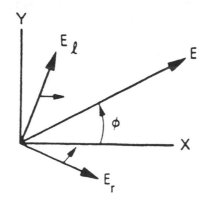

FIG. 6   Faraday rotation through an angle ϕ from the x axis.

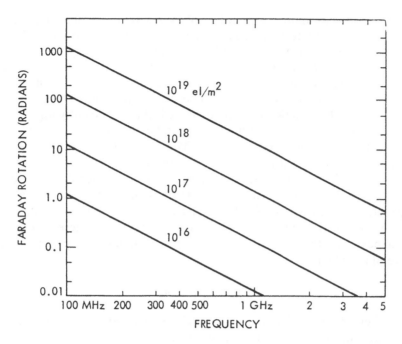

FIG. 7 Faraday rotation as a function of TEC and frequency. (From Ref. 18.)

where β_l and β_r are the phase constants for the left and right circularly polarized waves, respectively. The phenomenon of Faraday rotation may or may not present a problem, but frequently needs to be taken into account. Faraday rotation is shown as a function of total electron content (TEC) and frequency in Fig. 7.

It has been pointed out above that the polarization of a linearly polarized wave changes as it propagates in the ionosphere. (The polarization may stay linear, or nearly so, but the orientation of the **E** vector changes.) Characteristic waves are waves that do not change their polarization as they propagate, and a linearly polarized wave thus is not a characteristic wave. Left and right circularly polarized waves, however, do not change their polarization as they propagate and they are characteristic waves for propagation in the ionosphere.

Propagation of Electromagnetic Waves in the Ionosphere

The discussion following Eq. (16) is based on the assumption that K was less than unity but greater than zero. However, it is possible for K to be zero or even negative. The condition $K = 0$ occurs when $\omega = \omega_p$, and K is negative when ω is less than ω_p. When K is negative, the index of refraction n, which is the square root of K, is imaginary. Thus, it turns out that the condition $\omega = \omega_p$ plays the same type of role for propagation in a plasma as cutoff frequency plays for propagation in waveguides. Waves having angular frequencies ω greater than ω_p propagate in a plasma and, neglecting attenuation, their fields vary as $e^{-j\beta z}$ with z representing distance. Waves having angular frequencies ω less than ω_p do not propagate but are evanescent waves. Their fields decrease with distance as $e^{-\alpha z}$, even when no dissipative attenuation occurs. A propagating wave that impinges on a region or layer where K is negative experiences reflection.

The result of the situation described above is that sufficiently high-frequency waves pass through the ionosphere without being reflected, but sufficiently low-frequency waves may be reflected. The simplest case is that of vertical incidence. A linearly polarized wave that is unaffected by the earth's magnetic field will be reflected at vertical incidence from an ionospheric layer if the electron density in the layer reaches a value such that ω_p becomes equal to ω. In most practical applications of HF to long-distance communication, however, e-m waves are incident obliquely upon the ionosphere. In these cases, in terms of frequency f in Hz (or MHz, etc.) rather than angular frequency, reflection will take place when

$$f = f_p \sec \phi_o \qquad (20)$$

or less, where ϕ_o is the initial launch angle of the wave upon the ionospheric layer in question, measured from the vertical.

Certain basic aspects of ionospheric propagation are treated in this section, and some ionospheric effects that are especially pertinent to earth-space propagation, including ionospheric scintillation, are considered in the section, "Ionospheric Effects."

Terrestrial Line-of-Sight Transmission

Terrestrial microwave line-of-sight systems have been used extensively for telecommunications but now face stiff competition from satellite and fiber-optics systems. A number of propagation effects must be considered in the design of line-of-sight links, but we concentrate mainly on refraction effects in this section and give brief attention to effects associated with terrain also. Effects due to atmospheric gases, clouds, and precipitation are considered in a separate section. Terrestrial line-of-sight transmission effects, including those due to the refractive structure associated with atmospheric turbulence, are important to radar and remote sensing systems as well as to communications. Troposcatter systems rely, to a large extent at least, on scatter from refractive irregularities associated with turbulence. These irregularities are discussed briefly in this section, but we do not devote a separate section to troposcatter.

Index of Refraction of the Troposphere

The index of refraction is an important parameter affecting the propagation of e-m waves in the ionosphere, and also in the troposphere. The index of refraction of the troposphere is only slightly greater than unity, and it might appear that refractive effects would not be important, but such is not the case. Because the index of refraction is so close to unity it is common practice to use refractivity measured in N units rather than values of n itself when discussing the index of refraction of the troposphere. N units are given in terms of n by

$$N = (n - 1) \times 10^6. \tag{21}$$

If $n = 1.000300$, for example, the refractivity in N units is 300. The refractivity of the troposphere is a function of pressure, temperature, and water vapor (19) and is given by

$$N = \frac{77.6p}{T} + \frac{3.73 \times 10^5 e}{T^2}, \tag{22}$$

where p is atmospheric pressure in millibars (mb), T is temperature in kelvins, and e is water vapor pressure in mb. Note the major effect of e in determining N. The standard sea level pressure is 1013 mb. The water vapor pressure e is equal to the product of the saturation water vapor pressure and the relative humidity. For a temperature of 20°C and a relative humidity of 70%, for example, the saturation water vapor pressure is 23.4 mb, according to the 1984 *Smithsonian Meteorological Tables* (20), and e is 16.4 mb.

Propagation in the troposphere is influenced by the index of refraction profile with height. An exponential decrease of N with height often is assumed, in accordance with

$$N = N_s e^{-h/H}, \tag{23}$$

where N_s is the surface value of refractivity, h is height above the surface, and H is a constant. The average value of N_s in the United States has been taken to be 313 and a value of H of 7 km has been considered to be appropriate for the United States.

Ray Paths in the Troposphere

Because the index of refraction of the troposphere tends to vary with altitude, ray paths at elevation angles other than vertical tend to be curved rather than to follow straight lines. Curvature C at a point is defined mathematically as $1/\rho$ where ρ is radius of curvature. That is

$$C = 1/\rho. \tag{24}$$

It can be shown that, when the index of refraction n varies with altitude at a constant rate of dn/dh, the curvature of a ray is given by

$$C = \frac{dn}{dh} \cos \beta, \tag{25}$$

where h is height and β is the angle of the ray measured from the horizontal (21). The difference between the curvature of a ray and the curvature of the earth's surface is important when considering propagation over the surface of the earth. Taking the radius of the earth to be r_o and assuming a nearly horizontal path so that $\cos \beta$ is approximately unity, the difference between the earth's curvature and the curvature of the ray is

$$\frac{1}{r_o} - C = \frac{1}{r_o} + \frac{dn}{dh}. \tag{26}$$

Instead of using the actual earth curvature and the actual ray curvature, however, the same relative curvature can be maintained if one uses an effective earth radius kr_o and a ray of zero curvature such that

$$\frac{1}{r_o} + \frac{dn}{dh} = \frac{1}{kr_o} + 0. \tag{27}$$

The advantage of this approach is that, because the ray is taken to have zero curvature, the ray path can be drawn as a straight line. A typical value for dn/dh at the earth's surface is $-40\ N$ units/km, for which the corresponding value of k is 4/3. Assuming this value of k to be applicable, a radius of 4/3 times the radius of the earth can be used and the ray paths can be drawn as straight lines. However, 4/3 is only a typical value for k, and it generally is advisable to assume at least a limited set of values of k, say k equal to 4/3, 1, and 1/2. Another approach is to let the surface of the earth be flat and draw curved ray paths for different values of k (Fig. 8). A k value of $-157\ N$ units/ km is of interest because in this case a ray that is launched horizontally will remain indefinitely at the same fixed height above the earth. This condition, of course, is not met precisely in practice but an approach to this condition often is responsible for the ability of radar to detect targets at seemingly unusually large distances, as when radar monitors the movements of ships over the surface

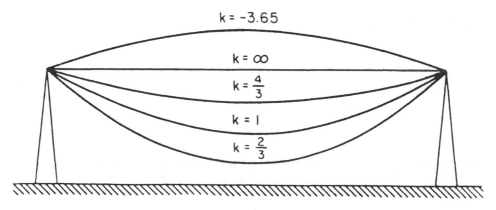

FIG. 8　Ray paths from a transmitter T to a receiver R for various values of k (exaggerated and not to scale).

of the ocean. This phenomenon is considered further in the section, "Propagation Effects on Interference."

In designing or analyzing possible or existing terrestrial microwave links, the earth's terrain needs to be taken into account in conjunction with ray paths corresponding to a set of k values as mentioned in the paragraph above. Possible reflections from such flat surfaces as bodies of water should be avoided and checks should be made of terrain irregularities and potential obstacles to see if sufficient path clearance is provided. The needed clearance commonly is expressed with relation to the first Fresnel-zone radius F_1. All rays passing through the first Fresnel zone add constructively, and the effect of obstructions within the first Fresnel zone may be important. The value of F_1 is given by

$$F_1 = 17.3\left[\frac{d_T d_R}{f_{\text{GHz}}d}\right]^{1/2},\tag{28}$$

where d_T is the distance along the path from the transmitter, d_R is the distance from the receiver, and $d = d_T + d_R$. The distances are in km and F_1 is in m. The criteria given by GTE Lenkurt (22) are at least $0.6\,F_1 + 3$m for light route or medium reliability systems' clearance, and at least $0.3\,F_1$ at $k = 2/3$ and at least $1.0\,F_1$ at $k = 4/3$ for heavy route or highest reliability systems' clearance.

A reason for avoiding reflections is to avoid potential interference between signals reaching the receiver by more than one path. In the absence of surface reflections, multipath conditions, however, can be encountered due to atmospheric conditions alone. Indeed a margin of 30 or 40 dB or even more must be supplied on most nearly horizontal terrestrial paths to overcome signal fading due to multipath effects. Multipath is most apt to occur under stable atmospheric conditions, especially in the presence of temperature inversions. All vertical motions are inhibited strongly in an inversion layer, in which the temperature increases with height, and both pollution and water vapor existing below the layer tend to be confined below it. The refractivity N may decrease rapidly and depart significantly from the exponential form of Eq. (23) in upward passage through an inversion layer. Inversions are a common occurrence, especially at night in arctic, subarctic, or desert conditions and in such regions of subsiding air as the Los Angeles basin. Another type of signal variation, known as *scintillation*, commonly due to atmospheric turbulence, involves more rapid variations of smaller magnitude than the signal fading caused by multipath effects (23).

In addition to the variation of index of refraction with height, the index also exhibits variations with atmospheric turbulence. The theory of turbulence indicates that it develops from wind shear, that turbulence is introduced in the form of large turbulent eddies or blobs of scale size L_o, and that energy is transferred from larger to smaller eddies throughout an inertial subrange corresponding to eddies of size l where $L_o \geq l \geq l_o$. For eddies of smaller size than l_o viscous effects dominate and turbulent energy is dissipated. The refractivity structure associated with atmospheric turbulence is considered to be the principal mechanism, or a principal mechanism, allowing communication over long non-line-of-sight paths by troposcatter.

Propagation through Earth's Atmosphere at Elevation Angles above 10°

This section deals with oblique radiowave propagation at microwave and millimeter wavelengths through the earth's atmosphere. Thus, the section includes effects on paths between the earth's surface and satellites in geostationary or lower-elevation orbits around the earth. It also applies to communication with spacecraft in deep space and may apply to communication with aircraft. We will refer to all of the paths as earth-space paths and assume that their elevation angles are 10° or greater, commonly 30° and greater. As a result, refractive fading effects and attenuation due to rain are minimized in comparison to the case of nearly horizontal paths. An obvious reason in the case of rain is that rain occurs only up to heights of several km above the earth's surface, and path lengths through rain therefore are lower on earth-space paths than on terrestrial paths. On the other hand, earth-space paths often operate with smaller margins than terrestrial paths and there presently is considerable interest in using higher frequencies (e.g., 30 and 20 GHz) on earth-space paths. These factors may cause rain to be very important in the consideration of earth-space paths. Ionospheric effects, including Faraday rotation, may be important on earth-space paths operating at frequencies below 10 GHz (especially below about 2 GHz) but ionospheric effects generally are negligible above 10 GHz. Attenuation and noise due to rain, clouds, and atmospheric gases, on the other hand, are considerably more serious above 10 GHz than below. It should not be assumed, however, that effects of rain are necessarily unimportant below 10 GHz. Because of the advantages of frequencies between about 1 and 10 GHz, early commercial-satellite communications systems operated with uplinks at frequencies near 6 GHz and downlink frequencies near 4 GHz. Other early satellite systems, military and otherwise, likewise have used frequencies in the 1 to 10 GHz band. The ACTS (Advanced Communications Technology Satellite), currently under development by NASA, however, will operate with frequencies near 30 GHz for the uplink and frequencies near 20 GHz for the downlink.

Before actually discussing propagation effects further, we now supply relations that show how propagation and noise affect the performance of telecommunications systems. It is necessary to have a sufficient carrier-power-to-noise-power ratio for telecommunications systems to operate satisfactorily; Eq. (29) shows the factors that determine this ratio. The quantity C/X is used for the ratio, with C standing for received carrier power and X standing for noise power. The expression is given in dB, the value of a power ratio P_2/P_1 being given in dB by $10 \log (P_2/P_1)$.

$$(C/X)_{\text{dB}} = (\text{EIRP})_{\text{dBW}} - (L_{FS})_{\text{dB}} - L_{\text{dB}} + (G_R/T_{\text{sys}})_{\text{dB}} - k_{\text{dBW}} - B_{\text{dB}} \quad (29)$$

Here, EIRP (effective isotropic radiated power) is the product of the transmitted power P_T and the gain of the transmitting antenna G_T; L_{FS} is the quantity $(4\pi d/\lambda)^2$ with d the length of the path and λ the wavelength; L_{dB} is the loss factor, which includes attenuation encountered along the path due to rain and the like; and G_R is the gain of the receiving antenna. Recalling that the Boltzmann's constant is 1.38×10^{-23} joules/kelvin (1.38×10^{-23} J/K), the k of Eq. (29) has the magnitude of the Boltzmann's constant but, unitwise, represents the product of Boltzmann's constant and a 1-K temperature interval and a 1-Hz bandwidth. Thus, k has units of J/second (s) as a Hz (cycle/second) has units of 1/s, and k also has units of power W (watts) as power is the rate at which energy is applied or utilized as a function of time and J/s represents energy divided by time. T_{sys} represents the magnitude only of the temperature on the kelvin scale, and B represents the magnitude only of the bandwidth in Hz, as the units of T_{sys} and B have been absorbed within k. The system noise temperature T_{sys} includes the noise emitted along the path when the path passes through such dissipative media as rain, clouds, and atmospheric gases. Noise is considered further in the section, "Atmospheric Radio Noise." Notice that L and T_{sys} are the two quantities of Eq. (29) of special interest from the viewpoint of the propagation of e-m waves.

Useful references having broad coverage of propagation effects on earth-space paths and/or radio wave propagation in general include texts by Allnutt (24), Hall (25), Hall and Barclay (26), Ippolito (27), and two NASA Reference Publications (28,29).

Ionospheric Effects

Earth-space paths that pass through the ionosphere experience ionospheric effects that tend to be proportional to the TEC along the path. TEC is the integral of electron density along the path, namely, $\int N \, dl$. It can be shown that the Faraday rotation of a linearly polarized wave traveling through the ionosphere at frequencies above about 100 MHz is given by

$$\phi = (2.36 \times 10^4/f^2)B_L \int N \, dl = (2.36 \times 10^4/f^2)B_L \text{ TEC}. \qquad (30)$$

Here ϕ is the angle of Faraday rotation in radians, f is frequency in Hz, N stands for electrons/m^3, and B_L is an average or effective value of the longitudinal component of the earth's magnetic field in Wb/m^2 or teslas. The longitudinal component refers to the component of the earth's field that is parallel to the path. It commonly is considered for paths through the entire ionosphere that it is sufficient to use the value of the longitudinal component of the earth's magnetic field at a height of 400 km, or at a height near 400 km such as 420 km, for B_L. A procedure for determining the value of B_L at the desired height is given in Flock (29).

A practical consequence of Faraday rotation is that, in the frequency range at which Faraday rotation is significant, one cannot transmit using one linear polarization and receive using an antenna with the same linear polarization without having a high probability of encountering a polarization mismatch. Among the techniques for avoiding or dealing with the problem are using a sufficiently high frequency that Faraday rotation is negligible, using a receiving antenna that can accept both orthogonal linear polarizations so that no polarization loss occurs, and using circular rather than linear polarization. The TEC along a path through the ionosphere varies from night to day, with the seasons, with the sunspot cycle, with traveling ionospheric disturbances, and with any other phenomenon that affects electron density and its distribution. Faraday rotation is discussed above as a nuisance or problem, but it can be used to monitor TEC and its variations, as indicated by Eq. (30).

Another effect of the TEC along a path is to produce excess group time and range delay. To consider this phenomenon, note that the integral $\int n \, dl$, where dl is an increment of length along a path, gives the true distance along the path if $n = 1$ but gives the value P, sometimes known as phase path length, which is different from the true distance if n does not equal unity. Thus ΔR, the difference between P and the true distance R, is given by

$$\Delta R = \int (n - 1) dl. \tag{31}$$

Neglecting refraction and considering that $f > 100$ MHz so that $n^2 \approx 1 - X$, where $X = 80.6 \, N/f^2$

$$n^2 = 1 - f_p^2 = 1 - 80.6 \, N/f^2, \tag{32}$$

where N is electron density (electrons/m^3) and f is frequency in Hz. Taking X as small compared to unity, as is the case for sufficiently high frequencies,

$$n = 1 - X/2 = 1 - 40.3 \, N/f^2. \tag{33}$$

For applying Eq. (33), however, one is concerned with group velocity rather than phase velocity. As $v_g v_p = c^2$ for ionospheric propagation when $v_p > c$, where v_g is group velocity and v_p is the phase velocity, the group refractive index is given by $n_g = 1 + X/2$. As a result

$$\Delta R = (40.3/f^2)\int N \, dl = (40.3/f^2) \text{ TEC}. \tag{34}$$

The excess range delay is seen to be inversely proportional to frequency squared and directly proportional to TEC (total electron content along the path). From Eq. (34), it appears that it is necessary to know or estimate the TEC in order to determine ΔR. However, a method for determining TEC by use of two different frequencies is described in the final section of this article, where atmospheric effects on range and velocity determination are considered more thoroughly.

Irregular variations or scintillations of the amplitude of radio stars were recorded first by Hey, Parsons, and Phillips (30), who reported variations in the amplitude of signals from Cygnus and Cassiopeia at 36 MHz. At first, it was thought that the emissions from the stars might be varying with time, but records obtained simultaneously from stations separated by 200 km shows no similarity; when the receiver separation was only about 1 km the records were closely similar (31,32). These results showed that the scintillations were not caused by the stars but were of local origin, and it was concluded that their source was in the ionosphere. The scintillations were attributed by Hewish to a diffraction pattern formed at the ground by a steadily drifting pattern of irregularities in the ionosphere at a height of about 400 km (33).

With the advent of satellites, scintillations of signals from such spacecraft also were observed (34). The signals from radio stars are incoherent and broadband and allow the recording of amplitude and angle-of-arrival scintillations but not phase scintillations. Coherent, monochromatic signals from spacecraft have the advantage of allowing the recording of phase scintillations (35–37). The early observations of scintillations were at comparatively low frequencies and, on the basis of the assumed form of decrease of scintillation intensity with frequency, it was expected that frequencies as high as those of the 6- and 4-GHz bands planned for the INTELSAT system would be free from scintillation effects. It developed, however, that scintillation does occur at frequencies of 6 and 4 GHz and somewhat higher at times (38–40). Peak-to-peak fluctuations as high as 7 dB have been observed.

Scintillation is caused by scattering from irregularities in electron density. The strongest scattering is observed in equatorial and auroral regions, especially the equatorial areas. The resulting scintillation is correspondingly intense and extends to higher frequencies than elsewhere. Scintillation tends to be weak at temperate latitudes. Maximum scintillation tends to occur at night in all regions. The pattern of occurrence is suggested in Fig. 9. Further information on scintillation can be found in CCIR Report 263-6 (41).

Effects of Precipitation, Clouds, and Atmospheric Gases

Precipitation

Much of the interest in effects of precipitation on telecommunications understandably has been directed to frequencies above 10 GHz, but the various

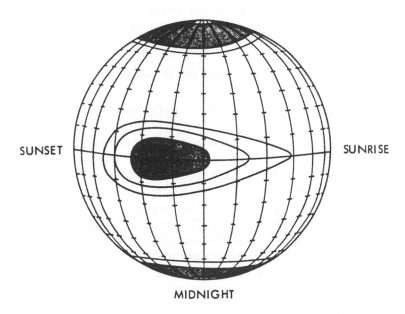

FIG. 9   Pattern of ionospheric scintillation. (From Ref. 41.)

models of attenuation due to rain generally are applicable below 10 GHz as well. Attenuation and noise due to precipitation may be important for frequencies as low as 8 GHz or lower and may need to be taken into account for frequencies as low as 4 GHz. Depolarization, the production of cross-polarized components that have polarizations orthogonal to the original polarizations, increases with attenuation above about 8 GHz. However, significant depolarization may take place for frequencies as low as 4 GHz. Backscatter from precipitation is important in radar observations at frequencies as low as those of the L band (e.g., around 1500 MHz), and bistatic scatter from rain is a potential source of interference for telecommunications system operations at frequencies this low as well as at higher frequencies.

Attenuation.   Theoretical treatments of the effects of precipitation on the propagation of e-m waves commonly have been based on the Mie theory for spherical drops (42,43), but more recent treatments take account of the departure of larger raindrops from the spherical shape. We emphasize here, however, empirical treatments of the effects of precipitation, as these are used widely for practical applications. These relations have the form of

$$\alpha_p = a(f)R^{b(f)} \qquad \text{dB/km}, \tag{35}$$

where α_p is the attenuation constant applying to power density for propagation of an e-m wave, R is rain rate in millimeters per hour (mm/h), and a and b are empirical constants that are functions of frequency f. Values of a and b for spherical drops have been provided by Olsen, Rogers, and Hodge (44). However, large raindrops are spheroidal in form rather than spherical, and the recent practice has been to use the following expressions for a and b (45). The spheroidal drops are larger in the horizontal direction than in the vertical direction, and thus attenuation coefficients for horizontally polarized waves (a_H) are greater than for vertically polarized waves (a_v).

$$a = [a_H + a_V + (a_H - a_V) \cos^2\theta \cos 2\tau]/2 \tag{36}$$

$$b = [a_H b_H + a_V b_V + (a_H b_H - a_V b_V) \cos^2\theta \cos 2\tau]/2a \tag{37}$$

The subscripts H and V refer to horizontal and vertical polarization, the angle θ is the elevation angle of the path, and τ is the tilt angle of the electric field from the horizontal. The angle τ can be taken to be 45° for circular polarization. Table 1 shows values of a and b for horizontal and vertical polarization and can be used directly for these polarizations and propagation over horizontal paths without using Eqs. (36) and (37). Or, for other conditions, the values of Table 1 can be substituted into Eqs. (36) and (37).

For estimating the attenuation constant to be encountered on a path, one needs a statistically based value for the rain rate R. The rain rate, exceeded for only 0.01% of the time, commonly is desired for commercial application, but values for other percentages may be needed in other cases. A number of models have been devised; we concentrate here on a CCIR model (46,47). Information

TABLE 1 Coefficients a and b for Calculating Horizontal and Vertical Polarization Attenuation

Frequency (GHz)	a_H	a_V	b_H	b_V
1	0.0000387	0.0000352	0.912	0.880
2	0.000154	0.000138	0.963	0.923
4	0.000650	0.000591	1.12	1.07
6	0.00175	0.00155	1.31	1.27
8	0.00454	0.00395	1.33	1.31
10	0.0101	0.00887	1.28	1.26
12	0.0188	0.0168	1.22	1.20
15	0.0367	0.0335	1.15	1.13
20	0.0751	0.0691	1.10	1.07
25	0.124	0.113	1.06	1.03
30	0.187	0.167	1.02	1.00
35	0.263	0.233	0.98	0.96
40	0.350	0.310	0.94	0.93

Source: Adapted from Ref. 45.

on rain rates R to be exceeded for various percentages of time is presented by a combination of maps showing regions of the earth labeled A through P (but with no I or O), and a table (Table 2). Figure 10 shows these rain rate regions for North America, and Table 2 shows rain rates exceeded as a function of percentage of time.

Having a value for the attenuation constant α_p from Eq. (35) is a first step in the estimation of the total attenuation to be expected along an earth-space path. One also needs a value for the path length L through rain, and a path reduction value r_p. The total attenuation A in dB then can be estimated from

$$A = \alpha_p L r_p. \tag{38}$$

TABLE 2 Rain Rates Exceeded as a Function of Year Percentage, Regions A to P

Percentage of Year	Rain Rates Exceeded (mm/h)													
	A	B	C	D	E	F	G	H	J	K	L	M	N	P
1.0	<0.5	1	2	3	1	2	3	2	8	2	2	4	5	12
0.3	1	2	3	5	3	4	7	4	13	6	7	11	15	34
0.1	2	3	5	8	6	8	12	10	20	12	15	22	35	65
0.03	5	6	9	13	12	15	20	18	28	23	33	40	65	105
0.01	8	12	15	19	22	28	30	32	35	42	60	63	95	145
0.003	14	21	26	29	41	54	45	55	45	70	105	95	140	200
0.001	22	32	42	42	70	78	65	83	55	100	150	120	180	250

Source: Adapted from Ref. 46.

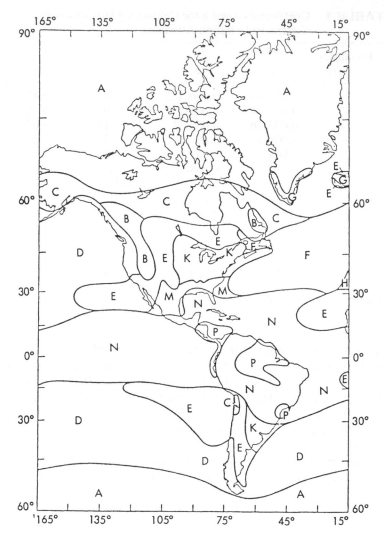

FIG. 10   Rain-rate regions of the Americas. (From Ref. 46.)

For determining the path length L, it is necessary to consider the vertical distribution of rainfall. Temperature tends to decrease with height and precipitation tends to occur as snow rather than rain above the 0°C isotherm. Snow causes considerably less attenuation than rain and it is the length of the path up to the 0°C isotherm that largely determines attenuation due to precipitation. Modeling the spatial distribution of rain is difficult, but several procedures have been proposed for determining a height extent of rain H that can be used for estimating attenuation. One recently recommended procedure is to take H as equal to 4 km for latitudes ϕ less than 36° and to use Eq. (39) to obtain a value for H for latitudes greater than 36°.

$$H = 4.0 - 0.075(\phi - 36) \quad \text{km.} \tag{39}$$

The path length L then is determined from

$$L = H/\sin \theta \quad \text{km.} \tag{40}$$

Intense rain tends to be localized, and the path reduction factor r_p takes account of the fact that the distribution of rainfall along the path tends not to be uniform. An empirical relation used for the path reduction factor is given by

$$r_p = 1/[1 + 0.045D], \tag{41}$$

where D is in km and is the horizontal projection of L. A recommended procedure for calculating attenuation A is first to make the calculation for the rain rate exceeded for 0.01% of the time $A_{0.01}$, and then to calculate attenuation A_p from

$$A_p = A_{0.01}0.12p^{-(0.546 + 0.043 \log p)}. \tag{42}$$

We now provide three estimates of the magnitude of $A_{0.01}$, the attenuation exceeded for 0.01% of the time. Consider first Rain Rate Region E of the CCIR model discussed above, a frequency of 10 GHz, circular polarization, a latitude of 40°, and an elevation angle of 42°. Using the Fig. 10, Tables 1 and 2, and relations given above, the attenuation on the earth-space path considered is estimated as 3.15 dB. Now considering Ka-band frequencies of 30 and 20 GHz, as will be used on the ACTS satellite (see the section, "Optical Fibers and Satellites"), the rain rate region K of the CCIR model, and other parameters as for the 10 GHz example, the attenuation at 30 GHz is estimated as 22.75 dB and the attenuation at 20 GHz is estimated as 16.97 dB. The higher frequency, 30 GHz, will be used for the uplink, and 20 GHz will be used for the downlink. These attenuation values, estimated for 30 and 20 GHz for a particular region and path, are substantial. Measures for countering the attenuation are mentioned in the section, "Propagation," in the discussion of optical fibers and satellites. The attenuation values are reduced significantly if attenuation exceeded for larger percentages of time is considered. For example, use of Eq. (42) shows that the attenuation of 22.75 dB estimated for 30 GHz is reduced to 8.69 dB in the case of the attenuation exceeded for 0.1% of the time.

A number of other models have been developed for estimating the attenuation expected due to rain. The 1980 global model (48) commonly is preferred for North America as it provides somewhat better detail for this region. The SAM (simple attenuation model) is described in Pratt and Bostian (49) and Stutzman and Yon (50). Several of the better-known models are described in Flock (29).

Depolarization. The term *depolarization* refers to a degradation or unwanted change in polarization, as from purely vertical linear polarization to linear polarization at an angle slightly different from vertical (51). This latter condition is equivalent to a combination of vertical and horizontal polarization. Such an effect can be caused by precipitation.

It is highly desirable in many circumstances to be able to use two orthogonal polarizations on the same signal path, but the ability to do so may be limited to some degree by antenna characteristics or depolarization caused by precipitation or some other phenomena. The two linear polarizations generally are referred to as vertical and horizontal, but the polarizations tend to be rotated somewhat from the local vertical and horizontal axes. The two circular polarizations are the right and left circular polarizations. Two orthogonal polarizations sometimes are referred to as cross-polarizations, and a wave of the opposite or orthogonal polarization that is produced by a process of depolarization is known as a *cross-polarized wave*. The production of a cross-polarized wave may result in unacceptable interference between orthogonally polarized channels of the same path.

In considering transmission through rain, the ratio of the power of the wanted or copolarized wave to the power of the unwanted wave is pertinent. Letting E_{11} and E_{22} represent the electric fields of copolarized waves and E_{12} and E_{21} represent the electric fields of cross-polarized waves and expressing the ratio in dB, the ratio may be written as $20 \log (E_{11}/E_{12})$, for example. This kind of ratio is referred to by the term *cross-polarization discrimination* (XPD). For the example mentioned above,

$$\text{XPD} = 20 \log \ (E_{11}/E_{12}) \tag{43}$$

Rather than using XPD to describe the state of polarization, use also can be made of its reciprocal, namely, D, which stands for depolarization. Thus,

$$\text{D} = 20 \log \ (E_{12}/E_{11}) \tag{44}$$

The rather high XPD value of 40 dB, for example, corresponds to the small depolarization of -40 dB; the quite low value of XPD of 10 dB corresponds to the large depolarization of -10 dB.

Depolarization due to precipitation is caused by the nonspherical shape of raindrops and ice crystals; spherical drops do not cause depolarization. Depolarization would not occur in the case of spheroidal drops either if the electric field vector of a linearly polarized wave were to lie strictly parallel to either the long or short axes of the drops. In the general case, however, the roughly spheroidal drops tend to be canted or tilted with respect to the electric field vectors. Wind contributes to canting and, even in the case of apparently vertical fall, the drops normally exhibit a distribution of canting angles. Differential attenuation and phase shift of field components parallel to the long and short axes of drops cause depolarization. The effect of differential attenuation is shown in a qualitative way in Fig. 11. A circularly polarized wave is equivalent to the combination of two linearly polarized waves that differ by 90° in both spatial configuration and electrical phase, and depolarization occurs for circularly polarized waves also. Indeed, it develops that depolarization tends to be worse for circularly polarized waves than for linearly polarized waves.

An expression of XPD given in CCIR Report 722-2 (52) is

$$\text{XPD}_{\text{dB}} = 30 \log f_{\text{GHz}} - 40 \log \cos \theta - 10 \log [1/2(-\cos 4\tau e^{-(\kappa_m)^2})]$$

$$- 20 \log A_{\text{dB}} \qquad (45)$$

where f is frequency. The expression relates XPD and attenuation A_{dB}. It shows that XPD decreases with increasing attenuation. The angle θ is the elevation angle of the path, and the equation shows that XPD increases with increasing elevation angle. The quantity τ is the tilt angle of the electric field from the horizontal in the case of a linearly polarized wave. It can be taken to be 45° for a circularly polarized wave. The quantity $(\kappa_m)^2 = 0.0024 (\sigma_m)^2$, where σ_m is the standard deviation in degrees of the mean raindrop canting angle from path to path and storm to storm, with 5° a suitable value for σ_m.

Equation (45), which is basically empirical, is stated to be valid for frequencies above 8 GHz. It is not obvious why one should use log A_{dB}, especially since A_{dB} is already a logarithmic function. Nor is it obvious why there should be a 30 log f term. Chu has treated depolarization in a manner that is related more closely to basic principles (52,53). He also has provided a justification for Eq. (45), except that he favors using $11.5 + 20 \log f_{\text{GHz}}$ instead of $30 \log f_{\text{GHz}}$.

Chu's analysis is based upon the use of the quantity

$$\sqrt{(\Delta\alpha_o)^2 + (\Delta\beta_o)^2} = \sqrt{x^2 + y^2}$$

where x is the differential attenuation constant/km and y is the differential phase constant/km. For frequencies below about 8 GHz, attenuation is small

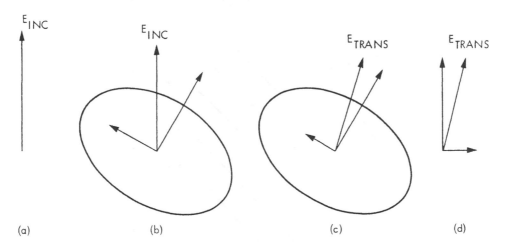

| (a) | (b) | (c) | (d) |

FIG. 11 An incident vertically polarized wave emerges from rain no longer strictly vertical after experiencing differential attenuation of components parallel to the major and minor axes of raindrops: a, E_{INC}, the field intensity of the incident wave; b, E_{INC} and its components are parallel to the major and minor axes of the drops; c, the effect of greater attenuation parallel to the major axis, resulting in E_{TRANS}, the field intensity of the transmitted wave; d, E_{TRANS} and its components in the vertical and horizontal directions. Note that E_{TRANS} is in a different direction from E_{INC}.

and depolarization is caused largely by differential phase rather than differential attenuation, whereas Eq. (45) involves only attenuation and not phase. Chu's treatment includes frequencies below as well as above 8 GHz. It is developed in terms of D rather than XPD and has the simplified form, for circular polarization, of

$$D_{cir}(dB) = 20 \log \left[\frac{1}{2} \sqrt{x^2 + y^2} L \cos^2\theta \right]. \tag{46}$$

An exponential term involving canting angle has been left out inside the bracket as the exponent can be set equal to zero as a conservative design procedure. The quantity L is the path length through rain, and θ is the elevation angle of the path. For linear polarization a term, like that of Eq. (45) involving τ and κ_m but with a positive sign, can be added. Figure 12 provides a plot of $[x^2 + y^2]^{1/2}$ as a function of frequency and rain rate.

Clouds

Rather than treating the complicated case of the effects of clouds on the entire electromagnetic spectrum, we confine attention here to the case of waves, such

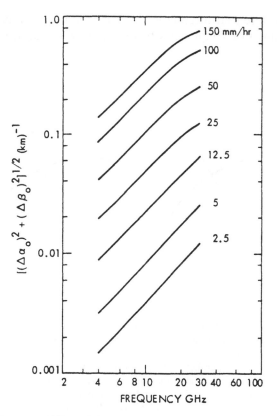

FIG. 12 Differential propagation constant. (From Ref. 53.)

TABLE 3 Attenuation and Noise Temperature Values of Cloud Models

ρ_ℓ g/m^3	Total Thickness km	S-Band (2.3 GHz) Zenith		X-Band (8.5 GHz) Zenith		X-Band (10 GHz) Zenith		Ka-Band (32 GHz) Zenith	
		T(K)	A(dB)	T	A	T	A	T	A
0.5	2	2.43	0.040	6.55	0.105	8.25	.133	61.00	1.083
0.7	2	2.54	0.042	8.04	0.130	10.31	.166	77.16	1.425
1.0	2	2.70	0.044	10.27	0.166	13.55	.216	99.05	1.939
1.0	3	3.06	0.050	14.89	0.245	19.66	.326	137.50	3.060
1.0	4	3.47	0.057	20.20	0.340	26.84	.457	171.38	4.407
.0	0	2.15	0.035	2.78	0.045	3.05	.049	14.29	0.228

Source: Adapted from Ref. 55.

as microwave and millimeter waves, for which the water droplets of clouds are small compared with wavelength. In this case, the theory of Rayleigh scattering applies, and this theory is considerably simpler than the complicated Mie or similar theories that must be used in theoretical analyses of propagation through rain at lower frequencies. Starting with Laplace's equation, $\nabla^2\phi = 0$ (3), the following equation for attenuation through uniform clouds can be derived.

$$\alpha_p = \left\{ 0.4343 \, \frac{6\pi}{\lambda} \, Im \left[- \frac{K_c - 1}{K_c + 1} \right] \right\} \rho \qquad \text{dB/km} \qquad (47)$$

Here, λ is wavelength in cm, Im indicates the imaginary part, K_c is the complex relative dielectric constant of water, and ρ is the water content of the cloud in grams per cubic meter (g/m^3). Table 3 shows values of attenuation A in dB for vertical paths through cloud models as a function of frequency, cloud thickness, and liquid water content ρ. For the worst case at 10 GHz, for example, the attenuation for a vertical path is 0.457 dB. For a path at an elevation angle of 10° or greater, however, the corresponding figure is found by dividing by the sine of the elevation angle. Thus, the attenuation for 10° is 2.63 dB. For 32 GHz, the corresponding attenuation at vertical incidence is 4.41 dB. Table 3 and consideration of elevation angle dependence illustrate the point that attenuation due to clouds increases with frequency but can be significant at a frequency as low as 10 GHz. The table also shows values of brightness temperature T. This quantity is treated in the section on atmospheric radio noise.

Attenuation values due to clouds generally are smaller than the values due to rain, but clouds generally are present for larger percentages of the time. For many commercial telecommunications systems, a high reliability, or system availability, of 99.99% is needed, but lower availabilities, from 95% to 99%, may be satisfactory in some other cases. The forthcoming new version of CCIR will say that attenuation due to clouds may be the most relevant impairment factor when such comparatively low availabilities are considered (45). It is stated that, "In that case, the propagation factors to be considered include only very light rain or no rain at all and cloud attenuation may determine the link margin" (45, p. 236) (for low elevation angles at frequencies above 20 GHz).

Atmospheric Gases

Absorption peaks due to water vapor occur at about 22.3, 183.3, and 323.8 GHz, and absorption peaks due to oxygen occur near 60 and 118 GHz. These peaks and the general increase in attenuation with frequency are illustrated in Fig. 13, which applies to total attenuation on a vertical path from sea level through the entire atmosphere for surface values of 1013 millibars (mb) for pressure, 20°C for temperature, and 7.5 g/m³ for water vapor content. For paths at an elevation angle of 10° or higher, the corresponding attenuation can be estimated by dividing by the sine of the elevation angle. Additional information about attenuation due to water vapor and oxygen is given in a CCIR report (56).

The CCIR reference includes plots of attenuation constant (specific attenuation) versus frequency at heights of 0, 5, 10, 15, and 20 km in the frequency range from 50 to 70 GHz (Fig. 14). The plot for the earth's surface (0 km) is a smooth curve due to pressure broadening, but for heights of 15 km and 20 km, especially the latter, significant cyclical variations are shown as a function of

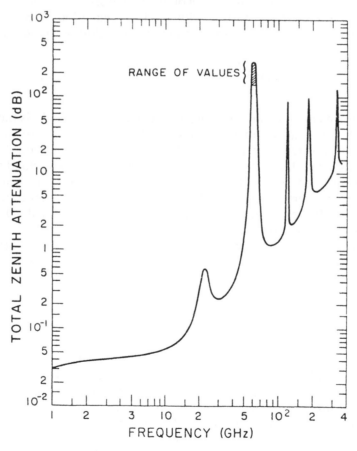

FIG. 13   Total zenith attenuation at ground level for $p = 1013$ mb, $T = 20$°C, and $\rho = 7.5$ g/m³. (From Ref. 56.)

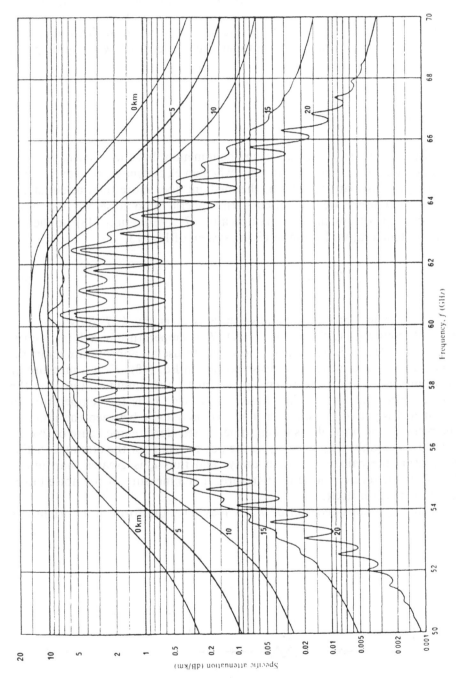

FIG. 14 Specific attenuation (attenuation constant) in the range 50–70 GHz at the altitudes indicated. (From Ref. 56.)

frequency. For most applications, it would be desirable to avoid the attenuation peaks near 22 and 60 GHz, but operation near 60 GHz, for example, might be desirable in such circumstances as remote sensing or where security is a consideration. During World War II, the attenuation peak near 22 GHz was overlooked and a radar was built to operate at a frequency near this value, with the result that the useful range of the radar was reduced below the anticipated value.

Atmospheric Radio Noise

Radio noise is not strictly a propagation phenomenon, but the subjects of propagation and noise are inseparable in that, when an electromagnetic wave propagates through a dissipative medium, the medium does more than attenuate; it generates radio noise as well. To discuss atmospheric radio noise, we first introduce the concept of system noise temperature T_{sys}. Equation (48) applies to system noise temperature for the receiving system of Fig. 15.

$$T_{sys} = T_A + (l_a - 1)T_o + l_a T_R \qquad (48)$$

The radio noise in watts in a bandwidth B is given by $k\,T\,B$, where k is Boltzmann's constant (1.381×10^{-23} J/K), T is temperature in kelvins, and B is bandwidth in Hz. It is convenient, however, to use an appropriate temperature, such as T_{sys}, as a measure of noise power. In Eq. (48), T_A is a measure of the noise introduced into the receiving system by the antenna, and it is the quantity of principal interest for present purposes. T_o is the temperature in kelvins of a dissipative transmission line that may exist between the antenna and the receiver and normally is taken to be the standard temperature of 290 K. The quantity l_a equals $1/g_a$, where g_a is the power "gain" of the attenuator. For an attenuator, the gain is less than unity, and thus l_a is greater than unity. Finally, T_R is the receiver noise temperature. It is a measure of the noise introduced by the receiver itself.

Returning to a consideration of T_A, it may have the form of a "brightness

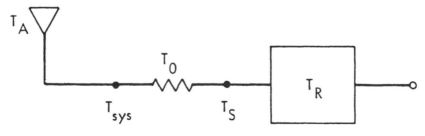

FIG. 15　Noise temperatures of receiving system.

temperature" T_b representing the noise temperature recorded when observing a noise source of noise temperature T_s through an absorbing region. An expression for T_b is

$$T_b = T_s e^{-\tau} + T_c(1 - e^{-\tau}). \tag{49}$$

Equation (49) shows that the source temperature T_s is reduced by the factor $e^{-\tau}$ or by A_{dB}, with $A_{dB} = 4.34\,\tau$. The term $T_c(1 - e^{-\tau})$ of Eq. (49) represents the noise introduced by an absorbing region of intrinsic temperature T_c. It is of interest to illustrate the degradation in signal-to-noise ratio due to an absorbing region along a path. Let the ratio be represented by C/X and the degradation in dB by $\Delta(C/X)_{dB}$, and take the system noise temperature T_1 in the absence of the absorbing region (for example, rain or cloud) to be 100 K. Consider an attenuation of 1 dB and let the intrinsic temperature T_c of the absorbing region be 280 K. Then, with the absorbing region present, the second term of Eq. (49) has the value 57.6 K. For this case,

$$\Delta(C/X)_{db} = 1 + 10 \log [(100 + 57.6)/100] = 1 + 1.98 = 2.98,$$

where $100 + 57.6 = 157.6$ K is T_2, the value of the system noise temperature when the absorbing region is present. The C/X ratio is reduced by 2.98 dB, but only 1 dB of this reduction is due to attenuation. Next, consider that T_1 is 25 K. Then $\Delta(C/X)_{dB} = 6.19$ and only 1 dB of this is due to attenuation, with 5.19 dB due to the noise emitted by the absorbing region.

Multipath and Shadowing in Mobile Communications

To this point, it generally has been assumed that an unobstructed free-space path exists for the signal, but, in some circumstances, especially in mobile communications, a signal may reach the receiving antenna over a direct path and a signal component (or components) may reach the antenna after reflection or scatter from nearby surfaces or objects. The case of reflection from a plane surface on an earth-space path is depicted in Fig. 16. Reflections generally are undesirable in such circumstances, as the relative phase of the direct signal and the reflected signal component varies, and this condition tends to result in signal fading. The magnitude of the wave reflected from a plane surface is determined by the reflection coefficient for the wave; the reflection coefficient is different for horizontal linear polarization and for "vertical" polarization (more properly, for polarization in the plane of incidence, namely, the plane shown in Fig. 15). The expression for the reflection coefficient for horizontal polarization ρ_h is given by

$$\rho_h = \frac{\sin\theta - [K - j\sigma/\omega\epsilon_o - \cos^2\theta]^{1/2}}{\sin\theta + [K - j\sigma/\omega\epsilon_o - \cos^2\theta]^{1/2}}. \tag{50}$$

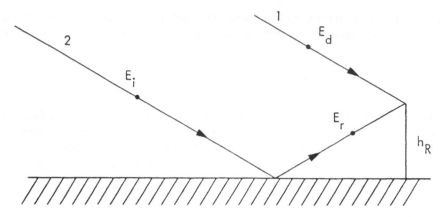

FIG. 16   Direct and reflected rays for an earth-space path employing horizontal polarization.

Here, θ is the elevation angle of the path; the wave is incident from Medium 1, which is air onto Medium 2; K is the relative dielectric constant of Medium 2; σ is the conductivity of Medium 2; ω is the angular frequency of the wave; and ϵ_o is the electric permittivity of free space (8.854×10^{-12} farad/meter). The corresponding reflection coefficient for vertical polarization is

$$\rho_v = \frac{[K - j\sigma/\omega\epsilon_o] \sin \theta - [K - j\sigma/\omega\epsilon_o - \cos^2\theta]^{1/2}}{[K - j\sigma/\omega\epsilon_o] \sin \theta + [K - j\sigma/\omega\epsilon_o - \cos^2\theta]^{1/2}}. \tag{51}$$

An interesting feature of the reflection coefficients is that the reflection coefficient for vertical polarization goes to zero for the lossless case ($\sigma = 0$) for the Brewster angle θ_p, but there is no elevation angle for which the reflection coefficient for horizontal polarization goes to zero. When the conductivity σ does not equal zero, the reflection coefficient for vertical polarization does not go to zero but has a minimum (Fig. 17). For the present case, when Medium 1 is air, having a relative dielectric that can be taken to be unity, the Brewster or polarizing angle is defined by

$$\theta_p = \tan^{-1}(1/K)^{1/2}. \tag{52}$$

The situation for circular polarization is interesting in that, upon reflection of a right circularly polarized wave, for example, both right and left circular components occur in the general case. It can be shown (29) that the reflection coefficient for the component of the same polarization as the incident waves ρ_c is given by

$$\rho_c = (\rho_h + \rho_v)/2. \tag{53}$$

The expression giving the cross-polarized or orthogonal component ρ_x is

$$\rho_x = (\rho_h - \rho_v)/2. \tag{54}$$

Another consideration is that the reflection coefficients given above may be reduced by the surface roughness factor ρ_s. This factor can be expressed as a function of $\Delta\phi = (4\pi h_s/\lambda) \sin \theta$, where h_s is the rms value of the terrain height irregularities, λ is the electromagnetic wavelength, and θ is the elevation angle of the path. One form for ρ_s is

$$\rho_s = e^{-(\Delta\phi)^2/2}I_0[(\Delta\phi)^2/2], \tag{55}$$

where I_0 indicates the modified Bessel function (of the quantity that follows it in brackets) (59). As h_s increases in magnitude, ρ_s becomes smaller, the amount of energy specularly reflected in the forward direction is reduced, and the condition of diffuse scatter (over a range of angular directions) tends to become dominant (60).

Multipath conditions and diffuse scatter may be somewhat of a problem in voice service in some rather extreme situations and are likely to become more

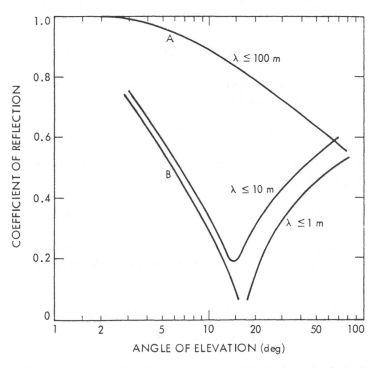

FIG. 17 Reflection coefficients for plane average ground (A, horizontal polarization; B, vertical polarization). (From Ref. 57. Also see Ref. 58.)

serious for data transmission, especially when high data rates are utilized. Multi-path can result from buildings, power poles, and structures and terrain of various kinds, as well as from reflections from flat surfaces. In practice, however, for mobile satellite voice service to vehicles on highways, shadowing due to roadside trees has become the subject of most concern. A number of experiments simulating mobile satellite service have been carried out (61,62).

Propagation Effects on Interference

In the section on ray paths in the troposphere, the possibility that radio waves that are launched essentially horizontally may propagate for unexpectedly long distances was mentioned. The atmospheric conditions responsible for this phenomena exist only over a limited range of altitude and commonly form what is referred to as a *duct*. Ducting generally does not occur consistently enough to be useful but can be a serious cause of interference, for example, between a terrestrial line-of-sight system and a satellite earth station. A second major cause of interference is scatter from rain cells, such as scatter from the beam of a powerful transmitter of an earth station to the receiver of a line-of-sight system. Analysis of interference is a specialty that cannot be covered fully here, but some attention is devoted to ducting.

First, we write the expression for L_{FS} (as in the section, "Propagation through Earth's Atmosphere at Elevation Angles above 10°") in decibel form and in terms of the factors that define it.

$$(L_{FS})_{dB} = 20 \log (4\pi) = 20 \log d - 20 \log \lambda \tag{56}$$

Commonly, however, L_{FS} is expressed in terms of frequency rather than wavelength. Replacing λ by c/f results in

$$(L_{FS})_{dB} = -147.55 + 20 \log f + 20 \log d \tag{57}$$

If f is expressed in GHz rather than Hz and d is expressed in km rather than m, the expression becomes

$$(L_{FS})_{dB} = 92.45 + 20 \log f_{GHz} + 20 \log d_{km} \tag{58}$$

If ducting is taking place, however, this expression for free-space path loss does not apply. Instead, one can use a loss factor L_b defined for transhorizon paths by

$$(L_b)_{dB} = 92.45 + 20 \log f_{GHz} + 10 \log d_{km} + A_c$$
$$+ (\alpha_d + \alpha_w + \alpha_o)d_{km} + A_s \tag{59}$$

An important point to notice is that variation with distance now is given by 10 log d rather than 20 log d as for free-space propagation. In a duct, the signal only spreads out in one direction instead of two to a first approximation. The quantity A_c is a coupling loss that takes account of the fact that not all of the closely spaced rays leaving the transmitting antenna are trapped within the duct, and A_s represents losses caused by obstacles along the path. The αs are attenuation constants, with α_d a duct attenuation constant having the theoretical minimum value of 0.03 dB/km. The constants α_w and α_o account for water vapor and oxygen, respectively, and may be important for long paths at high frequencies. In practice, one must determine what value of L_b is needed, and the distance d needed to obtain this value of loss then is calculated. This value of d is the "coordination distance" for great-circle propagation. All possible interfering systems within this distance then must be examined in detail to obtain an estimate of the actual interference level (63–65).

With respect to scatter from rain, most persons are familiar with the radar displays shown in TV reports on weather. In this case, energy is scattered by hydrometeors in the radar antenna beam back to the transmitting radar. But energy is scattered in other directions also, and the scattered energy has the potential to cause interference. Interference due to ducting and scatter from rain are discussed in Ref. 66 and in the CCIR references listed above. A coordination distance can be determined for rain scatter, and the largest of the two coordination distances (for ducting and rain scatter) is the one that should be used.

Optical Fibers and Satellites

During the first two decades of satellite operations since the launch of INTELSAT I in 1965, satellites supplied a substantial part of new long-distance telecommunications services. Impressive gains in satellite technology were made during this period. But, highly impressive advances in technology also were made in other transmission media, most notably in optical fibers. The installation of fiber-optical cables is proceeding rapidly at present. The next section treats the propagation characteristics of optical fiber and satellite systems, consistent with the emphasis of this section on propagation. Other factors are important, however, in comparing optical fiber and satellite systems, and brief consideration is given to switches in the following section.

Propagation

Attenuation

The decrease in attenuation of optical fibers to values around 0.2 dB/km is a major factor contributing to their rapid deployment. The received signal level in satellite communications may decrease with distance because of two different factors, namely spreading loss and attenuation. The L_{FS} term of Eq. (29) introduces a decrease in power density with distance d that is proportional to distance squared. L_{FS} commonly is referred to as the free-space loss factor. The physical basis for this "loss" is shown more simply by Eq. (60), which applies to "free" or empty space (a vacuum).

$$C = \frac{P_T G_T A_R}{4\pi d^2} \tag{60}$$

Here, C is received carrier power in watts, P_T is transmitted power, A_R is the effective area of the receiving antenna, and d is distance in meters. The decrease with distance is due to the fact that the antenna beam spreads over a larger area with increasing distance. This loss with distance squared is referred to as *spreading loss*. As a matter of semantics, spreading loss customarily is not referred to as attenuation, and we do not consider it to be attenuation.

By attenuation, we refer to such phenomena as dissipative processes involving the conversion of electromagnetic energy to thermal energy, the scattering of electromagnetic energy out of a transmitted antenna beam, and the radiation of electromagnetic energy out of an optical fiber at a bend in the fiber. Factors causing attenuation in optical fibers include bends in fibers, Rayleigh scattering, impurities, and absorption associated with molecular or lattice structure. Minimizing the effects of impurities has been a major factor in reducing attenuation in optical fibers. Factors causing attenuation of radio beams include precipitation, clouds, atmospheric gases, and dust. The general increase of attenuation due to rain and clouds with frequency in satellite communications requires consideration. The highly variable and potentially serious attenuation due to rain can be countered by adaptive compensation for signal level changes due to rain, including forward error correction and uplink power control to adjust automatically for fades, as developed for ACTS. Also, plans for ACTS involve reducing the burst rate of digital transmissions during fades due to rain to one-half the value normally used. It is stated that the throughput will not be reduced because increased dwell time will be utilized. In addition, signal regeneration, to be employed on ACTS, is reported to provide an improvement of 6 to 8 dB in signal-to-noise ratio as compared with the case for "bent-pipe" repeater satellites, which simply retransmit the received signal at a slightly different frequency, and have been the standard technology to now (67). As part of the ACTS program, plans are being formulated for scientific experiments to study such topics as the correlation between fading at 20 and 30 GHz at the same site, correlation of fades at spatially separated sites, and determination of correlation distance, fading rates, and probability distribution functions of fade depth.

While attenuation due to rain on earth-space paths is highly variable and additional research on the topics mentioned above will be valuable, the view is taken here that the attenuation problem is manageable for satellite communications as well as optical fibers.

Dispersion

Equation (34) shows that the excess time delay in ionospheric propagation is inversely proportional to frequency squared. If one determines the rate of change of delay with frequency, it is found that the difference $|\Delta t|$ in delay over the frequency range $|\Delta f|$ is given by

$$|\Delta| = \frac{2.68 \times 10^{-7}}{f^3} |\Delta f| \ \ (\text{TEC}).\tag{61}$$

This change of a propagation parameter with frequency is an example of the phenomenon of dispersion. If one inserts numbers into the equation, with $f = 870$ MHz, TEC $= 10^{18}$ electrons/m^2, and $\Delta f = 50$ MHz, and if one takes $\Delta f = 1/\tau$ where τ is the pulse width of digital transmissions, it appears that dispersion may limit the data rate for digital transmission to something less than 50 megabits per second (Mb/s). The frequency of 870 MHz, once considered for mobile satellite service, however, is low and Δt is inversely proportional to frequency cubed. For the 30- and 20-GHz frequencies of ACTS, ionospheric dispersion is negligible. As for the possibility of tropospheric dispersion at these frequencies, delay due to the troposphere is a very weak function of frequency. Measurements by Allen, Violette, and Espeland (68), using a 500-Mb/s biphase shift-keyed link on a 11.8-km path in a variety of atmospheric conditions, showed that dispersion will not limit the performance of wideband digital links near 30 GHz adversely.

Interest is being shown at present in the operation of digital communications systems at rates in the order of 1 gigabit per second (Gb/s). Transmission rates of several Gb/s have been demonstrated for single-mode optical fibers (6), whereas multimode fibers are limited to about 400 Mb/s. If transmission rates of 1 Gb/s, single-mode fibers systems, and satellite systems like NASA's ACTS are considered, it appears that both optical fibers and satellite systems can provide satisfactory service. The 1-Gb/s digital transmissions would involve many interleaved signals. A single HDTV (high-definition television) signal may involve a bandwidth of 155 MHz or higher, which could be transmitted satisfactorily by both optical fiber and satellite systems.

Delay

In the above section on dispersion, reference was made to differential time delay; we now consider total delay. The total time delay for a signal to travel from an earth station to a satellite in geostationary orbit and back to a second earth station is about 270 milliseconds (ms), or somewhat over a quarter of a second. This delay is quite obvious in voice communication, and some time is required for a person to become used to it. In large part for this reason and because of the advent of optical fibers that have very small delay, the U.S. domestic (but not international), public, high-density satellite telephone service is reported to have disappeared (67). The geostationary satellite delay is nearly 10 times as long as the longest delays encountered in terrestrial systems, and the echo cancelers, modems, and protocols developed for terrestrial systems cannot handle this long delay. In addition, there is a diurnal variation in delay due to the fact that satellite position is not perfectly stationary even in geostationary systems. The maximum total variation in total delay in such systems is about 1.1 ms. Delay is not so very important in such applications as the distribution of television signals to cable systems and local broadcast stations, and solutions to time-delay problems exist for data transmission and other applications. This subject is treated in Chapter 12 of Pritchard and Sciuli (69), and we will not discuss details of the solutions here. For satellites to operate satisfactorily in systems containing both satellites and terrestrial transmission systems (such as ISDNs [Integrated Systems Digital Networks]), standards should be formulated to accommodate satellites, fiber-optical systems, and other terrestrial systems. Use of a large number of low-orbit satellites has been proposed for satellite communications, and the low-orbit satellites would avoid the long delays associated with geostationary satellites but would have greater variations in delay.

Security, Privacy, and Interference

Fiber-optical systems have inherent advantages with respect to security, privacy, and interference, as satellite transmissions can be intercepted by parties other than the intended recipient. Security, however, can be obtained on satellite systems by encryption. Interference is a subject requiring careful attention in the case of radio-frequency transmissions, and careful attention serves to minimize the problem.

Discussion

Fiber-optical systems are well suited for high-capacity widebandwidth links, and such installations, including underwater transoceanic cables, are proceeding rapidly. Satellites also can provide high-capacity, widebandwidth service and the potential for competition exists between satellites and fiber-optical systems for this type of application. Fiber cables have advantages of small size and light weight and need not be restricted to high-capacity links. Fiber local-area networks (LANs) are being installed and may offer bandwidths sufficiently wide for data or video distribution. Optical fibers also have begun to be installed to a limited extent to provide ordinary telephone service. One problem about this application is that of providing power for telephones when the public power system fails. Presently, the needed power usually is supplied by the same copper wires that carry conversations, but optical fibers are not electrical conductors.

Satellites are recognized as being well suited for broadcast, point-to-multipoint service, mobile services, and medium and thin route multiple access communications in general. Maritime–mobile satellite service is well established, developments are proceeding on land–mobile satellite service, and activity on aeronautical–mobile satellite service is imminent. Whereas satellite earth stations tended originally to be large and expensive and satellite transmitted powers were low, satellite powers have increased, switching capability is being added, and small earth terminals, in particular VSATs (very-small-aperture terminals), now are used widely (70). The distribution of television signals to local broadcast stations and cable systems represents a major application of satellites at present. Satellite systems have been characterized by high long-term reliability, 99.99% in the case of the INTELSAT system. Optical fiber cables are subject to inadvertent or deliberate breaks, but the danger of breaks may diminish as redundancy increases. Satellite systems are capable of rapid reconfiguration and can provide restoration of service in emergencies when other systems experience outage. The INTELSAT system commonly has provided backup to terrestrial systems, including, recently, the TAT-7 and TAT-8 fiber transatlantic systems.

More than a satisfactory propagation medium is required for the telecommunications systems of the future. Another need is for high-speed, high-capacity switches. Two types of on-board switching modes will be available on NASA's experimental ACTS. The baseband processor mode is to be used with VSATs. The baseband LBR-2 system encompasses the 64-kilobit-per-second (kb/s) rate for a single telephone conversation and the 1.544-Mb/s rate of T1 systems. The baseband LBR-1 system encompasses the T1 frequency and the 45-Mb/s rate of T3 systems. The microwave switch mode accommodates the 2.4-kb/s to 64-kb/s frequencies that will be used for mobile services. It also serves the medium data rate band, including 45 mb/s to 220 mb/s, which could be used for HDTV, and the high data rate band, including 220 Mb/s to 1 Gb/s, which could be used for supercomputers. ACTS also will employ high EIRP (discussed in the section on propagation at angles above 10°), electronically hopping spot beams. Its multibeam antennas provide the capability for rapidly reconfigurable patterns of hopping spot beams and fixed spot beams. The ACTS beams will have the capability of hopping to as many as 20 locations during each 1-ms TDMA (time division multiple access) frame. INTELSAT VI, launched in 1989, was the first commercial communications satellite to have spot beams. ACTS will provide signal processing and regeneration as well as switching.

Although fiber-optical and satellite systems will be competitive in some situations, the view is taken here that emphasis should be given to their complementary nature and to the different types of applications to which they are best suited. Fiber-optical and satellite systems should be able to function together in hybrid systems, and standards should be formulated to allow such hybrid operation. Emphasis is given to technical considerations, especially to propagation considerations, in this chapter, but regulatory matters, economics, policy, and political and marketing skills will have major effects on how telecommunications systems of the future evolve. The July 1990 issue of the *Proceedings of the IEEE*, a special issue devoted to satellite communications, is a good source of information on that subject. A new quarterly, *IEEE LCS, The Magazine of Lightwave Communication Systems*, started publication in 1990 and promises to be a useful and interesting source of information on fiber-optical applications.

A most interesting development affecting propagation in optical fibers is reported in the March 1993 issue of *Spectrum* (71). Optical solitons, pulses of light whose shape and spectrum endure over vast distances, may allow repeaterless optical cable and simplify transoceanic communication systems.

Position Determination

This section deals with the determination of position, perhaps of a vehicle on the ground, an aircraft in the air, or a particular location on the ground, by the use of electromagnetic waves. To this end, we first discuss the excess range delay for propagation through the earth's ionosphere and troposphere with respect to the range delay for propagation in a vacuum. In the section on ionospheric effects, the excess range delay ΔR for an ionospheric path was stated to be $(40.3/f^2)$ TEC where f is frequency in Hz and TEC is total electron content along the path in electrons/meter^2. But to use this expression one needs to know the TEC. It develops, however, that the use of two frequencies allows determination of TEC and also allows bypassing TEC and determining ΔR and Δt, the corresponding excess time delay ($\Delta t = \Delta R/c$, where c is the velocity of an electromagnetic wave in a vacuum). Let f_1 be the frequency of primary interest and f_2 be a second lower frequency. Then, solving for the difference in time delay δt at the two frequencies gives

$$\delta t = \Delta t_2 - \Delta t_1 = \frac{40.3 \text{ TEC}}{c} \left[\frac{1}{f_2^2} - \frac{1}{f_1^2} \right]. \tag{62}$$

It now is possible to solve for Δt_1, which is found to be given by

$$\Delta t_1 = \frac{f_2^2}{f_1^2 - f_2^2} \delta t. \tag{63}$$

Thus, the excess time delay for the ionospheric path can be determined, to a first approximation at least. In cases in which the highest possible accuracy is desired, it may be necessary to look further into the justification for Eq. (33) and to consider a possible departure of the ionospheric ray from a straight line. As an example, in terms of $\Delta R = c \, \Delta t$ for $f = 2.3$ GHz and TEC $= 10^{18}$ electrons/m^2, the excess range delay is 7.62 m. That is, the true range is 7.62 m less than that which would be estimated by assuming a wave velocity of c.

The troposphere also contributes excess range delay. To estimate this excess range delay, we consider first the contribution of the first term, namely, (77.6 p/T) of Eq. (22), where p is pressure in mb and T is temperature in kelvins. If one assumes an exponential decrease in pressure p with height above the surface in accordance with $p = p_o e^{-h/H}$ where p_o is surface pressure, it develops that ΔR_1, the excess range delay from the first term of Eq. (22), is given for a vertical path by

$$\Delta R_1 = 2.2757 \times 10^{-3} p_o \qquad \text{m.} \tag{64}$$

Taking p_o to be 1013 mb, the nominal value for sea level, the excess range delay is 2.31 m. Obtaining Eq. (64) involves taking $10^{-6} \int (77.6 \, p/T) \, dh$ with $H = RT/Mg$. R is the gas constant [8.3143 $\times 10^3$ J / (K kilogram mole)], M = 28.9655 from Table 3 of the U.S. Standard Atmosphere, 1976 (72), T is treated as a constant, and the value of g, the acceleration of gravity, is taken as 9.7877 m/s^2. This value of g applies to the height where the pressure is 500 mb at 45° latitude. The value of ΔR_1 depends primarily on only the surface pressure and is independent of the temperature profile.

It remains to consider the excess range delay due to the second term of Eq. (22). The pressure p in the first term is the total pressure and includes the partial pressure of water vapor, but dry air is responsible for most of the excess range delay of about 2.3 m for a vertical path from sea level, as would be expected from the fact that the partial pressure of water vapor is only a small fraction of the total pressure. The second term of Eq. (22) gives the major contribution, ΔR_2, of water vapor to excess range delay. This delay caused by water vapor is highly variable and difficult to predict in advance, but a representative value can be calculated readily. To this end, it is convenient to replace the water vapor pressure in mb, namely e, by the water vapor density, ρ, in g/m^3. The needed relation is

$$e = \rho T/216.5 \qquad (65)$$

Substituting this expression for e into N_2, the second term of Eq. (22), taking ρ at the surface to be 7.5 g/m^3 and T at the surface to be 280 K, assuming that N_2 decreases exponentially with a scale height H of 2000 m, and integrating from $h = 0$ to $h = \infty$ gives a representative value of ΔR_2 or 9.23 centimeters (cm) at vertical incidence. The delay caused by water vapor is much smaller than that caused by the dry constituents of the atmosphere, but as it is highly variable and difficult to predict, it is responsible for a larger error in the estimate of range delay than is the dry air.

The GPS in its final form will use 18 satellites in orbits at an altitude of 20,183 km to provide three-dimensional position and velocity information to mobile receivers anywhere in the world, whether on land or sea or in the air. Signals are transmitted at two L-band frequencies, 1575.42 and 1227.60 MHz, in order to permit correction for ionospheric time delay. The excess range delay caused by the dry air can be corrected for readily. In some cases, the correction for water vapor may not be important and can be ignored, but, if highest accuracy is desired, an effort should be made to correct for it. The satellites of the GPS system carry precision cesium clocks and, if the user has a precision clock, signals from three satellites are sufficient to determine position. A fourth satellite is required for most users, however, who must have a clock of specified accuracy but not a truly precision clock. By making measurements of pseudo-range to the four satellites, four equations can be formulated and solved for four unknowns consisting of three position coordinates and the offset between precision GPS time and time as indicated by the user's clock. The term *pseudo-range* is used because the originally measured quantities are sums of true ranges and offsets due to user time error. Spread-spectrum techniques are used for separating the signals from a particular satellite from those of other satellites in the field of view. Signals are received at low level, usually well below the thermal noise level in the receiver. Each satellite operates with a unique P code that is the product of two PN (pseudonoise) codes that utilize the 10.23-MHz clock frequency. Each satellite also transmits a shorter C/A code based on a repetition rate of 1.023 MHz. This code is used for signal acquisition, and it also includes data on satellite position, errors of the satellite clock, and parameters for cor-

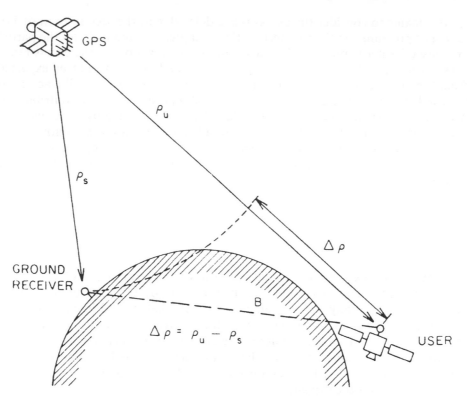

FIG. 18   Differential technique for utilizing Global Positioning System (GPS).

recting for ionospheric excess time delay. An introduction to GPS is included in Flock (28).

For highest accuracy, use can be made of carrier phase as well as the P code. One approach to using the GPS system for precision position determination is the differential technique (Fig. 18), in which position is determined with relation to a fixed reference point on the ground (73). For correcting for the delay caused by water vapor, use can be made of microwave radiometers that are passive receiving systems that record brightness temperature (see the section on atmospheric radio noise) on paths through the troposphere. By using two or more frequencies, such as 22.235 and 31.4 GHz, effects of water vapor, liquid water, and atmospheric gases can be separated and values for the water vapor contents along the paths can be determined. Another kind of system for precise position determination, the VLBI (very long baseline interferometer) system, also is affected by water vapor, and water vapor radiometers can be useful to VLBI systems. It is considered that it will be possible in some situations to determine distances to accuracies of a small number of cm (perhaps one or two cm in some cases) by use of the techniques mentioned here (74,75).

The Radio Determination Satellite Service (RDSS) for providing position determination and other services has been described by Briskman (76).

References

1. Johnk, C. T. A., *Engineering Electromagnetic Fields and Waves*, 2d ed., Wiley, New York, 1988.
2. Jordan, E. C., and Balmain, K. C., *Electromagnetic Waves and Radiating Systems*, Prentice-Hall, Englewood Cliffs, NJ, 1968.
3. Ramo, S. J., Whinnery, J. R., and Van Duzer, T., *Fields and Waves in Communication Electronics*, Wiley, New York, 1965.
4. Smith, E. K., and Kirby, R. C., The International Radio Consultative Committee (CCIR): Part 1, *Communications Society*, vol. 12, pp. 11–18, July 1974.
5. Codding, G. A., and Rutkowski, A. M., *The International Telecommunications Union in a Changing World*, Artech House, Dedham, MA, 1982.
6. Wilson, J., and Hawkes, J. F. B., *Optoelectronics, An Introduction*, 2d ed., Prentice-Hall, New York, 1989.
7. Miller, S. E., Marcatili, E. A. J., and Li, T., Research towards Optical Fiber Transmission Systems, *Proc. IEEE*, 61:1703–1751 (December 1973).
8. Cheo, P. K., *Fiber Optics and Optoelectronics*, Prentice-Hall, Englewood Cliffs, NJ, 1990.
9. Yariv, A., *Optical Electronics*, 3d ed., Holt, Rinehart & Winston, New York, 1985.
10. Yeh, C., *Handbook of Fiber Optics*, Academic Press, San Diego, 1990.
11. Flock, W. L., *Electromagnetics and the Environment*, Prentice-Hall, Englewood Cliffs, NJ, 1979.
12. Davies, K., *Ionospheric Radio*, IEE and Peter Peregrinus, London, 1990.
13. Budden, K. G., *Radio Waves in the Ionosphere*, Cambridge University Press, Cambridge, 1961.
14. Kelso, J. M., *Radio Ray Propagation in the Ionosphere*, McGraw-Hill, New York, 1964.
15. Ratcliffe, J. A., *An Introduction to the Ionosphere and Magnetosphere*, Cambridge University Press, Cambridge, 1972.
16. Rishbeth, H., and Garriott, O. K., *An Introduction to Ionospheric Physics*, Academic Press, New York, 1969.
17. Allis, W. P., Buchsbaum, S. J., and Bers, A., *Waves in Anisotropic Plasmas*, MIT Press, Cambridge, MA, 1963.
18. Klobuchar, J. A., Ionospheric Effects on Satellite Navigation and Air Traffic Control Systems, *Recent Advances in Radio and Optical Propagation for Modern Communication, Navigation, and Detection Systems*, AGARD Proceedings (April 1978).
19. Smith, E. K., and Weintraub, S., The Constants in the Equation for Atmospheric Refractive Index at Radio Frequencies, *Proc. IRE*, 41:1035–1037 (August 1953).
20. List, R. J., *Smithsonian Meteorological Tables*, 6th rev. ed., Smithsonian Institution, Washington, DC, 1984.
21. Bean, B. R., and Dutton, E. J., *Radio Meteorology*, Superintendent of Documents, U.S. Government Printing Office, Washington, DC, 1966.
22. GTE Lenkurt, *Engineering Considerations for Microwave Communication Systems*, GTE Lenkurt, San Carlos, CA, 1972.
23. Thompson, M. C., Wood, L. E., Janes, H. B., and Smith, D., Phase and Amplitude Scintillations in the 10 to 40 GHz Band, *IEEE Trans. Antennas Propagat.*, AP-23:792–797 (November 1975).
24. Allnutt, J. E., *Satellite-to-Ground Radiowave Propagation*, Peter Perigrinus, London, 1989.

25. Hall, M. P. M., *Effects of the Troposphere on Radio Communication*, IEE and Peter Peregrinus, London, 1979.

26. Hall, M. P. M., and Barclay, L. W., (eds.), *Radiowave Propagation*, IEE and Peter Perigrinus, London, 1989.

27. Ippolito, L. J., *Radiowave Propagation in Satellite Communications*, Van Nostrand, New York, 1986.

28. Ippolito, L. J., *Propagation Effects Handbook for Satellite Systems Design*, A Summary of Propagation Impairments on 10 to 100 GHz Satellite Links with Techniques for System Design, NASA Reference Publication 1082(04), NASA Headquarters, Washington, DC, 1989.

29. Flock, W. L., *Propagation Effects on Satellite Systems at Frequencies below 10 GHz*, NASA Reference Publication 1108 (02), NASA Headquarters, Washington, DC, 1987.

30. Hey, J. S., Parsons, S. J., and Phillips, J. W., Fluctuations in Cosmic Radiation at Radio Frequencies, *Nature*, 158:234 (August 17, 1946).

31. Smith, F. G., Origin of the Fluctuations in the Intensity of Radio Waves from Galactic Sources: Cambridge Observations, *Nature*, 165:422–423 (March 18, 1950).

32. Little, C. G., and Lovell, A. C. B., Origin of the Fluctuations in the Intensity of Radio Waves from Galactic Sources: Jodrell Bank Observations, *Nature*, 165:423–424 (March 18, 1950).

33. Hewish, A., The Diffraction of Galactic Radio Waves as a Method of Investigating the Irregular Structure of the Ionosphere, *Proc. Royal Soc. of London*, Series A, 228:238–251 (February 22, 1955).

34. Yeh, K. C., and Swenson, G. W., F-Region Irregularities Studies by Scintillation of Signal from Satellites, *Radio Science* (Sec. D, J. of Research, National Bureau of Standards), 68D:881–894 (August 1964).

35. Crane, R. K., Ionospheric Scintillation, *Proc. IEEE*, 65:180–199 (February 1977).

36. Woo, R., Multifrequency Techniques for Studying Interplanetary Scintillation, *Astrophys. J.*, 201:238–248 (October 1, 1977).

37. Smith, E. K., and Edelson, R. E., Radio Propagation through Solar and Other Extraterrestrial Ionized Media, JPL Pub. 79–117, Jet Propulsion Laboratory, Pasadena, CA, January 15, 1980.

38. Craft, H. D., and Westerlund, L. H., Scintillations at 4 and 6 GHz Caused by the Ionosphere, 10th Aerospace Science Meeting, San Diego, CA, AIAA Paper No. 72-179, Jan. 17–19, 1972.

39. Taur, R. R., Ionospheric Scintillation at 4 and 6 GHz, *COMSAT Tech. Rev.*, 3: 145–163 (Spring 1973).

40. Mullen, J. P., et al., UHF/GHz Scintillation Observed at Ascension Island from 1980 through 1982, *Radio Sci.*, 20:357–365 (May–June 1985).

41. International Radio Consultative Committee, Ionospheric Effects upon Earth-Space Propagation, Report 263-6, Vol. 6, *Propagation in Ionized Media, Recommendations and Reports of the CCIR, 1986*, International Telecommunication Union, Geneva, 1986.

42. Kerr, D. E. (ed.), *Propagation of Short Radio Waves*, Vol. 13, Radiation Laboratory Series, McGraw-Hill, New York, 1951.

43. Kerker, M., *The Scattering of Light and Other Electromagnetic Radiation*, Academic Press, New York, 1969.

44. Olsen, R. L., Rogers, D. V., and Hodge, D. B. The aR^b relation in the Calculation of Rain Attenuation, *IEEE Trans. Antennas Propagat.*, AP-26:318–329 (March 1978).

45. International Radio Consultative Committee, Attenuation and Scattering by Precipitation and Other Atmospheric Particles, Report 721-2, Vol. 5, *Propagation in*

Non-Ionized Media, Recommendations and Reports of the CCIR, 1986, International Telecommunication Union, Geneva, 1986.

46. International Radio Consultative Committee, Radio Meteorological Data, Report 563-3, Vol. 5, *Propagation in Non-Ionized Media, Recommendations and Reports of the CCIR, 1986*, International Telecommunication Union, Geneva, 1986.

47. International Radio Consultative Committee, Propagation Data Required for Space Telecommunication Systems, Report 564-3, Vol. 5, *Propagation in Non-Ionized Media, Recommendations and Reports of the CCIR, 1986*, International Telecommunication Union, Geneva, 1986.

48. Crane, R. K., Prediction of Attenuation by Rain, *IEEE Trans. Commun.*, COM-28:1717–1733 (September 1980).

49. Pratt, T., and Bostian, C. W., *Satellite Communications*, Wiley, New York, 1986.

50. Stutzman, W. L., and Yon, K. M., A Simple Rain Attenuation Model for Earth-Space Radio Links Operating at 10-35 GHz, *Radio Sci.*, 21:65–72 (January–February 1986).

51. Beckmann, P., *The Depolarization of Electromagnetic Waves*, Golem Press, Boulder, CO, 1968.

52. International Radio Consultative Committee, Cross-Polarization Due to the Atmosphere, Report 722-2, Vol. 5, *Propagation in Non-Ionized Media, Recommendations and Reports of the CCIR, 1986*, International Telecommunication Union, Geneva, 1986.

53. Chu, T. S., Microwave Depolarization of an Earth-Space Path, *Bell Sys. Tech. J.*, 59:987–1007 (July–August 1980).

54. Chu, T. S., A Semi-Empirical Formula for Microwave Depolarization on Earth-Space Paths, *IEEE Trans. Commun.*, COM-30:2550–2554 (September 1982).

55. Slobin, S. D., Microwave Noise Temperature and Attenuation of Clouds at Frequencies below 50 GHz, JPL Publication 81-46, Jet Propulsion Laboratory, Pasadena, CA, 1981.

56. International Radio Consultative Committee, Attenuation by Atmospheric Gases, Report 719-2, Vol. 5, *Propagation in Non-Ionized Media, Recommendations and Reports of the CCIR, 1986*, International Telecommunication Union, Geneva, 1986.

57. International Radio Consultative Committee, Propagation Data for Maritime and Land Mobile Satellite Systems Above 100 MHz, Report 884, Vol. 5, *Propagation in Non-Ionized Media, Recommendations and Reports of the CCIR, 1982*, International Telecommunication Union, Geneva, 1982.

58. International Radio Consultative Committee, Reflection from the Surface of the Earth, Report 1008, Vol. 5, *Propagation in Non-Ionized Media, Recommendations and Reports of the CCIR, 1986*, International Telecommunication Union, Geneva, 1986.

59. Miller, A. R., Brown, R. W., and Vegh, E., New Derivation for the Rough-Surface Reflection Coefficient and for the Distribution of Sea-Wave Elevations, *IEE Proc.*, Part H, 131:114–116 (April 1984).

60. Beckmann, P., and Spizzichino, A., *The Scattering of Electromagnetic Waves from Rough Surfaces*, Macmillan, New York, 1963.

61. Vogel, W. J., and Goldhirsh, J., Tree Attenuation at 869 MHz Derived from Remotely Piloted Aircraft Measurements, *IEEE Trans. Antennas Propagat.*, AP-34:1460–1464 (December 1986).

62. Goldhirsh, J., and Vogel, W. J., Mobile Satellite System Fade Statistics for Shadowing and Multipath from Roadside Trees at UHF and L-Band, *IEEE Trans. Antennas and Propagat.*, 37:489–498 (April 1989).

63. International Radio Consultative Committee, The Evaluation of Propagation Factors in Interference Problems between Stations on the Surface of the Earth at

Frequencies above about 0.5 GHz, Report 569-3, Vol. 5, *Propagation in Non-Ionized Media, Recommendations and Reports of the CCIR, 1986*, International Telecommunication Union, Geneva, 1986.

64. International Radio Consultative Committee, Propagation Data Required for the Evaluation of Coordination Distance in the Frequency Range 1 to 40 GHz, Report 724-2, Vol. 5, *Propagation in Non-Ionized Media, Recommendations and Reports of the CCIR, 1986*, International Telecommunication Union, Geneva, 1986.

65. International Radio Consultative Committee, Determination of Coordination Area, Report 382-5, Vols. 4 and 9 (Part 2), *Frequency Sharing and Coordination between Systems in the Fixed-Satellite Service and Radio-Relay Systems, Recommendations and Reports of the CCIR, 1986*, International Telecommunication Union, Geneva, 1986.

66. Crane, R. K., A Review of Transhorizon Propagation Phenomena, *Radio Sci.,* 16:649–669 (September–October 1981).

67. Campanella, S. J., Evans, J. V., Muratani, T., and Bartholome, P., Satellite Communications Systems and Technology, Circa 2000, *Proc. IEEE*, 78:1039–1056 (July 1990).

68. Allen, K. C., Violette, E. J., and Espeland, R. H., Observed Wide-Band Digital Performance at 30 GHz, *IEEE Trans. Commun.*, COM-34:733–736 (July 1986).

69. Pritchard, W. L., and Sciulli, J. A. *Satellite Communication Systems Engineering*, Prentice-Hall, Englewood Cliffs, NJ, 1986.

70. Rana, A. H., McCoskey, J., and Check, W., VSAT Technology, Trends and Applications, *Proc. IEEE*, 78:1087–1095 (July 1990).

71. Haus, H. A., Molding Light into Solitons, *Spectrum,* 30:48–53 (March 1993).

72. NOAA, NASA, United States Air Force, sponsors, *U.S. Standard Atmosphere, 1976*, Superintendent of Documents, U.S. Government Printing Office, Washington, DC, 1976.

73. Yunck, T. P., Wu, S.-C., and Lichten, S. M., A GPS Measurement System for Precise Satellite Tracking and Geodesy, *J. Astro. Sci.*, 33:367–380 (October–December 1985).

74. Resch, G. M., Water Vapor—The Wet Blanket of Microwave Interferometry. In: *Atmospheric Water Vapor* (A. Deepak, T. D. Wilkerson, L. H. Ruhnke, eds.), Academic Press, New York, 1980, pp. 265–282.

75. Gary, B. L., Keihm, S. J., and Janssen, M. A., Optimum Strategies and Performance for the Remote Sensing of Path Delay Using Ground-Based Microwave Radiometer, *IEEE Trans. Geosci. Remote Sensing*, GE-23:479–484 (July 1985).

76. Briskman, R. D., Radio Determination Satellite Systems, *Proc. IEEE*, 78:1096–1106 (July 1990).

WARREN L. FLOCK

Electron Beam Displays (Cathode-Ray Tubes)

Introduction

One of the major technological trends of our times has been the need to transfer information to the ultimate consumer, the human being. Entire industries have originated as a result, including branches of telecommunications, the computer industry, and the consumer sector.

Since human vision is the communication channel with the widest bandwidth among our senses, electronic displays have established a dominant position in relation to this area. When the objective is not so much the transfer of information but entertainment, once again displays play a vital role. In this context, display hardware is needed to generate pictures that include text, graphics (e.g., drawings or charts), and natural scenes of landscapes, people, and sporting events. The quality of the pictures must rival that of such other media as film; color, a technological breakthrough just a few decades ago, has become essential and the physical size of the picture must be commensurate with the properties and limitations of the human eye. Last, but certainly not least, cost is a major consideration, sometimes an overriding one.

These requirements represent an enormous technological and manufacturing challenge. To address them, a number of display technologies have emerged, particularly in the last two decades. Surprisingly, the cathode-ray tube (CRT) remains dominant although it is the oldest technology. The reasons are many, including an established manufacturing base that routinely produces tens of millions of the devices at a very high yield and the superior ergonomics of the device. Preeminent among all, at least to this writer, is the enormous dynamic range that CRTs offer. This has made them pervasive in all sorts of applications, including personal computers (PCs), television (TV), air-traffic control, and situation displays for command and control, to name just a few. This wide dynamic range of a CRT is summarized in Table 1.

Figure 1 shows the block diagram of a typical CRT monitor. The word *monitor* generally is used to describe an ensemble of waveform generators, power supplies, and amplifiers together with the display device and an overall mechanical structure or housing. A television set consists of a monitor with the addition of tuner, audio, and decoding functions. The heart of the monitor is of course the CRT device itself, which is shown diagrammatically in Fig. 2.

The CRT consists of an electron source (a thermionic cathode), some beam-forming and beam-focusing electrodes, a deflection system, and a screen. A glass envelope is required in order to sustain the vacuum and is a major cost factor for most CRTs.

In synthesis, electrons are emitted by the cathode, focused into a small bundle by some lenses, and accelerated to several kilovolts (kV) before hitting the phosphor screen. There the beam is converted into a luminous spot. The electron beam is modulated in space by a deflection field. In addition, the beam

TABLE 1　Cathode-Ray Tube Performance

Size	1-cm to 1-m direct view
	1- to 50-m projection
Resolution	200^2 to $10,000^2$
Luminance	<1 to $>10,000$ cd/m^2
Contrast	Ambient dependent (30 easy in the office)
Gray levels	Yes
Color	Yes
Cost	$20 monochrome, $50 Color*
Data rates	A few MHz to 1 GHz depending on system requirements
Image retention time	μs to ms (allowing fast updating)
Interface	Simple serial
Power	30 W to 200 W typical for a complete monitor
Status of technology	Mature but constantly improving
Status of manufacturing	Approximately 100 million devices produced in 1991

Source: From Ref. 1.
*Entertainment-grade devices. Industrial grade are 2× higher and up.

current can be modulated in intensity by applying a voltage between the CRT grid and cathode. As a result, a picture appears on the screen and is viewed by the observer. There are two popular ways of generating a picture: calligraphic and raster.

The most intuitive manner of picture generation is called *calligraphic*, *directed beam*, or *stroke writing* (2). The electron beam is deflected in such a way as to produce directly the picture of interest (see Fig. 3). After the picture is completed, the process is repeated at a rate high enough to avoid flicker. Until recently, calligraphic displays were quite common in such specialized applications as air-traffic control and displays for the cockpit.

By far the most popular way of addressing a CRT to generate a picture is to cover the CRT screen repetitively with a "striped pattern" or raster (see Fig. 4) (3). Linear deflection waveforms are applied along each axis at frequencies that are fixed and multiples of one another. The electron beam is turned on or off at the right time so that a picture is formed (see Fig. 5).

Another major trend is color. A few years ago high (spatial) resolution and color were mutually exclusive. Today, thanks to the introduction of fine-pitched masks and screens, color displays with good resolution are commonplace. A lot of work remains to be done, however, in both the hardware and the perceptual or human interface side of color displays in order to achieve the quality produced by photographic film.

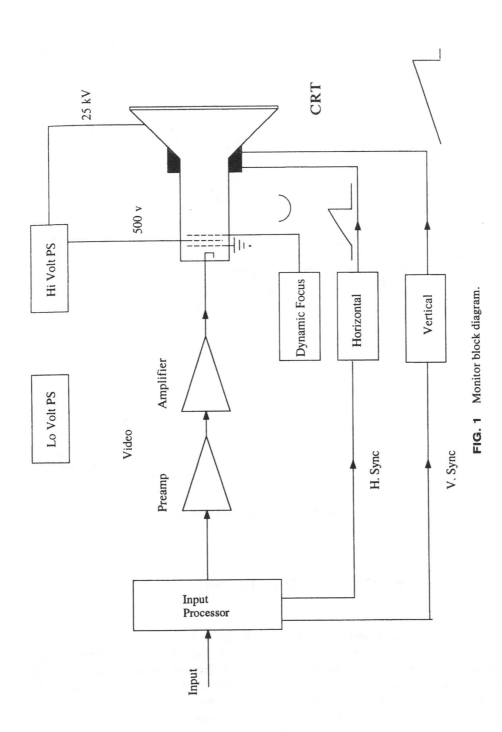

FIG. 1 Monitor block diagram.

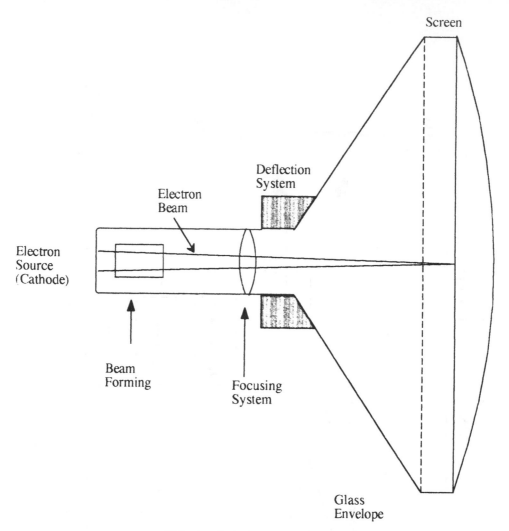

FIG. 2   Cathode-ray tube diagram.

Monochrome Cathode-Ray Tubes

The structure of a monochrome CRT is shown in Fig. 6. A good way to study the essential characteristics of a CRT is to subdivide it. One considers a beam-forming region, a beam-shaping region, a deflection region, and finally the light-producing region or screen. The beam-forming region is made up of the filament, the cathode, the control grid, and the first anode, G_2. The beam-forming region often is referred to as the *triode*; together with the beam-shaping region, it forms the CRT electron gun. Electrons leave the cathode, form into a bundle called the *crossover* (ideally, a point source), and are accelerated and

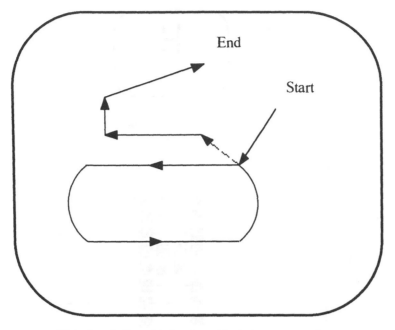

FIG. 3 Calligraphic (stroke writing) picture generation.

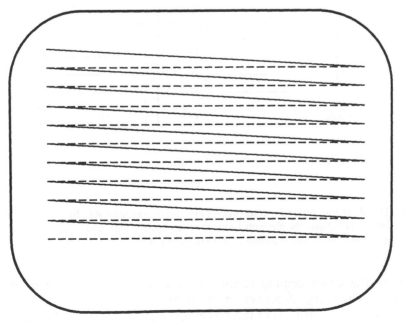

FIG. 4 Raster scanning.

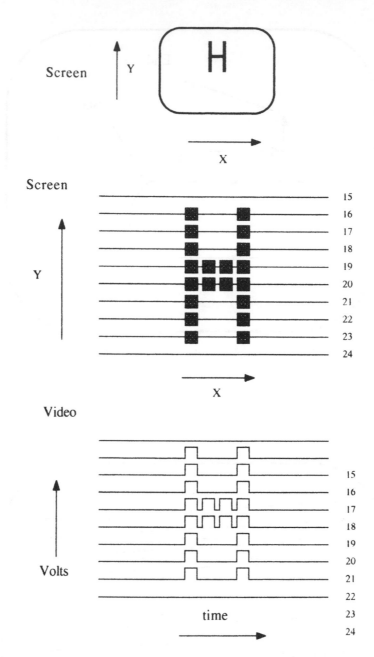

FIG. 5 Details of the picture forming process in a raster-scanned CRT. In order to generate the letter "H" (*top*), the electron beam is swept horizontally and vertically (*middle*). The grid-cathode voltage is pulsed positive (*bottom*) at the appropriate time.

focused in the beam-shaping region. They finally impinge on the phosphor screen where the energy is converted into light.

The cathode of most CRTs is a nickel cylinder heated by a filament, usually of tungsten. The emitting portion of the cathode is coated with an oxide. At the

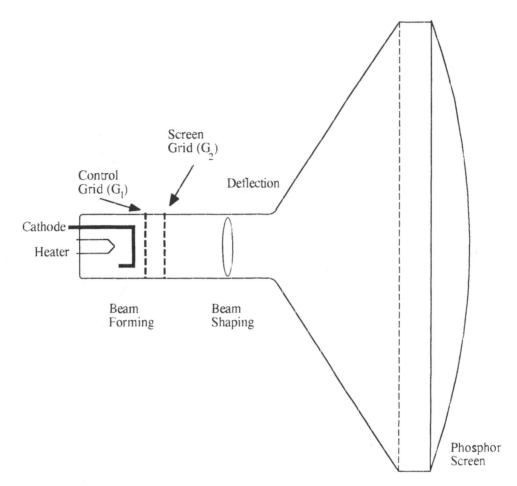

FIG. 6 A monochrome cathode-ray tube. Beam-forming, beam-shaping, and deflection regions are seen. The phosphor screen or light-emitting region is where the electron energy is converted to light.

typical operating temperature of the cathode (1000°K), there are enough energetic free electrons available that some will be emitted and form a beam. To make a cathode, a mixture of strontium and barium carbonate is applied to a nickel base. After pumping, the cathode is heated to about 1300°K for about 1 minute. The temperature then is reduced to about 1100°K and a voltage applied. The voltage is increased slowly and current is drawn for about 10 to 20 minutes. This constitutes the so-called activation of the cathode.

Building, processing, and activation of cathodes is a technology in itself, with substantial variations depending on the manufacturer and on the intended use of the CRT. The technology is quite mature (about 100 million CRTs are produced routinely each year). In actual use, the cathode current is a strong function of the grid-to-cathode voltage. A typical dependency is shown in Fig. 7. If the grid voltage is more negative than a threshold value (usually called *cutoff voltage*, E_{co}), essentially no current flows. As the grid is made more positive

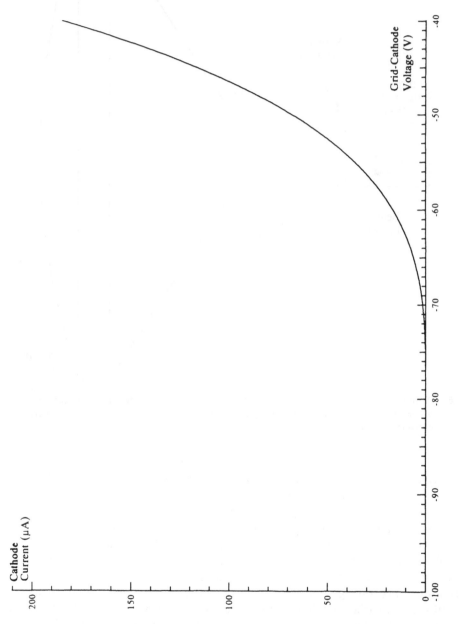

FIG. 7   Typical current-voltage relationship for a cathode-ray tube.

than E_{CO}, current flows more quickly with voltage. An approximate mathematical relationship is given by Refs. 2, 4, and 5.

$$I_k = 3 \frac{E_d^{3.5}}{E_{CO}^2}, \tag{1}$$

where E_d is the so-called grid drive ($E_{GK} + E_{CO}$) and I_k is expressed in microamperes (μA). This relationship is plotted in Fig. 8. In a given CRT, E_{CO} can be changed by operating the G_2 grid at a different voltage. As the voltage on G_2 is increased, the electric field in the vicinity of G_1 is increased, thus increasing E_{CO}.

If one calculates the current density at the cathode and averages it (spatially) over the whole cathode area, one finds (1)

$$\overline{\rho_c} = \frac{1.2 \times 10^{-3}}{\pi} \frac{E_D^{1.5}}{D^2}, \tag{2}$$

where D is the diameter of the grid aperture in millimeters (mm). The current density has a peak value given by

$$\rho_p = 1.5 \times 10^{-3} \frac{E_D^{1.5}}{D^2}. \tag{3}$$

These current densities are referred to as *cathode loading* and are expressed in amperes/centimeter^2 (A/cm^2). If long cathode life is important and a conventional cathode is used, peak cathode loading should be limited to 1 A/cm^2 or less. Recently, a number of alternative cathode materials and structures have been proposed that allow considerably higher values of cathode loading (up to 4 A/cm^2).

We now consider the process that converts the cloud of electrons emitted at the cathode into a focused spot. As we discussed above, the electron gun produces an electron beam concentrated into a crossover. Subsequently, a system of lenses will focus the crossover into a narrow spot on the screen. Fortunately, electron paths can be described by the same sort of equations that describe light rays (6). This means that such concepts as index of refraction, focal lengths, aberrations, and magnification all have an exact analogy in electron optics. More importantly, one can show that some simple structures (e.g., a set of coaxial cylinders at different voltages) have properties very similar to those of an optical lens. Some of the key relationships for both kind of lenses are shown in Fig. 9 (4).

The analogy extends to formal properties of both systems. The principal equation for optical systems

$$(1/p) + (1/q) = 1/f \tag{4}$$

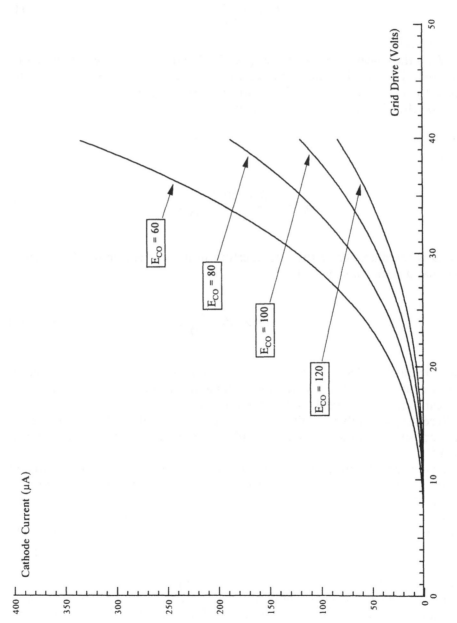

FIG. 8   Cathode-ray tube cathode current as a function of grid drive for various cutoff voltages.

that relates object distance p and image distance q to focal length f has an identical electron-optical counterpart. Similarly, the magnification of an optical system

$$m = q/p \qquad (5)$$

has a counterpart.

In optical systems, most real lenses are more complex than the simple "thin" lens shown in Fig. 9. Similarly, in electron-optical systems, real lenses are more complex. The general analogy still holds, however. Just like in optics, real electron lenses depart somewhat from the ideal, and suffer from various imperfections commonly referred to as *aberrations*. This is detailed briefly in Table 2. In practice, spherical aberrations are often the limiting factor in the resolution of a display system.

Let us now consider an important characteristic of CRTs, that is, the size of the spot on the screen. In any practical CRT, the beam size at the center is finite due to a number of causes. First, electrons leaving the cathode do not

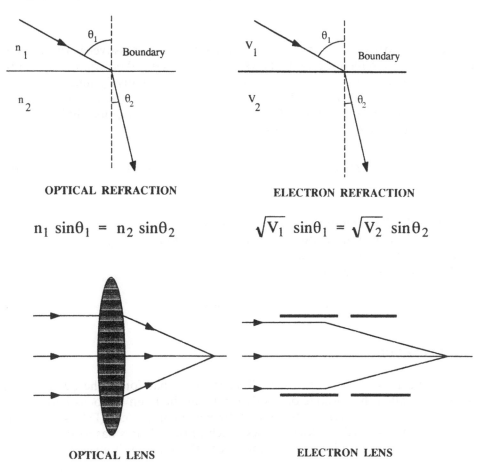

OPTICAL REFRACTION

$$n_1 \sin\theta_1 = n_2 \sin\theta_2$$

ELECTRON REFRACTION

$$\sqrt{V_1}\, \sin\theta_1 = \sqrt{V_2}\, \sin\theta_2$$

OPTICAL LENS

ELECTRON LENS

FIG. 9 Optical/electron-optical analogy.

TABLE 2 Aberrations

Type	Optical Effect	Electron-Optical Effect
Chromatic	Rays of different wavelengths have different focal lengths	Electrons of different energies have different focal lengths
Spherical	Rays parallel to the axis have a focal length that depends on the distance from the axis	Electron rays parallel to the axis have a focal length that depends on the distance from the axis
Astigmatism	Off-axis rays have a focal length dependent on their angle	Off-axis electrons have a focal length dependent on their angle
Coma	A round object is focused into a tear-shaped image	A round crossover is focused into a tear-shaped spot
Curvature, distortion of field	Rectangular grids become curved, pincushioned, barreled, and so on	Rectangular grids become curved, pincushioned, barreled, and so on

have equal velocity, but a spread. Thus, the size of the crossover is finite, as is its image on the screen. Second, the various lenses have aberrations, especially spherical aberrations. Third, electrons are charged negatively and thus repel each other. This is known as the *space-charge effect* and will spread the beam out as the current is increased. As a result of these factors, one finds that the spot size at the screen can be expressed as (7)

$$D = \sqrt{D_{sa}^2 + (0.3D_{th})^2} + 0.7D_{th} + D_{sc} \qquad (6)$$

where
D = the diameter of the spot at the screen
D_{sa} = the contribution due to spherical aberration
D_{th} = the contribution due to thermal velocity
D_{sc} = the contribution due to space charge

The effect of each of these is discussed next.

Spherical Aberration

There are a number of lenses in a CRT: the one that forms the beam into a crossover (called the cathode or triode lens) and the main or focus lens. Some electron gun designs such as Dynamic Astigmatism and Focus (DAFs) or Trinitrons employ so-called pre-focus lenses. Each of these lenses has its own spherical aberration. One finds that the total spherical aberration is given by (7)

$$D_{sa} = M_p M_L D_{sac} + M_L D_{sap} + D_{saL}, \qquad (7)$$

where M_p, M_L are the magnification of the prefocus and main lens, respectively, and D_{sac}, D_{sap}, and D_{saL} are the contribution of the cathode, prefocus lens, and main lens, respectively.

The main lens is often the leading contributor to spot size. One finds that D_{saL} is given by

$$D_{saL} = \text{const.} \frac{\Phi^3}{D_L^2}, \tag{8}$$

where Φ is the diameter of the beam in the lens and D_L is the diameter of the lens. It is important to remember that we can expect a smaller and crisper spot as we make the lens larger. The importance of this will become clearer in the discussion of color systems, in which three guns must coexist in one CRT envelope.

Thermal Velocity

Even if we could build perfect lenses, we still would wind up with a finite spot due to the spread of velocities. One finds that the current distribution on the screen is given by a Gaussian distribution (2)

$$j(r) = j_0 e^{-\frac{r^2}{\rho^2}} \tag{9}$$

where $j(r)$ and j_0 are the current densities at a distance r and at the center, respectively, and ρ is the parameter that describes the "width" of the distribution. It is given by (2)

$$\rho = \frac{2r_0 L}{\Phi} \sqrt{(kT/qV)} \tag{10}$$

where
   r_0 = the radius of the emitting area on the cathode
   L = the distance from gun to screen
   Φ = the diameter of the beam in the main lens
   k = Boltzmann's constant (1.38×10^{-23} J/°K)
   T = the absolute temperature of the cathode (°K)
   q = the charge of the electron (1.6×10^{-19} coul)
   V = the screen voltage

The importance of these relationships is due to the fact that the resolution of a CRT is limited principally by the size of the electron beam at the screen. As more current is drawn from a smaller cathode area, cathode loading increases at a potential cost in CRT life. Equation (10) also reveals that as the distance to the screen L is made smaller, the size of the beam improves but the deflection angle increases and thus deflection power decreases. Similar considerations apply to the screen voltage and the cathode temperature.

Space-Charge Effects

Space-charge effects are important at large beam currents, usually well past 200 μA or so. One finds

$$D_{sc} = \frac{i_b L^2}{2\pi\epsilon_0 \sqrt{\dfrac{q}{m}}\ \Phi V^{1.5}} \tag{11}$$

where m is the mass of the electron.

From all this, one sees that electron guns are complex. Thus, no simple method exists for analyzing, much less synthesizing, an arbitrary electron-optical system. With the prevalence of powerful computers, simulation techniques have become available that offer an invaluable aid to gun design. A number of structures can be simulated and performance predicted without resorting to the "build it first and see if it works" methodology prevalent years ago.

In Fig. 10, electron trajectories are shown for the so-called unipotential gun, one of the more popular electron lens structures in use for monochrome CRTs. The control grid G_1 is shown together with some of the other elements.

In Fig. 11, the unipotential gun is compared to the bipotential gun. The first system usually involves focus elements at higher voltages and is preferred when both good resolution and high beam currents are required. The second is the simplest and has found wide application in both moderate- and high-resolution systems and often is referred to as a low-voltage lens.

Deflection

We now consider the techniques for deflecting the electron beam. In general, one wants to move the beam as rapidly as possible over as wide an angle as

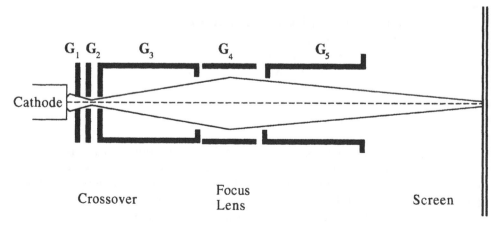

FIG. 10 Schematic diagram of the unipotential gun. The electron cloud is formed into a small crossover, which is focused onto the screen.

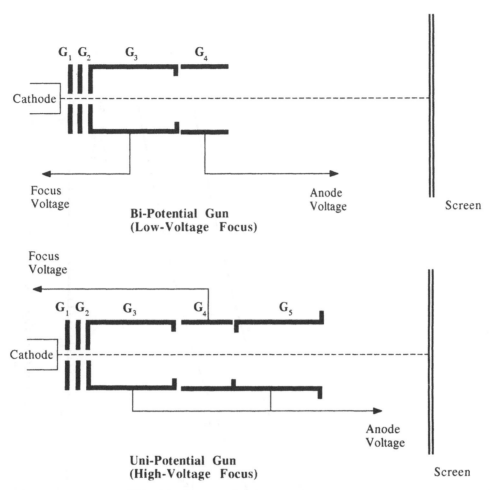

FIG. 11 Two popular electron gun structures.

possible and with as little power as possible. The *deflection angle* is defined as twice the angle from the deflection center to the corner of the screen; its effect on overall CRT length is shown in Fig. 12, in which a 15-inch (in) diagonal CRT is assumed (just large enough to display an 8½ × 11-in area).

There are two known ways to deflect a beam: electrostatic and magnetic. In electrostatic deflection, the beam is made to pass between two sets of deflection plates that are orthogonal to one another. The electric field resulting from voltages applied to the plates allows positioning of the beam anywhere on the screen. This technique generally is used in oscilloscopes in which deflection speed is a major requirement. In applications in which deflection angle and brightness are more important, magnetic deflection is preferred. Instead of deflection plates, one uses deflection coils, wound orthogonally to the CRT axis in order to take advantage of the well-known Lorentz equation

$$\vec{F} = q(\vec{v} \times \vec{B}). \tag{12}$$

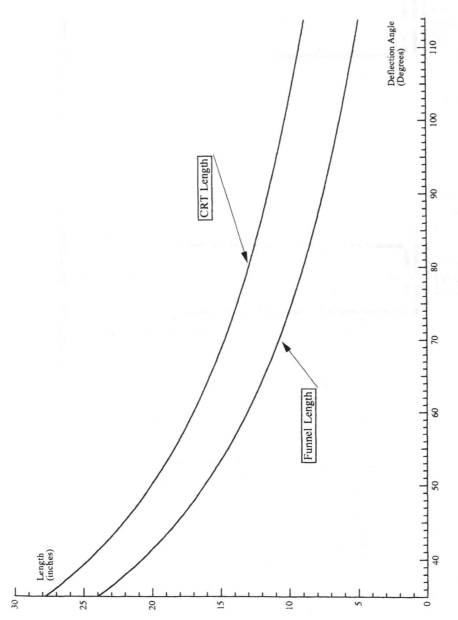

FIG. 12 Effect of deflection angle on cathode-ray tube length (15-in diagonal). As the deflection angle is made larger, the overall length of the cathode-ray tube becomes smaller.

The relevant equations governing deflection angle θ and on-screen deflection y are

$$\sin \theta = 3 \times 10^5 \frac{B\ell}{\sqrt{E_b}} \tag{13}$$

and

$$y = L \tan \theta, \tag{14}$$

where
   B = the magnetic flux density
   ℓ = the length of the coil
   L = the distance from the deflection center to the screen(meter-kilogram-second [mks] units)

It is important to note that while the deflection voltage increases linearly with the screen voltage for electrostatic deflection, in magnetic deflection the required field B, hence the deflection current, increases with the square root of E_b. Since display luminance is proportional to E_b, as is shown by Eq. (20), magnetically deflected systems are preferred in many applications.

One of the important characteristics of a deflection yoke is its effect on spot focus as the beam is deflected. If the magnetic field within the yoke is not uniform, different electrons within the beam will be deflected differently and an astigmatic spot results (see Fig. 13). The optimum turns distribution required to generate a uniform field has been determined and is a cosine function of the angle between the CRT axis and the coil loop (3).

Empirically, one finds that the growth of the spot with deflection is given by

$$\delta = K \left(\frac{d\theta}{D}\right)^2 \tag{15}$$

where
   K = a nondimensional yoke quality factor
   d = the size of the beam bundle within the deflection region
   D = the CRT neck diameter

In summary, a good monochrome yoke is wound with the proper distribution in such a way as to guarantee stability and consistency. This is achieved by using magnetic materials in which slots or teeth are provided and by carefully controlling manufacturing techniques. Any real deflection yoke has another, separate, effect on the spot focus. As the beam is deflected away from the axis, it encounters some fringing field lines and must travel a greater distance before reaching the screen. Both of these effects cause the beam to come to a focus before the screen (see Fig. 14). So-called dynamic focus techniques for compensating this effect usually involve changing the focus voltage as a function of the beam position. Usually parabolic or second-order functions of beam position are involved.

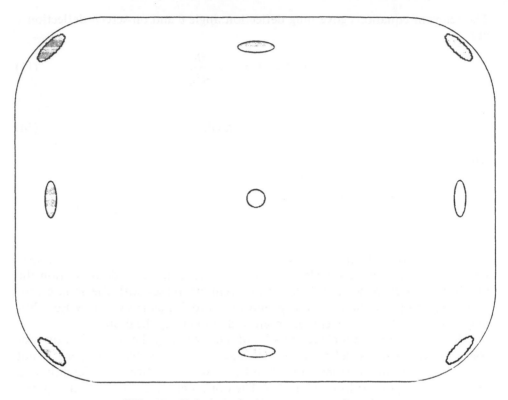

FIG. 13 Yoke imperfections may cause astigmatism.

Another important consideration is pattern distortion. Deflection current is related linearly to the sine of the deflection angle.

$$\sin \theta = ki \tag{16}$$

The on-screen deflection y, however, is proportional to the tangent of the deflection current

$$y = L \tan \theta \tag{17}$$

Consequently, the relationship between y and the deflection current is linear only for very small angles (Fig. 15). One finds

$$y \cong \frac{Lki}{\sqrt{1 - k^2 i^2}} \tag{18}$$

If one carries out a similar analysis in both x and y rather than in just one dimension, one finds that linear current steps are displayed as a grid with pronounced pincushion distortion (see Fig. 16). A number of compensating techniques have been developed, including magnetic (the addition of fixed magnets), hybrid (electronic/magnetic), and all electronic techniques.

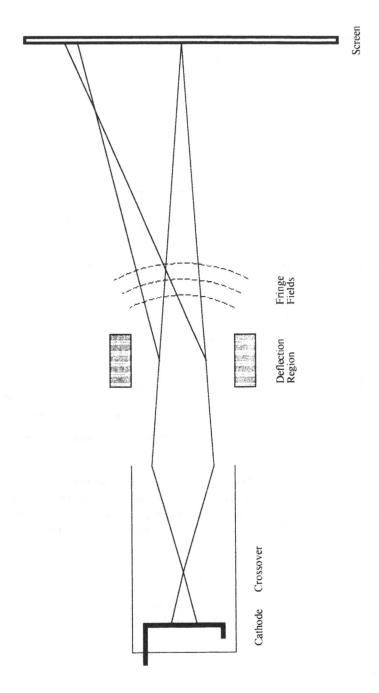

FIG. 14 The causes of deflection defocusing.

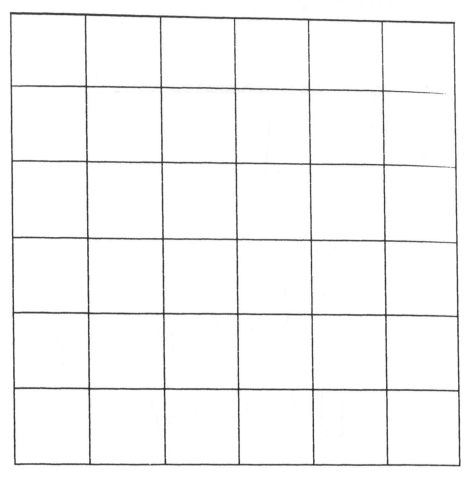

FIG. 15   An ideal grid. Lines are straight and spaced equally; they form right angles.

Clearly, for a given CRT and yoke form factor, the magnetic energy $Li^2/2$ is constant. One therefore can trade off the yoke inductance against the current required to deflect a given distance on screen by employing a yoke with more or fewer turns. Furthermore, the energy constant of the yoke Li^2 is related to the CRT neck diameter by

$$Li^2 \propto D \qquad\qquad (19)$$

Recalling the discussion about spherical aberrations and lens size, one can understand how deflection power versus beam size can be traded off as the neck diameter is varied.

Phosphor Screens

After the beam has been deflected, it lands on the screen where the electron energy is converted into light. The screen consists of a glass faceplate, with the appropriate optical properties, on which a suitable phosphor material has been deposited. Deposition methods include dusting, spraying, and settling through liquid means. For satisfactory system operation, the right choice of phosphor material and deposition techniques must be made. The characteristics one usually looks for in a phosphor are efficiency, resolution, and good life, together with the appropriate color and persistence.

The phenomenon by which a phosphor emits light under electron bombardment is called *cathodoluminescence* and is a quantum phenomenon. Energetic electrons from the electron beam penetrate deep into the phosphor, with a mean penetration on the order of 14 μm (micrometers) at 20 kV. As the beam is absorbed, it raises some of the phosphor electrons to an excited state. The excited electrons then return to a stable state, releasing the excess energy in the form of light with a wavelength and a persistence dependent on the particular phosphor and its preparation. Phosphors operated above a few kV usually are coated with an aluminum film that improves the luminous efficiency by both

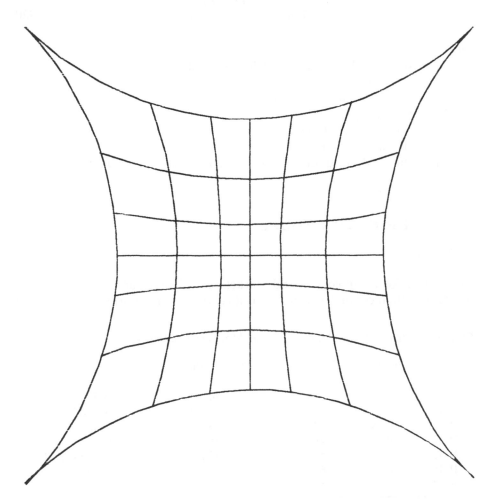

FIG. 16 Pattern distortion in an uncorrected display. The grid lines are curved and are closer to one another in the center.

stabilizing the operating voltage and reflecting more of the emitted light toward the observer. The essential properties of some of the more popular phosphors are summarized in Table 3. (A more comprehensive listing of phosphors and their properties can be found in Ref. 8.)

An important consideration in the choice of a phosphor is life. To a first order, phosphor life can be correlated to the total amount of charge per unit area falling on the phosphor. As an example, zinc sulphide (ZnS) phosphors such as P-4 and its derivatives are used widely in monochrome television and computer displays. The initial efficiency is about 35 radiated lumens/watts excitation. This will decrease by 50% after the phosphor has absorbed about 35 coulomb/centimeter^2 (1,9).

A very important relationship exists between the CRT operating parameters and the luminance of the resulting image (see Fig. 17). For a raster-scanned system, one finds, using metric units,

$$\text{Lum} = \frac{\eta T_G}{\pi} \frac{i_b Eb}{A} \delta \tag{20}$$

where

$$\eta = \text{the phosphor efficiency in lumens/watt}$$
$$T_G = \text{the glass transmission}$$
$$i_b = \text{the beam current in amps}$$
$$E_b = \text{the voltage in volts}$$
$$A = \text{the raster area (meters}^2)$$
$$\delta = \text{the raster duty cycle}$$
$$\text{Lum} = \text{the resulting luminance in nits (candela/meter}^2 \text{ or cd/m}^2)$$

Color Cathode-Ray Tubes

Introduction

Color is a major technological and marketing challenge for displays. Color is important both in computer displays, as it represents an additional information

TABLE 3   Properties of Commonly Used Monochrome Phosphors

Phosphor	Color	Peak (nm)	Decay Time (to 10%)	Usage
P-1	G	525	24 ms	DVST
P-4	W	450	24/60 μs	TV, WP, CG
P-7	B-W	440	4 s	Radar
P-11	B	460	35 μs	OSC
P-31	G	525	38 μs	WP, OSC, CG
P-35	O	600	1.1 s	Radar
P-44	G	550	1.7 ms	WP, CG

WP = word processing; CG = computer graphics; OSC = oscillography; DVST = direct view storage tubes.

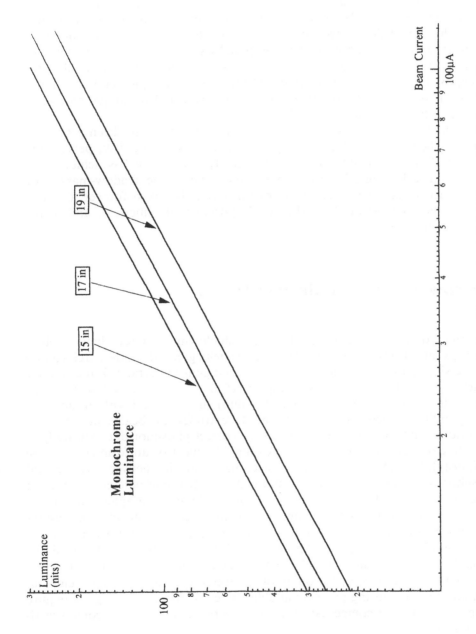

FIG. 17 Luminance of CRTs of different sizes per Eq. (20). Phosphor is P-4 (40 lumens/watt). Anode voltages are 15, 17, and 20 kV, respectively.

channel, and in the entertainment area, as color displays are much more natural and appealing than monochrome.

To display color, one takes advantage of Helmholtz's trichromaticity theory. Thus, each color can be represented by the addition of three independent primaries. Consequently, much more information must be processed in order to display color. Practical devices achieve color by some form of sharing or separation mechanism in either *space* or in *time*. Picture elements in color are obtained by synthesizing either subpixels that are spatially separated or individual monochrome picture frames that are temporally distinct. Each subpixel or subframe is composed of a single primary color. The subpixel technique implies an inherent loss in spatial resolution, while the frame sequential technique requires a faster refresh rate in order to avoid flicker.

For color CRTs, a number of technological solutions have been proposed, including the Columbia Broadcasting System (CBS) spinning disk, penetration CRTs, and the liquid crystal color shutter (1,3,10). None has achieved the market and technological success approaching that of the shadow-mask CRT and its successors. The development of this device is linked forever to David Sarnoff, the founder of RCA (Radio Corporation of America), and to the laboratory that bears his name.

Shadow-Mask Cathode-Ray Tubes

A number of alternative architectures for shadow-mask CRTs have evolved over the years (11) (see Fig. 18). The so-called delta gun is the oldest architecture and the simplest conceptually. Three individual guns are arranged in a triangle or delta inside the tube neck and angled (1° or so) with respect to the axis. The shadow mask consists of a thin (0.1 mm) sheet of steel with an aperture corresponding to each phosphor dot triad. It "extrudes" the beams into beamlets and insures that each beamlet lands on the correct phosphor dot. Naturally, a full frame of a single primary (e.g., red) should appear of uniform color to the eye (purity). For satisfactory performance, the registration between beamlet and phosphor dot must be held to no more than a few micrometers. A further requirement on a color display device is that small images (graphic lines or text characters) appear without fringing colors (convergence). This in turn requires that the centroid of each beam coincide with the other two over the entire screen to within 100 microns or so. The delta architecture makes good use of the available neck space, allowing relatively large electron-optical lenses. However, maintenance of accurate positioning of the electron guns over time, temperature, and manufacturing process variables is very difficult. Delta guns require complex circuitry to compensate for these tolerances and have become practically obsolete.

In the in-line structure (12,13), the guns are arranged horizontally with the center (green) gun parallel to the axis. The outer guns are tilted inward by about 0.5°. In high-resolution applications, a dot screen similar to the delta dot screen is used, while for lower-resolution applications a slotted mask is preferred for

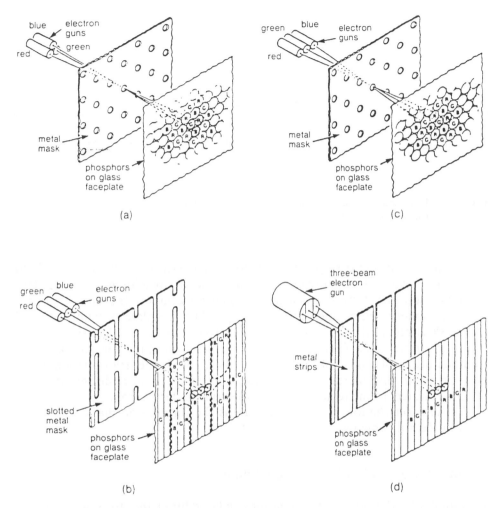

FIG. 18 Shadow-mask architectures: *a*, delta gun/dot mask; *b*, in-line gun/slot mask; *c*, in-line gun/dot mask; *d*, in-line/strip mask color cathode-ray tube (Trinitron). (Used by permission of Ref. 11.)

increased brightness. Eliminating the vertical tilt angle between guns and adopting a unitized mechanical structure greatly reduces purity and convergence difficulties to the point that these corrections routinely are built into the deflection yoke. A detailed description of self-converging yokes is beyond the scope of this work. The interested reader should consult the excellent literature on the subject (14). The detail of the electron-optical lens arrangement in an in-line gun is shown in Fig. 19-*a* (15). Shown in Fig. 19-*b* is a more modern implementation concept in which the lenses are extended (overlapping lens field) in order to reduce aberrations (16).

A different approach to the problem was developed by Sony with the Trinitron (17). The guns again are arranged horizontally, but the beams go through a common focus lens. This achieves an even better space utilization factor than

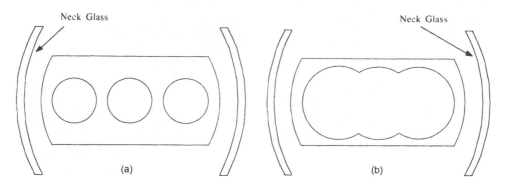

FIG. 19 Lens arrangement in in-line guns; *a*, conventional in-line gun; *b*, improved in-line gun with overlapping field (OLF).

delta guns by having a large common focus lens (see Fig. 18-*d*). For this to work, Sony had to use a more complex gun structure by adding a prefocus lens so that most of the electron rays will go through the center of the lens as opposed to the periphery, where aberrations are high. The faceplate is cylindrical and the mask is replaced by a grilled aperture. The phosphor is deposited on the screen in stripes. This eliminates the loss in resolution in the vertical direction. In moderate-resolution applications, the grille allows higher transmission of the electron beam, thus increasing luminance. This comes at the added expense and weight of a steel frame required to hold the grille under tension.

Yet another shadow-mask architecture is the flat-tension mask (18) developed by Zenith, which is shown in Fig. 21.

Effects of the Shadow Mask on Luminance

In color CRTs, the key equation relating displayed luminance to system parameters (Eq. [20])

$$\text{Lum} = \frac{\eta T_G}{\pi} \frac{I_b Eb}{A} \delta$$

is replaced by

$$\text{Lum} = \frac{\eta T_G T_M}{\pi} \frac{i_b Eb}{A} \delta \tag{21}$$

The shadow mask intercepts a considerable fraction of the electron beam. This is taken into account by T_m, the effective mask transmission (typically 0.13). η is the effective phosphor luminous efficiency. Typical values for η are 40, 16, and 8 lumens/watt for green, red, and blue, respectively. Equation (21) is plotted in Fig. 20.

A comparison of Figs. 17 and 20 shows that color CRTs are not nearly as bright as their monochrome counterparts. Nonetheless, satisfactory brightness

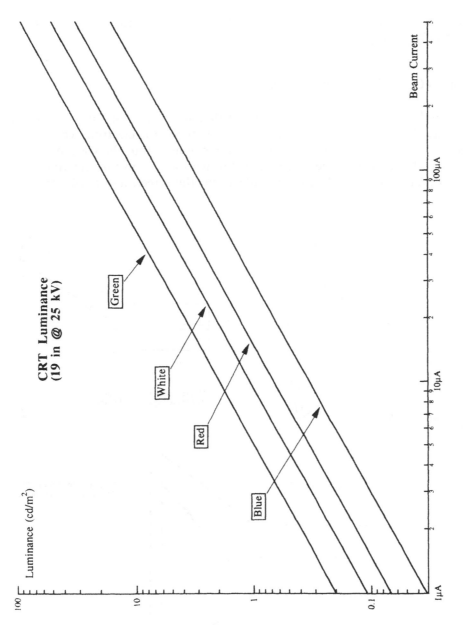

FIG. 20 Luminance of a color cathode-ray tube.

is achieved routinely by operating color CRTs at substantially higher accelerating voltages, 20 to 30 kV versus 12 to 18 kV for monochrome. The beam power intercepted by the mask causes thermal expansion (doming) that can result in loss of purity. This effect can be corrected, at additional cost, either by using such low-expansion materials as invar in the mask or by keeping the mask under tension in one direction (Trinitron) or both vertically and horizontally as in the Zenith (18) (see Fig. 21).

Electron Guns: Moiré Considerations

As a result of these factors, electron gun design for a color tube is a major challenge. In order to achieve satisfactory brightness, the electron beam power must increase. In order to accommodate three guns within the CRT neck, the diameter of each is reduced, increasing lens aberrations, in turn making for a larger beam. There is a further consideration: the screen and mask structure

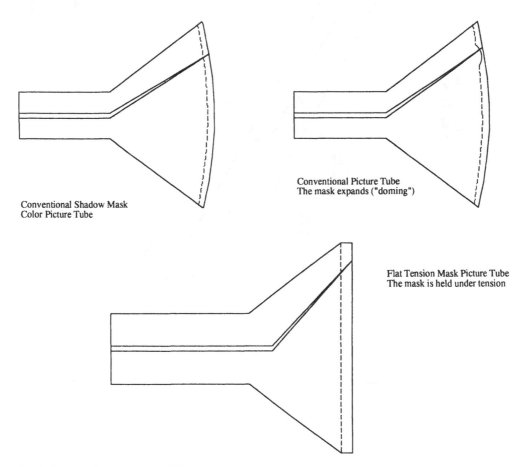

Conventional Shadow Mask
Color Picture Tube

Conventional Picture Tube
The mask expands ("doming")

Flat Tension Mask Picture Tube
The mask is held under tension

FIG. 21   At large values of beam power, the shadow mask expands (doming). This effect is minimized in the flat-tension mask.

are not continuous. This introduces a sampling effect on the original picture information. The raster process is also a sampling mechanism. In the presence of uncorrelated sampling processes, aliasing phenomena or Moiré effects may take place. This is exactly what happens in color systems, discussed below (12,19).

One assumes a dot mask with apertures of diameter D and pitch p scanned by a beam of width s; the scan line spacing is h (see Figs. 22 and 23). In order to see how moiré can arise, assume one line is centered on a row of dots, so that the output is maximum. If the line spacing h is slightly greater than the vertical separation of the dots $p/2$, then

$$h \geq \frac{p}{2}. \tag{22}$$

After a sufficient number of scan lines, a scan line will lie right in the middle of the dots and the output will be minimum. This will be reached after n scan lines, where

$$n\left(h - \frac{p}{2}\right) = \frac{h}{2}, \tag{23}$$

that is, on each scan line the process "gains" one difference, until, by definition, after n it has gained enough to be at a minimum output or at a distance $h/2$ less than the starting position.

From Eq. (23),

$$n = \frac{1}{2 - \dfrac{p}{h}}. \tag{24}$$

In other words, a brightness fluctuation in the picture is obtained with a period of $(2n - 1)$ lines. In order to minimize moiré, the proper relationship must be maintained between the scanning pattern, the mask parameters, and the electron beam characteristics (12,19). The same general considerations apply when the

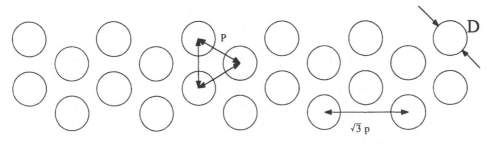

FIG. 22 Relationships between the apertures in a shadow mask.

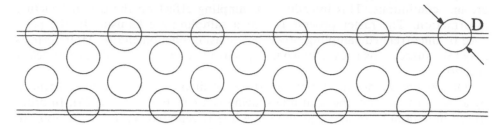

FIG. 23   Cause of moiré.

beam scanning pattern is quantized in the horizontal direction by a dot clock in a computer display.

Shadow Masks and Screens for Color Cathode-Ray Tubes

The engineering and manufacturing techniques involved in making shadow masks and color screens have been some of the great technological advances of our century, rivaled only by the evolution of the integrated circuit. The shadow mask itself is made of extremely thin (0.15 mm) steel in which the aperture pattern has been obtained by photo etching. The apertures themselves are not cylindrical but tapered, with the narrow part of the aperture facing the gun. This is to reduce the amount of electrons generated by secondary emission. Since the registration requirement between mask aperture and phosphor dots is extremely small, it is impossible to fabricate masks and screens separately and join them together at some point in the manufacturing process. One takes advantage of the fact that both electrons and light rays travel in straight lines and one uses a given mask to prepare the phosphor screen with which it will be used. A source of light (lighthouse) is placed at a point corresponding to the center of deflection. Light rays pass through the mask apertures and strike the screen where the corresponding phosphor dots are to be placed. The screen is made by applying a color phosphor mixed with a photosensitive binder to the faceplate. The mixture is exposed through the mask with ultraviolet light. The exposed binder hardens, while the unexposed portion is washed away later. The process is repeated three times, at the end of which a mask and a screen that will work together are obtained.

The description above, while accurate, is highly simplified. On one hand, the (electron) deflection center is not fixed but changes with deflection. Furthermore, the deflection field causes a distortion in the triad pattern. These effects are compensated with the aid of appropriate (optical) lenses. While the details of screen printing, mask making, and lighthouse correction techniques are beyond the scope of this work, one can appreciate the breadth of the technology. Missing dots, especially at the center of the screen, are unacceptable. During the whole exposure process, the mask must be removed and put back a number of times with a high degree of repeatability to allow phosphor deposition and washing. The thickness of the phosphor layer and the size of the dots must be

TABLE 4   Cathode-Ray Tube Technology Comparison

Parameter	Technology			
	Delta	In-Line	Trinitron	FTM
Luminance	Good	Good	Best	Highest
Resolution Mask (mm)	0.3	0.15	0.26	0.05
Resolution (gun)	Best	Good	Best	Good
Number of manufacturers	2–3	Many	1	1–2
Mechanical ruggedness	Good	Good	Worst	Best
Cost/power	Worst	Best	Good	?
Esthetics	Good	Good	Best	Best[+]
Research & Development activity	None	Lots	Good	Fair
Prospects for higher resolution	None	Excellent	Good	Good

kept within close tolerances for good color balance and purity. That the industry has been able to solve all these problems with a high degree of consistency (85% to 95% yields are routine) so as to produce millions of color CRTs, each with millions of mask holes and phosphor dots, is a tremendous achievement.

Phosphors

The discussion of phosphor properties for monochrome CRTs is applicable to color. Due to the low transmission of the shadow mask, phosphor efficiency is very important. A considerable amount of research and development effort was expended on the problem during the early years of color. This resulted in an order of magnitude improvement in efficiency between 1958 and 1972 (12,20). Subsequently, developments in screens have taken a different direction, that of improving contrast. One uses pigmented screens in which phosphor particles are coated in order to increase the reflection of ambient light selectively and thus the contrast. This process is used only on the red and blue phosphors. Matrix screens in which the screen area between phosphor dots has been printed with graphite are commonplace. This reduces the reflectivity of the background and thus improves contrast.

At one point, there was considerable effort in synthesizing phosphors with long persistence in order to reduce refresh rates, especially in high-resolution applications. Currently, the demands for fast image update and the popularity of such input devices as mice and trackballs, have made this need obsolete. This trend has been reinforced by the development of faster deflection systems, video amplifiers, and memories.

Summary

The properties of each of the shadow-mask structures discussed above are summarized in Table 4.

System and Ergonomic Considerations

Introduction

A description, however condensed, of the key aspects of CRT technology would not be complete without a discussion of those factors that make a display system useful. Furthermore, in the modern business environment, products must comply with a number of mandatory requirements that depend on the targeted market. For instance, displays aimed at the military market must meet applicable military (MIL) standards. For civilian markets, some of the most important standards have been issued by such agencies as Underwriters Laboratory (UL), Deutsche Industrie Norm (DIN), the Federal Communications Commission (FCC), and the International Organization for Standardization (ISO). It is beyond the scope of this article to examine each standard in detail, but an overview and a discussion of some important issues may prove useful. The limitations placed by such organizations as ISO fall into one or more of the following categories.

- *Ergonomics.* Insures that the display will be of sufficient quality so as not to cause unnecessary operator fatigue or eyestrain.
- *Electrical Safety.* Insures that the operator of the equipment will not be subject to electrical shock as a result of inadvertence or equipment malfunction.
- *Radio-Frequency Interference/Electromagnetic Interference (RFI/EMI).* Insures that the equipment will not interfere and will not be susceptible to interference from similar equipment.
- *Radiation Hazard.* Limitations on radiation hazards insure that the operator will not be subjected to excessive levels of radiation.

In the United States, the American National Standards Institute (ANSI) ANSI/HFS 100 standard was published in 1988 and has served as a foundation for a series of European requirements (21). The ANSI/HFS standard contains a number of useful recommendations. The most important are summarized in Table 5.

Flicker

Flicker is the sensation of pulsation or intermittency we perceive when looking at an image, such as that generated by a CRT, whose repetition frequency is insufficient. The *critical fusion frequency* (CFF) is defined as the frequency at which the image just fuses. While some of the fundamental principles governing flicker have been known for some time (23), a complete description of the phenomenon still is lacking. What is known is that CFF depends on a number of factors such as the image luminance, the ambient illumination, the angle the image forms with the eye, and the age and possibly even the gender of the observer. Confounding the problem is the fact that not all individuals are equally sensitive to flicker so that any recommendation relating to flicker of

TABLE 5 ANSI/HFS 100-1988

Characteristic	Specification
Ambient illuminance	500 lux nominal
Display luminance	At least 35 cd/m^2
Luminance adjustment	Mandatory
Viewing distance	500 mm nominal
Character height	2.3 to 6.5 mm at nominal view distance, 3.1 mm preferred
Character width	92%–93% of height
Image polarity	Both acceptable (black on white and white on black)
Character format	7 × 9 minimum
Character modulation	$(LM - Lm)/(LM + Lm) > 0.75$
Luminance uniformity	Better than 50%
Jitter	Less than 0.1 mm at nominal viewing distance
Linearity (integral)	Better than 2%
Linearity (differential)	Better than 10%
Image quality	Use modulation transfer function area (MTFA) metric*
Flicker	Flicker free for 90% of viewers

*The MTFA formula given in the standard is in error (22).

necessity must be statistical. The most comprehensive work to date on the subject is that of Farrell, who published analytical equations that allow prediction of flicker performance under a variety of conditions (24,25). These equations are at the basis of recommendations as to minimum refresh rates for CRT displays used in the office environment and are plotted in Fig. 24.

Since television displays generally are viewed at greater distances than computer monitors, viewing angles are much smaller and the flicker requirements are not nearly as severe. Thus, in television displays, the practice has been to choose the refresh rate to be equal to the power line frequency (60 hertz [Hz] in the United States and 50 Hz in Europe) so as to minimize interference. In order to minimize the bandwidth over which the signal is transmitted, television systems generally employ interlace. This means that the picture is not scanned progressively from top to bottom. Rather, half the picture is scanned first (e.g., with 250 scanning lines) in 1/30 of a second (3). During the next "pass" or field, the remaining half of the picture is scanned, with the scanning lines halfway between the old ones (see Fig. 25).

Image Quality—Resolution

One of the most complex and controversial subjects in the ergonomics of displays in general and of CRT displays in particular is that of image quality. While this may be surprising, especially in view of the many excellent displays currently on the market, there are good reasons for it. First, the shape of the scanning aperture is not square, as was shown in the discussion of electron-optics. This means that it is not at all obvious a priori just what is meant by "resolution." We look into the issue more deeply and invoke complex criteria

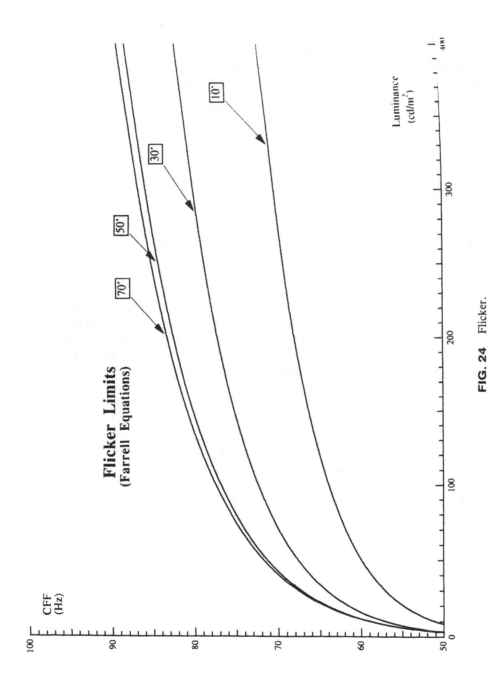

FIG. 24　Flicker.

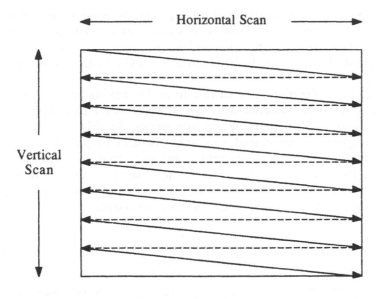

Progressive Scan

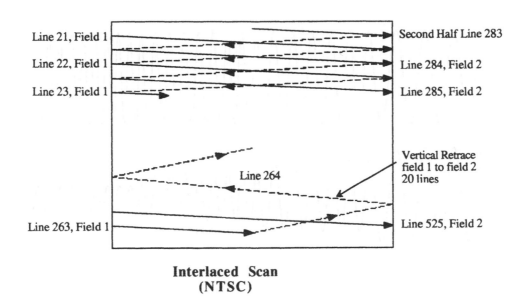

**Interlaced Scan
(NTSC)**

FIG. 25 Raster scanning.

such as Rayleigh's. Second, in an optimum display system, we expect the human eye to be the limiting factor, or at least be a significant contributor to the resolution of the system. Thus, even a first-order resolution theory must start with a description that encompasses the behavior of both Gaussian spot profiles and the human eye.

It is tempting to begin the discussion of resolution by seeking help from the theory of resolution of optical systems, since optical systems are so well known and were of significant help in defining electron optics. In this context, the notion of spatial frequency is introduced, similar to ordinary frequency. Spatial frequency characterizes sinusoidal excitations in intensity that are periodic in either the space on the screen or at the angle of the observer's eye. We define the modulation transfer function (MTF) of an electron optical system as the behavior with spatial frequency of the modulation of the system

$$\text{MTF}\ (u)\ =\ \frac{L_{\text{MAX}}\ -\ L_{\text{MIN}}}{L_{\text{MAX}}\ +\ L_{\text{MIN}}} \tag{25}$$

where L_{MAX} and L_{MIN} are the peak and minimum luminance, respectively. The MTF definition is illustrated in Fig. 26 (26). For a CRT at moderate beam currents, the spot profile can be closely represented by a Gaussian function (2,27) and an analytic expression of the MTF is available. One finds (28)

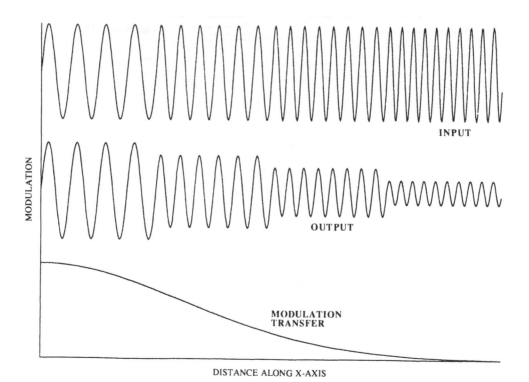

FIG. 26   Definition of modulation transfer function (MTF).

$$\text{MTF}(v) = e^{-\frac{\pi^2}{4\ell n 2} s_{50}^2 v^2} \tag{26}$$

where v is spatial frequency and s_{50} is the beam width measured between the 50% points (see Fig. 27). The behavior of the MTF is shown in Fig. 28.

MTF is a powerful concept because one can describe a complex system once the individual MTFs of each of its components are known. For instance, in a system with optical lenses, photographic film, CRTs, and phosphors, each of these components can be characterized by its MTF. The MTF of the system is the product of the individual functions.

The human eye, of course, has its own resolution limitations that depend on a number of factors, including luminance. While the MTF concept can be used (29), the contrast threshold function (CTF) is preferred. The inverse of MTF, it can be thought of as the minimum contrast required by the eye to perceive a given spatial frequency. A number of analytic formulations have been proposed; the one that gives the best fit to experimental data has been proposed by Barten (30):

$$\text{CTF}(v) = \frac{1}{b_0 v \exp(-b_1 v) \sqrt{1 + b_2 \exp(b_1 v)}} \tag{27}$$

where

$$b_0 = 440 \left(1 + \frac{0.7}{\text{Lum}}\right)^{-0.2}$$

$$b_1 = 0.30 \left(1 + \frac{100}{\text{Lum}}\right)^{0.15} \quad \text{and}$$

$$b_2 = 0.06 \tag{28}$$

where Lum is the display luminance in cd/m^2. Equation (27) is plotted in Fig. 29. The eye has an optimum acuity (minimum CTF) at about 3–5 cycles/degree and becomes progressively worse outside this range.

The perceived sharpness of an image is a function of the parameters of both the display (MTF) and the eye (CTF). In Fig. 30, the concept of modulation transfer function area (MTFA) is presented, the area between the MTF and the CTF curves.

Mathematically,

$$\text{MTFA} = \int_0^{v_0} [\text{MTF}(v) - \text{CTF}(v)] dv \tag{29}$$

v_0 is the "crossover" spatial frequency for which

$$\text{MTF}(v) = \text{CTF}(v) \tag{30}$$

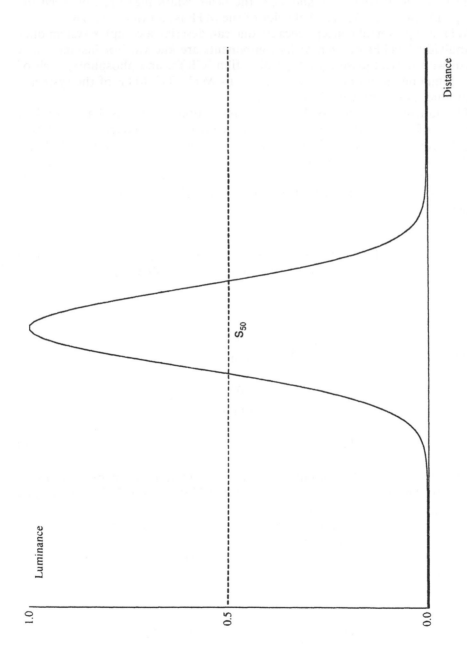

FIG. 27   The luminance profile of a cathode-ray tube spot. The "width" is generally defined as the distance between the 50% points.

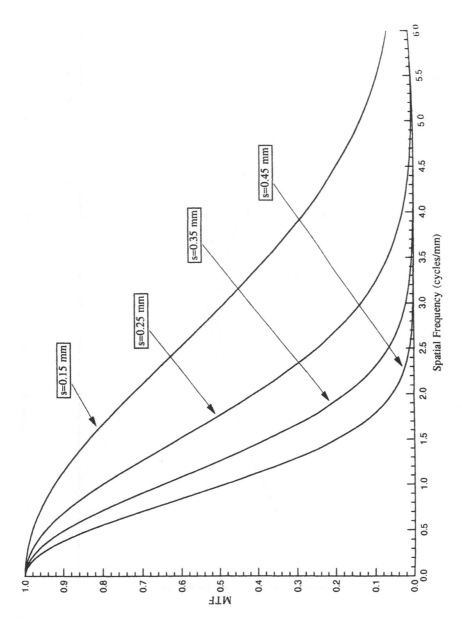

FIG. 28 Modulation transfer function of Gaussian spots.

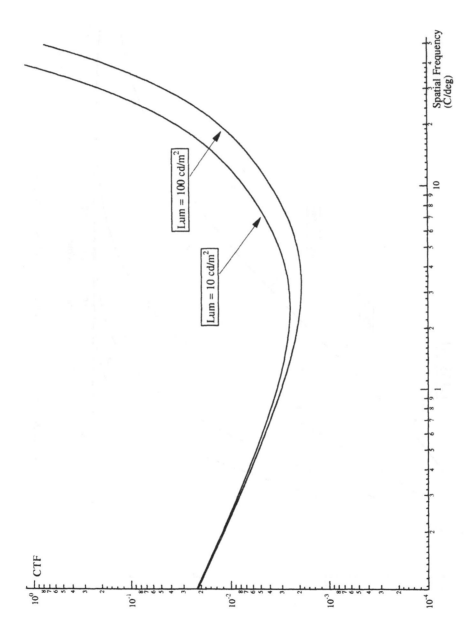

FIG. 29 The contrast threshold function of the human eye described by Eq. (27).

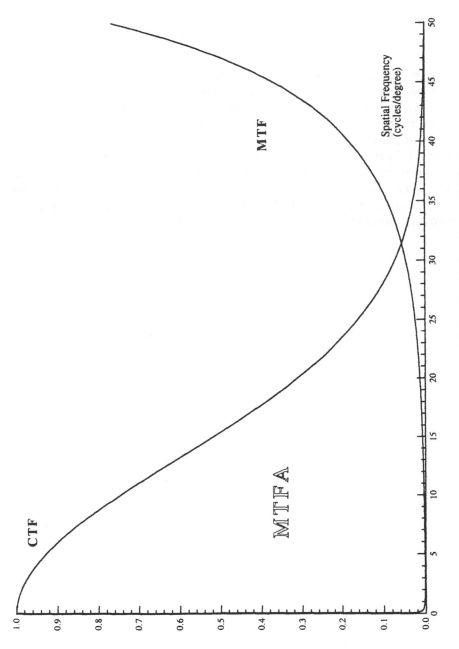

FIG. 30 Definition of modulation transfer function area (MTFA).

The MTFA, a quantity derived from physical measurements, correlates fairly well with the subjective sensation of sharpness (23). A metric that shows correlation with subjective judgments is the SQRI (square root integer) (30).

$$\text{SQRI} = \frac{1}{\ell n 2} \int_{v_{\min}}^{\infty} \sqrt{\frac{\text{MTF}(v)}{\text{CSF}(v)}} \frac{dv}{v} \tag{31}$$

SQRI is expressed in just noticeable differences (JNDs). Thus, different displays may be simulated by computing the SQRI and the effects of changes assessed. For example, Fig. 31 shows the effect of the beam size on the perceived quality of a monochrome display. Such graphs as Fig. 31 can be very useful in assessing the effects of parameter tolerances.

Color CRTs also can be accounted for by quantifying the effect on the MTF of the shadow-mask structure (28,31). Similarly, the effect of ambient lighting can be studied. One finds that the light reflected from the display reduces the available modulation and thus the MTF (32). This explains the common practice of improving image quality by reducing the ambient illumination.

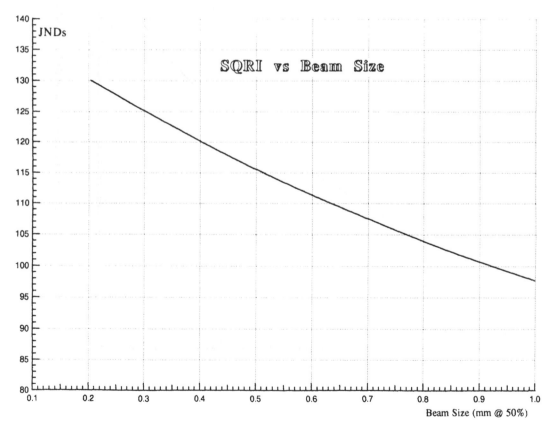

FIG. 31   Effects of variations in beam size on the resultant image quality.

Radiation: X-Rays

A form of radiation that has been of some concern is x-rays. They are emitted when energetic electrons strike a material. The wavelengths involved are between 0.01 and 10 nanometers (nm). X-rays penetrate quite deeply into matter, especially at the shorter wavelengths, and in excessive doses pose a health hazard. The phenomenon by which the energy of the electrons can produce x-rays is known as *Brehmsstrahlung* ("braking radiation" in German). A typical spectrum is shown in Fig. 32, which shows that the available energy of the electron beam limits the shortest wavelengths that can be produced.

Only rays with a wavelength greater than λ_0 can be created where

$$\lambda_0 = \frac{hc}{eV_b}, \tag{32}$$

where *h*, *c*, and *e* are Planck's constant, the speed of light, and the charge of the electron, respectively. Since CRTs, especially color CRTs, are operated at elevated voltages, x-ray emission can and does occur, in fact. Regulatory agencies limit the amount of x-rays that may be emitted by a commercial product to

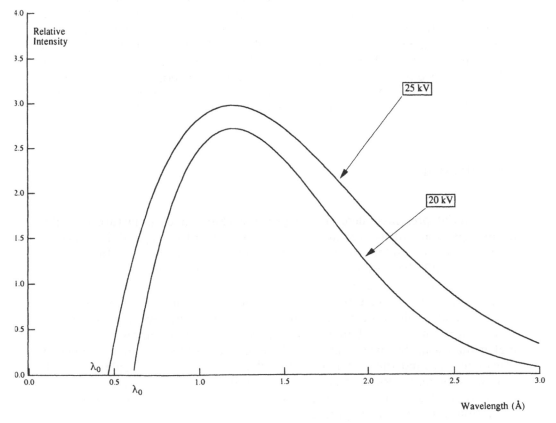

FIG. 32 X-ray spectrum.

0.5 milliroentgen (mR) or less, which is less than the amount of the background radiation on earth. In practice, most of the x-rays generated at the screen are absorbed by either the phosphor itself or the glass so no problem exists for CRTs operated at about 18 kV or below. For higher voltages, some precautions must be used, usually involving lead-enriched glass. CRT manufacturers furnish a so-called isodose curve from which one determines appropriate operating limits. For example, if one designs the high-voltage supply so that it cannot exceed about 30 kV under any condition (including malfunction), the user can be considered safe from x-ray hazards.

Radiation: Low-Frequency Electric and Magnetic Fields

Recently, a number of articles in the popular media have expressed concerns over potential health effects of the weak electric and magnetic fields produced by CRT monitors. At the time of this writing, no conclusive evidence exists that these weak fields do in fact cause adverse health effects. To the contrary, a major experiment conducted by the National Institute of Safety and Health (NIOSH) found no increase in miscarriages as a result of exposure to the weak electromagnetic fields emanating from CRT displays. Naturally, a variety of techniques exist that allow the reduction of these fields. The techniques include coating the front glass surface of the CRT with a conductive layer to eliminate the electrostatic charging of the front surface. Techniques to reduce dynamic electric and magnetic fields include canceling coils and electrodes, driven in synchronism with the horizontal sweep and arranged to cancel the stray flux (33,34). Irrespective of the physics of the situation, concerns have developed in the user community. As these market concerns coalesce, some form of regulation similar in intent to that in effect for x-rays undoubtedly will emerge.

Conclusions

The material presented here should give the reader a clear picture of CRT technology, some of its accomplishments, and some of its capabilities together with some of its limitations. While their demise was predicted confidently decades ago, CRTs have established themselves in achieving low-cost implementations of superior image quality.

For the technology, extending performance in terms of brightness, size, contrast, and resolution currently is not so much a technological challenge as a cost/performance issue. Laboratory displays with performances significantly higher than what has been mentioned here have been reported by a number of researchers around the world. For example, 19-in, 2000-line displays in both monochrome and color have been reported in the literature in some detail (35). Thus, there does not appear to be a significant technological barrier to extending resolution well past current practice.

In August 1992, the general consensus among the display technical community is that the position of dominance that CRTs enjoy will not be seriously challenged for at least another decade. Time alone will tell whether these predictions are accurate.

References

1. Infante, C., Introduction to CRT Systems, *SID Seminar Lecture Notes*, 1:1–106 (1986).
2. Soller, T., Staff, M. A., and Valley, G. E., *Cathode Ray Tube Displays*, McGraw-Hill, New York, 1948.
3. Zworykin, V. K., and Morton, G. A., *Television*, Wiley, New York, 1954.
4. Spangenberg, K. R., *Vacuum Tubes*, McGraw-Hill, New York, 1948.
5. Moss, H., *Narrow Angle Electron Guns and Cathode Ray Tubes*, Academic Press, New York, 1968.
6. Klemperer, O., *Electron Optics*, Cambridge University Press, Cambridge, England, 1953.
7. Barten, P., CRT: Present and Future, *SID Seminar Lecture Notes*, 2.8:1–43 (1988).
8. Electronic Industries Association, *Optical Characteristics of Cathode Ray Tubes*, Vol. TEPAC No. 116-A, EIA, Washington, DC, 1985.
9. Sherr, S., *Electronic Displays*, Wiley, New York, 1979.
10. Infante, C., Advances in CRT Displays, *Proc. Intl. Display Res. Conf., San Diego, CA*, 1–5 (1988).
11. Holmes, R. D., High Resolution CRT Displays, *SID Seminar Lecture Notes*, 2: 1–27 (1983).
12. Morrell, A. M., Law, H. B., Ramberg, E. G., and Herold, E. W., *Color Television Picture Tubes*, Academic Press, New York, 1974.
13. Hedler, R. A., Maninger, L. L., Momberger, R. A., and Robbins, J. D., Monochrome and Color Image Display Devices. In: *Television Engineering Handbook* (K. B. Benson, ed.), McGraw-Hill, New York, 1985.
14. Dasgupta, B. B., Deflection and Convergence Technology, *SID Seminar Lecture Notes*, 1:1–43 (1991).
15. Hirabayashi, K., Kitegawa, O., and Natsuhara, M., *In-line Type High Resolution Color Display Tube*, NHK, Japan, 1982, pp. 12.1–12.53.
16. Hosokoshi, K. S., Ashizaki, S., and Suzuki, H., Improved OLF In-line Gun System, *Japan Display Proceedings*, 272 (1983) (in English).
17. Yoshida, S., Ohkoshi, A., and Miyaoka, S., The Trinitron: A New Color Tube, *IEEE Trans. Consumer Electronics,* CE-28:56 (1982).
18. Dietch, L., Palac, K., and Chiodi, W., Performance of High-Resolution Flat Tension Mask Color CRTs, *SID Symposium Digest*, 17:324–326 (1986).
19. Ramberg, E. G., Elimination of Moiré Effects in Tri-Color Kinescopes, *Proc. IRE*, 40(N8):916 (1952).
20. Takano, Y., Recent Advances in Color Picture Tubes. In: *Advances in Image Pickup and Display* (B. Kazan, ed.), Vol. 6, Academic Press, New York, 1973.
21. American National Standards Institute, *American National Standard for Human Factors Engineering of Visual Display Terminal Workstation—1988*, ANSI/FS-100, Human Factors Society, Santa Monica, CA, 1988.
22. Infante, C., Numerical Methods for Computing MTFA, *Displays*, 12(2):80 (1991).

23. Snyder, H., Image Quality Measures and Visual Performance. In: Flat-Panel Displays and CRTs (L. Tannas, ed.), Van Nostrand Reinhold, New York, 1985.

24. Farrell, J. E., Benson, B. L., and Haynie, C. R., Predicting Flicker Thresholds for Video Display Terminals, *Proc. SID*, 28(4):449–453 (1987).

25. Farrell, J. E, Casson, E. J., Haynie, C. R., and Benson, B. L., Designing Flicker-Free Video Display Terminals, *Displays*, 115–122 (July 1988).

26. Jansen, L. A., Display Image Measurement and Characterization, *SID Seminar Lecture Notes*, 2:10/1–73 (1983).

27. Infante, C., The Use of Lorentzian Functions in the Analysis of Non-Gaussian CRT Spots, *SID Journal*, March 1992.

28. Infante, C., On the Resolution of Raster-Scanned CRT Displays, *Proc. SID*, 26(1): 23 (1985).

29. Schade, O. H., Optical and Photoelectric Analog of the Eye, *J. Optical Society of America*, 46:721 (1956).

30. Barten, P., The SQRI as a Measure for VDU Image Quality, *SID Intl. Symposium Digest*, 23:867 (1992).

31. Ohishi, I., *Spatial Frequency Characteristic of Picture Devices with a Mosaic Screen Structure*, NHK, Japan, 1971 (in Japanese).

32. Farley, W. W., Determination of Monochrome Display MTFs in the Presence of Glare, *SID 89 Digest*, 212 (1989).

33. Sluyterman, S. A., The Radiating Fields of Magnetic Deflection Systems and Their Compensation, *Proc. SID*, 29/3:207–211 (1988).

34. Cappels, R., Cancellation of ELF/VLF Electric Fields from CRT Displays, *Proc. SID*, 23:137–140 (1992).

35. Yoshida, H., and Yamazaki, E., Superhigh-Resolution Color Display Tubes, *Hitachi Review*, 32(4):199 (1983) (in English).

CARLO INFANTE

Electronic Displays—Human Factors

Introduction

An information display provides the critical link between a human mind and an electronic database. To be effective, the display must portray information in a form compatible with the mind's processing capabilities. This means that the information must be formated to utilize fully the intrinsic properties of the user's sensory and mental abilities. The initial element in this link is the data display system, which must be matched to the human visual system in order to facilitate information exchange. In a sentence, the human sensory system should be considered as a design criterion for an electronic information display.

The effective transfer of information from an electronic database to a human mind includes both the physical characteristics of the hardware upon which the information is generated as well as the underlying software that formats the information in a manner compatible with the user's mental capabilities. In this article, primary emphasis is on the design of display hardware mated to human sensory capacity. The human-factors requirements for manipulating the information to be displayed on that hardware lies outside the domain of this article.

Human factors, or *ergonomics*, is concerned with the forging of a productive and healthy link between a human operator and a machine. The discipline of human factors is relatively new, owing its growth to the increasing complexity of the machinery under human control. In fact, the design of aircraft during World War II, spurred by the inordinately large number of training crashes, marked the emergence of human factors as an independent discipline. A key element in the need for human factors in cockpit-control layout arose from the training of pilots who had little or no aeronautical engineering experience. That is, the new generation of pilots did not have an integral understanding of the functional relationships that governed the behavior of the complex machine — they had to rely on the interface alone. A similar trend has occurred in the automotive- and consumer-electronics fields and is evident in informational displays. One no longer can assume that the operator of an informational display either understands or cares to understand the underlying principles of the instrument.

As a relatively immature discipline, human factors has drawn on a number of established areas in its development. Most notably, modern human factors is a combination of engineering (mechanical, industrial, and electrical), experimental psychology, physics, anatomy, neurophysiology, medicine, and even law. Although our concern here is with the human factors of displays, similar needs exist in such diverse undertakings as tools, workplaces, living spaces, public facilities, transportation equipment, and information transfer.

Human factors approaches the problem of the human interface to a machine by viewing the human as an instrument with a complex design that is complete

I would like to express my gratitude to Robert Beaton and Gene Lynch for the comments, suggestions, and input to this chapter.

243

and unalterable—a design to which any product must link in order to effect an optimal exchange of information. Human-factors technology seeks to amass the known and to uncover the unknown attributes of the human side of the link in order to improve the design of any product. As applied to electronic displays, emphasis is placed upon three areas within the broader discipline of human factors: anthropometric, the physical configuration of systems to match the physical capabilities of the user (i.e., keyboard layout, panel and switch design, workstation layout, etc.); sensory, the design of systems that input/output (I/O) information to the human sensory systems, with concentration on the visual, auditory, and tactile interface (examples range from the reduction of flicker, effective use of color, voice I/O, touch panels, and hard-copy quality to the evaluation of the perceptual aspects of emerging technologies); and cognitive, the utilization of human mental capability and capacity in the design of systems (i.e., menu configuration and layout, error handling and messages, operating instructions, training, and mental manipulation of large color palettes).

Anthropometrics is a relatively mature discipline within human factors and has a large base of scientific data that can be applied to products. This database is available in the form of detailed tables and charts that indicate the range of human-body dimensions from the small female (5th percentile) to the large male (95th percentile). Naturally, new approaches and innovation in the interface to devices will require the development of new anthropometric solutions and the generation of new data.

The available information on *sensory* capacity, which reflects the state of scientific understanding of the sensory processes, is much less complete than that of anthropometrics. It is in this area that an increasing level of research activity can be seen throughout the electronics industry. To a great extent, this is due to the central role afforded display systems in all electronic instruments.

With an eye toward the future, there is an obvious need to develop an understanding of the mental processes of the human brain in order to build products effectively that have increased intelligence and sophistication. This *cognitive* area has the weakest database and can be expected to increase in importance over the next 5 to 10 years. The need is evident from the disturbing trend that, as the intelligence of electronics instrumentation increases, so do the demands placed upon the operator by the instrument. Reversing this trend is the major challenge for cognitive human factors.

The process of matching an instrument to user capabilities (see Fig. 1) begins with a careful and elaborate description of the instrument's objectives (in essence, a philosophy of the instrument's application). From the tasks that an instrument must perform, the engineer determines the required functions (input and output) and judiciously assigns some functions to the electronics and some to the user. For instance, humans excel over machines in detecting critical signals against high-noise backgrounds, extracting complex and varying patterns, or remembering solution strategies and "shortcuts." On the other hand, instruments can store large amounts of information, perform repetitive functions, and measure events beyond the sensory range of human experience. Many other functions are not assigned so easily to either user or machine due to limits in the available technology and to gaps in the understanding of human behavior. In addition, cost and engineering-development time impose limitations.

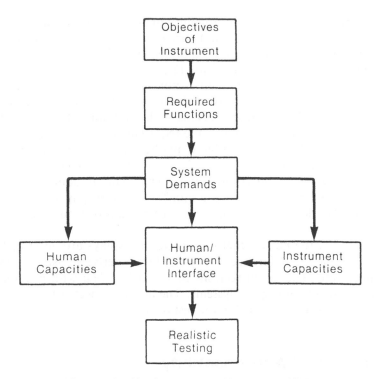

FIG. 1 Matching instrument and human capabilities.

After allocating functions, the engineer must perfect the interface between the system and the user. This task requires a consideration of all elements: anthropometric, sensory, and cognitive. Finally, testing the complete system under realistic conditions can ensure that its analytically determined attributes meet any empirically substantiated expectations.

Clearly, human factors needs to be an integral part of the design process. It cannot be "tacked on" at the end of the design. Perhaps the greatest task for human factors is to gain the prerequisite respect within the display community to be perceived as an equal partner in the design of display systems.

The database of human factors consists of the accumulated knowledge of human and machine capabilities as well as the sophisticated research methodology that allows an accurate evaluation of any proposed system. For the information display, human-factors specialists rely most heavily on two major resources: sensory neurophysiology and visual psychophysics.

Sensory neurophysiology provides the basic understanding of how the human senses function. Obviously, any system design that is to interface through the human senses must be compatible with the input needs of those senses. For example, a knowledge of the mechanism of visual accommodation allows the design of visual displays for which the lens of the eye is able to maintain a comfortable and accurate focal length (1). Visual psychophysics provides both a database and a methodology for assessing the functional link between the display and the viewer. An example is the contrast-sensitivity function, which is a psychophysical measure of the capability of the human visual system to process complex images in terms of the Fourier components of the image. Understanding this function allows the design of displays with resolution that matches the resolving capability of human vision (2).

The utilization of sensory human factors in the design of display systems has led to the approach known as *engineering to human specifications*. This implies that the display engineer working with the human-factors professional can design a display system to match the requirements of human sensory physiology. This approach has led to a large number of improvements in the classic cathode-ray tube (CRT) and has opened the door for new technologies that may be better matched to human visual requirements. The fundamental database of human vision has been described in a number of recent papers in a form from which the display designer can apply the information to future display systems (3–5).

In this article, the human-factors aspects of displays are described as engineering requirements. Reference is made to the underlying visual system needs when appropriate. The initial sections of the article deal with general human-factors requirements of display systems. The unique aspects of color CRTs and flat-panel displays are discussed in the rest of the article.

Resolution and Addressability

Conceptually, *resolution* refers to the capability of a display to present fine detail. In essence, resolution is dependent on the design of the CRT. It is derived from the width (line-spread function) of a line drawn on the screen; the narrower the line is, the higher the resolution will be. Although a number of definitions for line width are used, here we will consider the line width at 50% of the maximum luminance. The critical aspect of resolution is that it is a property of the CRT design.

For a monochrome CRT having a continuous phosphor, the line width and, hence, resolution, is a function of a number of factors such as phosphor, deflection fields, and electron-gun design. In theory, if one could make resolution infinitely high, then one would produce infinitely narrow lines on the display.

For shadow-mask color displays, the obtainable resolution is limited by the pitch of the shadow mask (the distance in millimeters between vertically adjacent mask-hole centers). In order to avoid moiré patterns, the minimum line width must be greater than the mask pitch (6).

The *addressability* of a display is a characteristic of the display controller and represents the ability to select and address a specific point (pixel) or x,y coordinate on the screen. For rastered displays, addressability usually is stated in terms of the number of lines scanned from the top to the bottom of the display and the number of points along a raster line. Since addressability is determined by the hardware driving the CRT and resolution is determined by the design of the CRT, they are independent of one another. However, in order to produce high display quality, certain relationships need to be maintained between resolution and addressability. For example, if the resolution is greater than the addressability, the raster lines will not merge; that is, the pixels will appear as separate dots. Conversely, if the resolution is too low (large spot size) relative to addressability, successive lines will overwrite adjacent lines, which may render critical details of the image invisible.

Given a particular resolution as an aspect of the CRT, an addressability is desired that allows adjacent lines to merge. This combination produces a full field (all pixels on) that, when viewed from the nominal viewing distance, has an imperceptible raster structure. This combination also allows the pixels forming alphanumeric characters to merge, which enhances their legibility. At the same time, the addressability of a display must allow individual lines or pixels to be visible when alternate lines or pixels are active. Whether a given addressability for a CRT of a fixed resolution will be visible is determined by the contrast sensitivity of the human eye. Figure 2 plots this function. (Figure 2 assumes an intermediate luminance in the 10–15 foot-lambert [fl] range. Contrast sensitivity increases marginally for high luminances and decreases markedly for lower luminances.) The graph indicates the necessary modulation between adjacent lines required to detect that particular line frequency. Modulation is defined as

$$M = \frac{L_{max} - L_{min}}{L_{max} + L_{min}} \qquad (1)$$

where L_{max} and L_{min} refer to the maximum luminance of the line (pixels) and the minimum luminance between lines (pixels), respectively. A line frequency with a modulation falling in the shaded portion of Fig. 2 is not visible. For example, a 30 cycles/degree raster structure would be just visible at about 0.13 modula-

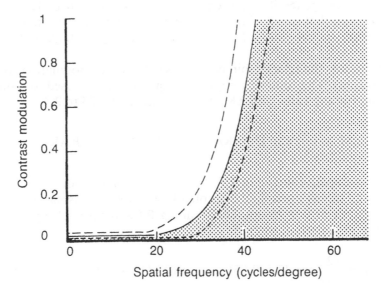

FIG. 2 Contrast sensitivity function of human vision with 90% population limits (broken lines). The graph indicates the necessary contrast between line gratings required to detect a grating of a given line frequency. (Data from Ref. 7.)

tion. A 35 cycles/degree structure, conversely, would be detectable at about 0.33 contrast modulation. The design goal, then, is to select an addressability that will result in a raster structure falling just outside the detection limit (i.e., the shaded portion of Fig. 2).

Obviously, one critical determinant for the visibility of a given line frequency is the distance between the observer and the display screen. Therefore, in Fig. 2, line frequency is defined in terms of the number of cycles/degree (number of lines or pixels) of visual angle. A display designed for close viewing will have a corresponding lower spatial frequency. It follows that a design viewing distance range should be specified for any display system.

The modulation between adjacent horizontal lines will depend upon both resolution and addressability. For a fixed spot size, as the addressability increases, the modulation between the lines decreases. As spot size increases with a fixed addressability, contrast modulation also is reduced. This relationship can be indexed by the resolution/addressability ratio (RAR) (8). As mentioned above, *resolution* is defined as the width of the spot at one-half the maximum luminance, while *addressability* is defined as the peak-to-peak separation between adjacent lines (pixels). This definition of addressability in terms of line (pixel) separation is equivalent to the reciprocal of the more familiar definition of addressability as the maximum number of displayable lines divided by the total raster height. Thus, for example, a 19-inch (in) diagonal display with a height of 10.8 in and an addressability of 1024 lines would have a peak-to-peak separation of $1/(1024/10.8) = 0.0105$ in. Assuming this display to have a 0.015-inch spot, the RAR value would be 1.43.

For RAR values greater than 1.0, the overlap between adjacent pixels will be great enough that their energies will sum. Thus, the contrast modulation between adjacent horizontal pixels for an RAR of 1 approaches 0. This is the reason behind the generally accepted design goal of an RAR of 1. It should be noted, however, that this goal fails to take into account the sensitivity of vision and, therefore, unnecessarily limits the addressability of high-resolution systems.

Figure 3 plots the relationship between RAR and the modulation for Gaussian lines (pixels). The following regression equation can be used to estimate the modulation for any value of RAR within the range of 0.35 to 2.4:

$$M = \left(\frac{2}{\pi}\right) \exp\left[3.6\,(\text{RAR}) - 7.0\,(\text{RAR})^2 + (\text{RAR})^3\right]. \qquad (2)$$

As an example, consider a CRT having a resolution of 0.015 in and an addressability of 480 horizontal lines on a 10.8-in vertical display area (peak-to-peak separation of 0.0225 in). The resulting RAR is 0.66. Using the estimate for contrast modulation, we find

$$M = \left(\frac{2}{\pi}\right) \exp\left[3.6(0.066) - 7.0(0.66)^2 + (0.66)^3\right] = 0.43.$$

Although we have considered the case of adjacent pixels along adjacent horizontal lines in which no appreciable decay of the phosphor occurs, this

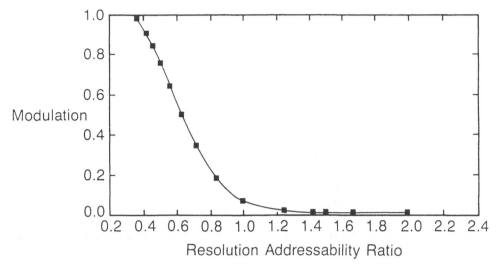

FIG. 3 Modulation between adjacent Gaussian pixels as a function of resolution/addressability ratio. A least-squares polynomial approximation of the function is shown as a solid line.

analysis will extend to situations in which an appreciable decay does occur. Because the visual system has its own intrinsic persistence, in which the image is maintained for a period of 20–50 milliseconds (ms), the decay due to the display phosphor can be disregarded. If the display is perceived as flicker free, the phosphor persistence is long enough so that the influence of phosphor decay can be considered negligible.

Consider the example of a 19-in display (10.8-in × 13.5-in display area) having a resolution of 0.020 in and an addressability of 1024 × 1280. The addressability of 1024 is equal to 94.8 lines/in or a separation of 0.0105. The RAR is 1.9, which, according to Fig. 2, means that the modulation can be taken to be zero. Obviously, the raster structure of the display will not be visible. To test that lines separated by an inactive pixel will be visible, apply the same calculations. Now, the addressability is taken as one-half the actual because only one-half (94.8/2 = 47.4) the lines or pixels are active. In the current example, the RAR becomes 0.95 (0.020/0.021). The modulation is 0.08. Assuming a viewing distance of 18 in and taking the spatial frequency as one-half the full-field value of 29.8 cycles/degree, or 14.9 cycles, we can see from Fig. 2 that the lines separated by a pixel will be just barely distinguishable. In practice, a larger modulation would be desirable. This can be accomplished by increasing the resolution (i.e., to 0.018 in) or by decreasing the addressability. Obviously, a much higher resolution (small spot size) at the same addressability would result in a visible raster structure.

Knox (9) has tested the subjective impression of the quality of displays that either fail or meet the adjacent and alternate line criterion proposed in the RAR metric. A modified high-resolution Tektronix GMA-201 monochrome monitor was used in which the resolution (spot size) or addressability (lines per inch) could be varied independently. Knox tested four spot sizes (0.005, 0.010, 0.015,

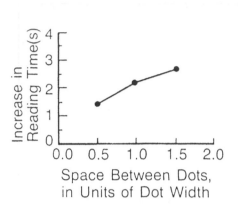 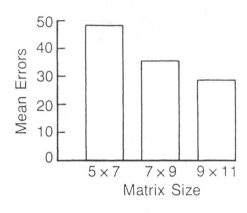

FIG. 4   Character legibility is influenced by: (*a*), interpixel distance; and (*b*), dot matrix size. (Data from Ref. 7.)

and 0.120 in) and four levels or addressability (100, 120, 150, and 200 lines per in). In one segment of his study, secretaries conducted a word-processing task, while in a second phase engineers corrected errors in schematic drawings. For both groups of subjects, resolution had a significant impact on judgments of image quality. Spot sizes in excess of 0.010 in produced a significant reduction in reported image quality. Addressability, on the other hand, had an impact on image quality assessments to the degree that a specific addressability was appropriate to the display resolution. That is, highest ratings went to images with RAR values in the range of 1.0 to 2.0, which are combinations in which both the adjacent pixel criterion (no visible raster) and alternative pixel criterion (alternative lines are discriminable) are satisfied.

An inappropriate addressability or inadequate resolution also can lead to a reduction in human performance with a display. Snyder and Maddox (10) found marked increases in reading time as the space between pixels within a character was increased (Fig. 4-*a*). Note that only the 0.5 spacing would produce an RAR value in which the adjacent and alternative criteria were met, while the 1.0 and 1.5 spacing would fail the adjacent criterion: The space between adjacent pixels would be visible, which gives the characters a discontinuous appearance. Similarly, Fig. 4-*b*, shows the impact of dot matrix size on symbol search times. Increasing the size of the character matrix from 5 × 7 to 7 × 9 reduced search time by 24%, while a further increase to 9 × 11 reduced time by 32%.

Luminance and Illuminance

The amount of light emitted or reflected from a display is specified as *luminance*. Similarly, the volume of light falling on a surface is known as *illuminance*. Both units of measurement are photometric (as opposed to radiometric) in that they correct the physical energy in terms of the spectral sensitivity of the human eye. The correction curve V_λ was adopted by the CIE (Commission Internationale de L'Eclairage) in its 1931 chromaticity-specification system. (The CIE is an international standards body for specifying light and color.)

Luminance and illuminance values are expressed in a variety of units. Most areas of science use the metric units candela/meter^2 (cd/m^2), or *nits*, for luminance and lumen/meter^2 (l/m^2), or *lux*, for illuminance. The display community has been slow to switch from the familiar foot-lambert (fl) and foot-candle (fc) specifications. Table 1 provides the conversion factors for the various measurement units.

The lower limit for the ideal luminance level for a display is given by the limits of high-acuity vision. The cones of the eye's central fovea need a minimum of about 5 cd/m^2 to function, and they demonstrate increasing sensitivity with increasing luminance up to about 1500 cd/m^2. While the visual system shows increases in acuity with increasing luminance, the result of increased display luminance is usually an increase in spot size. Thus, the gain afforded the eye is offset by the resolution loss in the display.

The target luminance for a display is more often a function of a contrast afforded between the written and unwritten portions of the display. Contrast, in turn, depends upon the level and type of ambient illumination. In general, it is wise to consider the levels of illumination present in the working environment of the display and, in particular, the reflectivity of surfaces in the immediate field of view. Figure 5 plots the recommended light levels for office work over the past 75 years (11). Note the steady increase until about 1960. Current recommendations are for about 500 lux. If the display is used extensively with white source documents, the luminance of the paper typically will be between 300 and 400 cd/m^2. To avoid the visual requirement of continuous readaptation of the eye, one should remain within the operating range of the visual system, which covers, at these levels, 1 to 2 log units. Thus, a practical luminance goal would be a minimum of 30 to 40 cd/m^2 for devices used in an office environment. Naturally, other ambient light levels would demand alternative luminance limits. The science of illumination engineering and architecture generally still are coming to grips with the demands of display technology, so the future requirements for lighting will see a matching of the environment to the display technology in order to satisfy the visual needs of the user better.

TABLE 1 Conversion Factors for Luminance and Illuminance Measurement Units

Luminance	
1 foot-lambert	= 3.4260 candela/m^2
1 foot-lambert	= 1.0760 millilamberts
1 candela/m^2	= 0.3142 millilamberts
Illuminance	
1 foot-candle	= 10.7600 lux
1 foot-candle	= 1.0760 milliphot
1 lux	= 0.01 milliphot

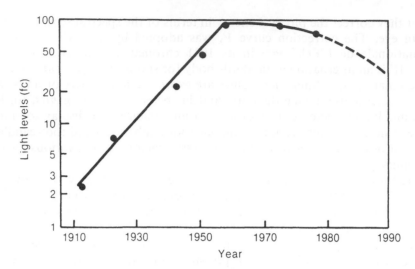

FIG. 5   Recommended light levels for visual characteristics of office work. (Adapted from Ref. 11.)

The use of luminance and illuminance measures for displays is made more difficult by the fact that the V_λ scale was derived under very specific measurement conditions that, by and large, do not apply to visual displays. This means that the perceptual experience of brightness only generally correlates with the measured luminance (12). In addition, the relationship between brightness and light intensity is nonlinear. That is, at low light-intensity levels, small increases produce large perceptual increases. At higher levels, an increasingly larger increment in light intensity is required to produce a perceptible change. Unfortunately, a CRT behaves in the same manner: small voltage increases produce large intensity increases.

The result of this property of vision and of CRTs is a loss in the number of discernible gray-level steps. Consider the curve in Fig. 6 marked "Gamma uncorrected." For a CRT with 4 bits of gray level, shown on the abscissa, the curve shows the number of just noticeable brightness steps encompassed by each increase in intensity. For example, an intensity increase of 0 to 1 covers three gray-level steps while, at the upper end, an increase from 11 to 16 is needed to cover one step. Obviously, a large number of perceivable levels of gray are lost. By introducing an offsetting gamma correction, one can obtain a linear-perceived gray-level curve. Here, the maximal number of visible gray levels results.

Visual scientists long have been aware that for cone vision or photopic vision a relatively constant relationship exists between the absolute luminance (L) and the increment or decrement in luminance (ΔL), which can just be detected (13). The ratio of ΔL/L, known as the Weber–Fechner ratio, takes the value of about .02 for spatially adjacent areas subtending 2° or more of visual angle. Thus, for the minimum discernible difference in luminance—one gray-scale step—a difference of about 2% is required. Naturally, for areas that are proximal but not adjacent, the size of the ratio increases dramatically. A single pixel separation (assuming an RAR of 1) raises the value to .07, indicating a 7% difference for the threshold of detection of a luminance difference. Widely separated areas may differ as much as 50% before a luminance difference can be seen.

Contrast

In simplest terms, in order for information to be visible on any display, the information must be of a higher or lower intensity than the noninformation areas of the screen. This difference is referred to as *contrast* and relates to the amount of light emitted by the informational and noninformational areas of the screen plus the ambient light added to each area from the environment. Thus,

$$\text{Contrast ratio} = CR = \frac{B_s + [R_p T^2{}_F E/\pi]}{B_b + [R_p T^2{}_F E/\pi]} \tag{3}$$

where
B_s = the light emitted by the area of greatest intensity in cd/m^2
B_b = the light emitted by the area of least intensity in cd/m^2
R_p = the reflectivity of the phosphor layer
T^2 = the transmittance of the faceplate and filter
E = the illumination on the surface of the display in lux.

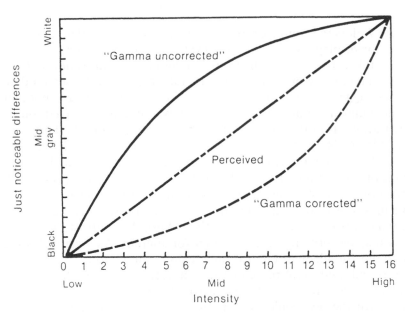

FIG. 6 Maximizing the number of visual gray levels on a display through gamma correction.

Consider a negative-image display (light characters on a dark background) in which a written area, with all pixels active, measured in a dark room with a spot photometer capable of measuring just the luminous area of active pixels, emits 50 cd/m^2 and the nonwritten background emits 4 cd/m^2. With 500 lux falling on the screen surface, a phosphor reflectivity of 0.33, and a transmittance of filter and faceplate for the ambient light of 0.32, we obtain

$$CR = \frac{50 + 0.33\,(0.32)^2 500/\pi}{4 + 0.33\,(0.32)^2 500/\pi} = 5.9 \text{ to } 1$$

Although several formulas for specifying contrast are in use, the above expression of contrast ratio and the measure of contrast percentage are most common:

$$\text{Contrast modulation} = C = \frac{(B_s + R_p T^2{}_F E/\pi) - (B_b + R_p T^2{}_F E/\pi)}{(B_s + R_p T^2{}_F E/\pi) + (B_b + R_p T^2{}_F E/\pi)} \quad (4)$$

For the above example, the contrast would be 69%.

When direct luminance measurements under normal ambient illumination levels are made, these rather complex appearing formulas reduce to the simple form of

$$\text{Contrast ratio} = \frac{\text{Luminance Maximum}\,(L_{max})}{\text{Luminance Minimum}\,(L_{min})} \quad (5)$$

$$\text{Contrast modulation} = \frac{L_{max} - L_{min}}{L_{max} + L_{min}} \quad (6)$$

Figure 7 plots the contrast ratio as a function of the ambient light level for a typical color display. Note the ranges evaluating the adequacy of the contrast. These are taken from such numerous human-factors guidelines as DIN 66234 (14) and represent the opinions of experts rather than solid research data. Such guidelines suggest a minimum contrast ratio of 3 : 1 (contrast modulation of 50%) and a maximum of 10 : 1 (82%). The lower limit has received some empirical verification, as in the study reported by Timmers, van Nes, and Blommaert (15) (see Fig. 8). Here, response latencies for identifying three-letter words were measured for targets presented directly to the fovea (0°) or slightly to the left and right of the fovea (−1.5°, +1.5°). Performance does appear to fall off at a contrast of 3 : 1 (50%) but seems unaffected at the 90% level.

One must exert care in interpreting studies of contrast, as they usually track performance for very short periods of time. Also, the contrast requirements for one task may be very different from that for another. As a general principle, it is wise to test for adequate contrast in a realistic simulation.

One recent study (16) presented text at various contrast levels on a CRT. The technique was to scroll the text across the screen to determine the fastest rate at which the text could be read accurately. Measurement sessions lasted about one hour so that sustained performance could be assessed. In one of the

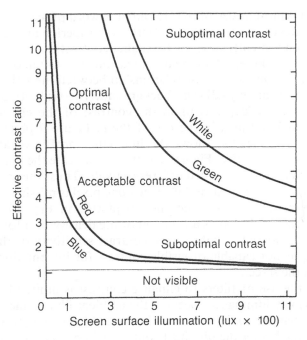

FIG. 7 The effective contrast ratio of colors on a cathode-ray tube with increasing levels of ambient light.

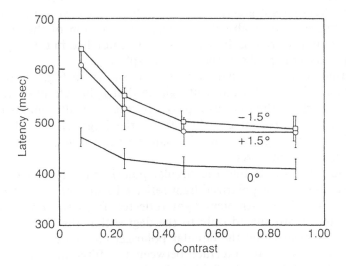

FIG. 8 Verbal response latencies (milliseconds) for Dutch three-letter words as a function of text-background contrast for three retinal locations. Vertical extenders indicate the standard deviation of the means. (Adapted from Ref. 5.)

conditions, text typical in size of that found on a word processor (15 minutes of arc) was used. Reading performance clearly was superior above a contrast of 3 : 1 (50%) but only slightly superior for higher contrasts.

Contrast, as used here, refers to large-area contrast. Often of greater importance is *small-area contrast*, or the relationship between the written area of the screen and the immediately adjacent background. The visibility of such a character as a lowercase *e* depends upon the contrast of the character elements relative to the enclosed area in the eye of the *e*. Figure 9 plots a measure of small-area contrast; Fig. 9-*a* shows the luminance profile of an area of the screen having all pixels active. Figures 9-*b* and 9-*c* indicate the luminance profile for one pixel and two pixels turned off, respectively. Thus, the contrast between the written and unwritten areas is 45% when one pixel is missing and 60% when two pixels are missing. These relations are plotted in Fig. 10, in which the amount of contrast is plotted as a function of the density of information on the screen. When considered in light of the contrast sensitivity of the eye, we see that the display is not capable of visible contrast for information greater than 80 cycles/in on the screen.

Obviously, the use of filters to enhance contrast is limited by the absolute reduction in display luminance. Because the ambient illuminance is reduced by a factor of two relative to the reduction in emitted light, the greater the density of the filter is, the higher the obtained contrast will be. Yet, at some point, the absolute amount of light emitted is so low that the display is useless. As noted in the section on luminance, a minimum display luminance of about 35 cd/m^2 for a monochrome device seems reasonable. Given a range of ambient light levels and this minimum luminance goal, the display designer can calculate a filter that is optimized to the display.

Thus far, a neutral density has been assumed — one that attenuates ambient and emitted light in equal amounts. For monochrome devices that emit light in a restricted portion of the visible spectrum, it often is possible to use a notched filter. Such filters pass a greater amount of light in the spectral region of the phosphor's emission while absorbing a larger proportion of the ambient light. For multichromatic displays, these filters must be notched in three spectral areas that correspond to the peaks of the three phosphors.

An alternative is the circular polarizer. As shown in Fig. 11, the polarizer functions as a light trap for ambient light in that light falling on the screen is polarized circularly to the right as it passes through the filter and is reflected as a left circularly polarized image. The left-polarized light then is blocked by the filter. Obviously, the light passing through the filter is attenuated only by the transmittance of the polarizer.

It is important to note that the circular polarizer is superior to the neutral density filter only in that polarized light reflected from the surface of the CRT is trapped by the filter. Ambient light reflected from the phosphor layer is depolarized and is attenuated only to the degree that the filter absorbs light. Practically, this means that the circular polarizer eliminates the double image created by the glass-to-air interface between the filter and the CRT's front surface. A potential drawback of this technology is that the maximum transmittance, at best, can be 50%. For many applications, this may reduce the absolute luminance level below an acceptable level.

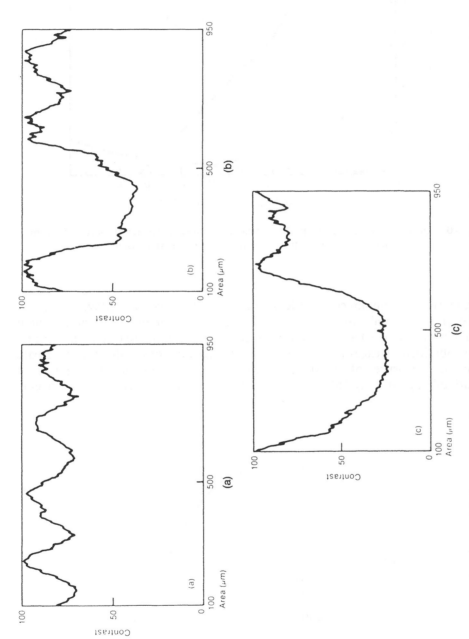

FIG. 9 Contrast modulation on a monochrome display: *a*, with all pixels on; *b*, one pixel on adjacent to an inactive pixel; and *c*, one pixel on adjacent to two inactive pixels.

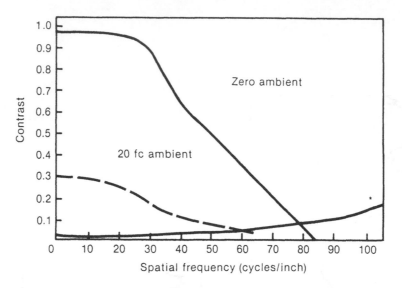

FIG. 10 Display contrast under 0 and 20 foot-candles ambient for increasing levels of information density (* = visual contrast sensitivity, 18-inch viewing distance).

Clearly, a minimum contrast level is required for a good display. There are, however, upper limits to the amount of contrast. Although the previously cited guidelines of 10 to 1 have received little empirical support, overly high contrast levels obtained by excessively reducing the background of the display can create problems. Measures of the ability of the eye to focus on the characters of a negative-image display indicate reduced accommodative accuracy for extreme

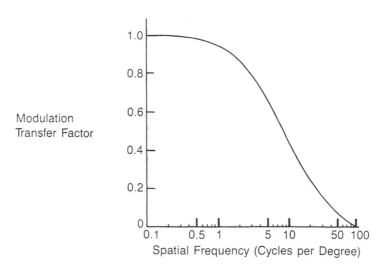

FIG. 11 A typical modulation transfer function.

contrast (17). While no concise data allow the exact specification of the limit, it is clear that one needs a minimum background level to avoid the condition known as *empty-field myopia*. When the eye is given no surface upon which to focus, it reverts to a resting position that usually is inappropriate for the viewing distance. Such a display gives the impression of a black hole in which elements on the screen appear to float in space.

Viewability: Image-Quality Metrics

Having considered a number of such attributes of CRT images as resolution, addressability, luminance, and contrast, it is obvious that a general measure of image quality is desirable. This metric should permit the designer to trade off display system characteristics in order to obtain the highest possible level of display viewability. In this context, then, *viewability* can be conceptualized as the degree of match between the physical output characteristics of an imaging system and the sensory input processing capabilities of a user of that system. The highest degree of viewability would result from a system perfectly attuned to the visual system of the observer such that all signals would be noiseless and maximized in information-carrying capacity. Clearly, then, we need a complete physical description of a given imaging system that can be matched to the functional properties of vision.

Applied to image quality, two elements are required: an accurate means of describing the relevant physical characteristics of the imaging system, and a description of the salient mechanisms of sensory functioning. Obviously, neither element is available as a complete entity. We are forced to use what is known currently about sensory functioning to select what appears to be relevant physical characteristics for the attempted match. We do not need to engineer characteristics into an imaging system for which the sensory system of the user is insensitive. For example, applying the existing knowledge of the relationship between visual acuity, luminance, and contrast allows us to set a limit for optimal contrast for displays rather than continually striving for higher and higher contrast.

In most approaches to image quality, a global measure of the operating characteristics of the imaging system is related to a global measure of visual sensitivity. Historically, most physical measures of image quality have followed the lead of Otto Schade (18), who proposed a figure of merit calculation based on the integration of the modulation transfer function (MTF) of the imaging system [MTF = $M(K)$] and the modulation transfer function of the human visual system [$O(k)$].

$$N_e = \int M^2(K)O(k) \tag{7}$$

With the availability of more accurate measures of both the system and human MTF, several alternative models have been developed.

The MTF of an imaging system can be conceived as a measure of the capability of the system to transfer the information contained in an object to an image of that object. Usually, the function relates the ratio of a sinusoidal modulation in the imaging system (M_{out}) to the object modulation (M_{in}). This modulation transfer ratio (M_{out}/M_{in}) can be calculated for the entire spatial frequency range of the object to be imaged. *Spatial frequency* refers to the number of modulations per unit area. To facilitate comparisons with vision, spatial frequency is expressed in cycles per degree of visual angle. Figure 11 shows a hypothetical modulation transfer function derived in this way.

Over a limited range of frequencies, the imaging system transfers the modulation perfectly (Modulation Transfer Ratio = 1.0, or $M_{out} = M_{in}$). As spatial frequency increases, the transfer is reduced such that the M_{out} reaches one-half (0.5) at a spatial frequency of 20 cycles/degree. A similar measure can be taken for the visual system's MTF, which is the reciprocal of the contrast sensitivity function (CSF). The spatial frequency of a sinusoidal grating is varied, and the observer adjusts the contrast (maximum luminance − minimum luminance/ maximum luminance + minimum luminance) to a point at which the grating is just visible. This function was shown in Fig. 3.

Taking in relation the contrast sensitivity function and MTF for vision, a plot of both functions on the same graph (see Fig. 12) provides the measure of image quality developed by Snyder (4) termed modulation transfer function area (MTFA). The limiting resolution of the imaging system is given as the crossover point of the two functions (lower right of Fig. 12). In the example, this occurs at 41 cycles/degree. That is, the amplitude of the M_{out} is lower than the required contrast modulation of the visual system. No information falling outside the shaded portion will be transmitted by the imaging system to the viewer.

As with Schade's original concept, a single measure of image quality can be obtained as the integrated difference between the two curves.

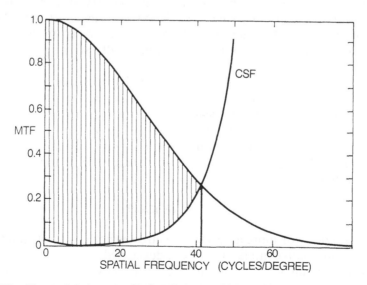

FIG. 12   The modulation transfer function area (CSF = contrast sensitivity function).

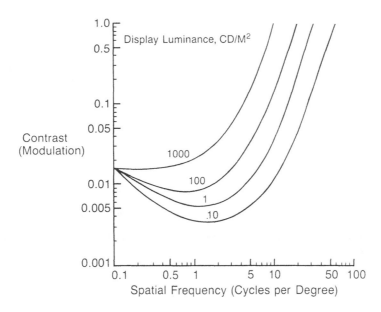

FIG. 13   The influence of luminance on the contrast sensitivity function. (After Ref. 19.)

$$\text{MTFA} = \int \text{MTF}_{(K)} - \text{CSF}_K \qquad (8)$$

Snyder's measure can be applied to complex images by performing a two-dimensional Fourier analysis of the object to be imaged (M_{in}) and the imaging system's transfer of the object (M_{out}). As demonstrated by Beaton (20), the degree to which the fundamental amplitude of the signal and the higher harmonics can be transferred to the viewer can be calculated as the MTFA.

The usefulness of the MTFA comes from examining the impact of a variation in the parameters of the imaging system on the system and observer MTF. Figure 13 shows the effect of display luminance on the CSF. As luminance is decreased from 1000 cd/m² to 0.10 cd/m² the system becomes more and more insensitive to higher frequencies—those responsible for sharp images. Further, the minimum contrast required increases.

Figure 14 shows the increase in resolving power of the human eye as a function of viewing distance. For distances below 50 centimeters (cm) a marked decrease in the crossover point of limiting resolution occurs. Focusing at a closer distance reduces the MTF of the lens. To capitalize on this characteristic of the visual system, such imaging systems as high-resolution displays should be viewed at distances greater than 38 cm.

The major weakness of MTFA analysis lies in the generalized assumption that the CSF can be used as an adequate description of the visual system's image-processing characteristics. An alternative concept, also based on the MTF, is that of Carlson and Cohen (21).

Carlson and Cohen developed a signal-detection model of the process of imaging and image perception. Based on the psychophysical research literature, they assume that the visual CSF represents the integrated output of a finite

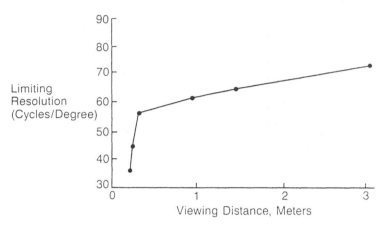

FIG. 14 The point of limiting resolution as a function of observer viewing distance. (After Ref. 7.)

number of spatial frequency channels in the visual system; a total of 7 channels with peak sensitivities at 0.5, 1.5, 3.0, 6.0, 12, 24, and 48 cycles/degree is assumed. The concept is similar to that of the visual sensitivity function (V_λ) as the summed output of three component color channels. Here, as with color, the responses of each channel are not identical with differences characterized by variation in the magnitude of change in modulation transfer ratio, which can just be detected. For example, the 12 cycles/degree channel can discriminate 28.3 changes in modulation transfer ratio over the modulation transfer range of 0 to 1.0 while the 24 cycles/degree channel discriminates only 12.9.

The size, and hence number, of just noticeable difference steps is developed from analytic considerations of

1. intrinsic background "noise" level of the visual system
2. signal strength capacity proportional to noise
3. a nonlinear square law model describing the contrast sensitivities of each channel

The final response is that of a spatial integrator that sums the output of each channel.

Although Carlson and Cohen developed the mathematical version of their model for direct calculation of image quality, they also prepared a large number of discriminable difference diagrams (DDD) that allow one to plot the MTF of an imaging system against visual sensitivity in each channel. Figure 15 shows one such diagram to be used in evaluating the quality of a large, high-luminance analog display; it plots the MTFs of two imaging systems (A and B). System A is clearly superior to System B; the "perfect system" would have a constant modulation transfer ratio of 1.0. However, System A transmits information in the 80 cycles/degree range, which is beyond the visual limit and hence unnecessary. System B, on the other hand, fails to transmit information above 24 cycles/degree and would produce a poorer image. The two plots for System B show the amount of increase or decrease in modulation transfer ratio that could just be detected in one channel (12 cycles/degree). Practically, Carlson and Cohen show that a trained observer can detect a 1-JND (just noticeable difference) change. But that significant difference requires a 3-JND change (total across all 7 channels). A change of 10 JNDs or more is substantial. Carlson and Cohen have developed over 100 DDDs to be used with a wide variety of analog, sampled, and raster displays. Choice of appropriate diagram is based on image size, viewing distance, mean luminance, and intrinsic noise power.

Given that one is able to measure the MTF of an imaging system, the analysis of image quality is very simple. Plotting the function on the appropriate diagram allows one to count the number of steps on each channel between the measured system and the ideal system. For the imaging systems shown in Fig. 15, then, System A is 2 units from perfect while System B is 23.6 units.

As with MTFA, a means of direct measurement of the image's MTF is required to apply the signal detection model. Sherr (22) and Keller (23) describe several techniques for measuring the MTF with CRT displays.

Although a number of other physical measures of image quality have been proposed, MTFA and signal-detection modeling appear to be the best candidates for the assessment of image quality. However, both models have shortcomings. Three distinct assumptions underlie the two models.

1. Visual data is based on threshold measures of contrast sensitivity. These data then are extrapolated to suprathreshold values. That the visual system functions at threshold and suprathreshold in an identical manner is questionable.

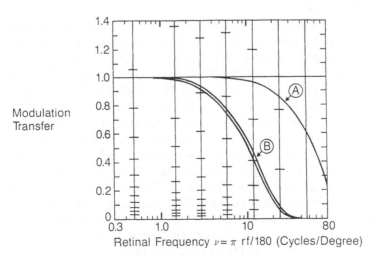

FIG. 15 The discriminable difference diagram for estimating image quality. (From Ref. 21. Permission for reprint courtesy of Society for Information Display.)

2. A ramification of Assumption 1 is the assumption that image quality increases with increased contrast. However, psychophysical research has shown that visual acuity is reduced at very high contrast levels.

3. The signal-detection model (DDD diagram) assumes the subjective size of the JNDs to be the same irrespective of contrast or frequency. Substantial psychophysical research renders this a questionable assumption.

Despite the limitations, both MTFA and signal-detection models have developed a sizeable body of research literature to support their utility to assessing global display viewability. The recently adopted American National Standards Institute (ANSI) Standard for Visual Display Human Factors, for example, contains the specification that a minimum MTFA of 5 is required for a display to meet the requirements of the standard.

Flicker

The visual system is responsive to the temporal luminance modulation of a light. At low frequencies of modulation, the light appears to flicker. If the modulation frequency increases and/or the modulation depth decreases, the flickering light changes to an apparently steady light. This transition is called the *critical flicker frequency* (CFF). Any refreshed-display devise must seek an optimum match of refresh rate and trace decay. Obviously, the decay rate of the phosphor trace is related directly to the required rate of refresh: the faster the phosphor decays, the higher the refresh rate must be. Yet other factors will contribute to the perception of flicker on a display. These include the waveform of the trace, luminance, contrast, ambient illuminance, display size, position in the visual field, and other elements. A number of such individual factors as age, fatigue level, health, and so on will contribute to the rate required to eliminate flicker (24).

The complexity of analyzing the elements of flicker has led most human-factors engineers to conclude that it is much easier to measure the necessary refresh rate for a given display than it is to develop a general model that would predict display flicker accurately. An example is provided in Ref. 25. Subjects viewed a monochrome AEG-Telefunken Model DCM 38 monitor from a distance of 50 cm. The screen showed a full white luminance of 100 cd/m^2. The visible area of the screen was masked to provide several sizes of circular viewing areas ranging from 0.8° to 30° (0.7–26.2 cm). The CRT had a P-4 phosphor (short persistence) and was driven in a noninterlaced mode. Figure 16 shows the results in terms of the flicker thresholds of the six subjects for each size viewing area as a function of fixation point. Thus, a 15° (13.1-cm) area centrally fixated had to be refreshed at 60 hertz (Hz) in order to reach the flicker threshold. Remember that threshold means a rate at which 50% of the judgments are for no flicker and 50% are for flicker. The larger area (corresponding roughly to a 12-in screen) required 68 Hz to reach the threshold. Note also that flicker sensitivity is greater when the device is viewed in the periphery of vision—a situation that frequently occurs when devices are left on but are not in use.

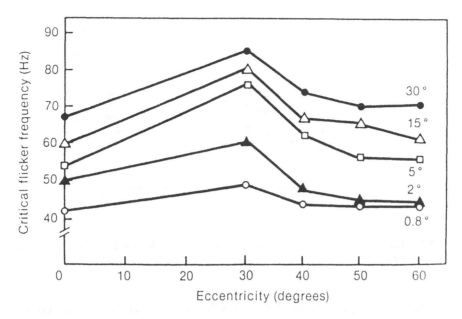

FIG. 16 Mean critical flicker frequency for six observers as a function of field size and retinal eccentricity. (After Ref. 25.)

Beyond the obvious contribution to flicker of the decay rate of the phosphor, image size, and position of the display in the visual field, the luminance of the display long has been known to be a critical determinant of flicker. Vision research dating to the turn of the century shows that the flicker perception threshold increases proportionately with luminance (26). This relation, known as the Ferry–Porter law, states that the flicker threshold increases as a function of the logarithm of the retinal illuminance. (Retinal illumination in trolands indicates the amount of light reaching the retina by calculating the area of the pupil and multiplying by the light intensity.)

The luminance dependency in flicker perception applied to the perception of flicker on a CRT has been studied extensively by Eriksson and Bächström in Sweden (27). Using a Tektronix 608 monitor with a P-31 phosphor, subjects viewed a 7×7-cm square area through a viewing hood from a distance of 5 cm. In total, flicker thresholds were obtained for four display sizes (10°, 30°, 50°, and 70° of visual angle) and five luminance levels (25, 50, 100, 200, and 400 cd/m^2). The results indicate a clear linear relationship between flicker threshold and luminance (see Fig. 17). Similar to the data of Ref. 25 shown in Fig. 16, decreasing the size of the display area resulted in a higher threshold for flicker. Note that if we compare the flicker thresholds for a 30° display area with a luminance of 100 cd/m^2 in the two studies, a threshold difference of about 20 Hz results (85 Hz for Ref. 25 and 65 Hz for Ref. 27). This large difference is presumed to be the result of the difference in phosphor decay rate between P-31 and P-4. Although a color difference exists (green versus white), previous studies have shown little differences in flicker thresholds due to color (24).

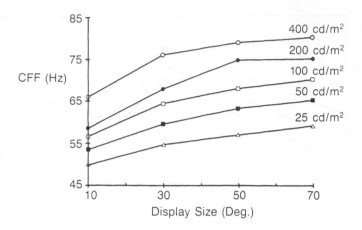

FIG. 17   The influence of luminance and display size on the perception of flicker. (After Ref. 27.)

Naturally, the polarity of the display has an impact on the perception of flicker. Grimes (28) showed that a display with a P-4 phosphor in a positive image mode (dark characters on a light background) required a 10-Hz higher refresh rate to be flicker free than the same display operated in a negative image mode.

An analytical model for flicker in a color display is even more difficult to develop as the problem is multiplied by the existence of three phosphors with different persistence and temporal wave characteristics. Figure 18 shows the data for a Tektronix 690SR monitor (short-persistence P-22 phosphor) for which subjects rendered a judgment of flicker or no flicker for randomly selected refresh rates. Judgments were made at four fixation points (29). The display showed a full white field with a series of crossed "HH" sections. The white field was adjusted to 75 cd/m² and viewed from 50 cm. The 19-in monitor subtended a visual angle of 55°. Interestingly, the 60-Hz refresh rate again fell right at threshold — 50% of the subjects detected no flicker with central fixation. For peripherally viewed display, however, 70 Hz reached the 50% threshold. Similar results were reported by McCulloch (30), who noted the expected increase in flicker threshold with increasing luminance.

The frequently encountered claim of "60 Hz noninterlaced flicker-free" rests on a statistical interpretation of the CFF. Recall that centrally viewed P-31 monochrome and P-22 color displays refreshed at 60 Hz reach the 50% CFF. This average means that half the observers will not see flicker but the other half will. To be truly flicker free for the majority of viewers, a higher criterion is needed. Such ergonomic standards as the ANSI Video Display Terminal ergonomic standard suggest a criterion of 90% judgments of no flicker.

Clearly, rates higher than 60 Hz are required to reduce the amount of perceived flicker. There is also a correlation between flicker perception and the observer's age. As a person grows older, flicker sensitivity falls off. Figure 19 shows data for the flicker threshold of 23 observers as a function of their age (29).

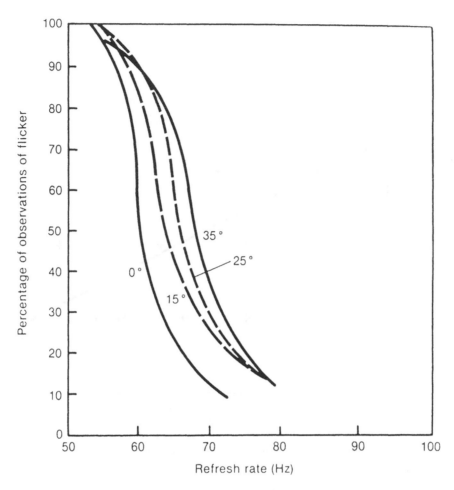

FIG. 18   The perception of flicker on a color monitor as a function of viewing angle.

Recently, several attempts have been made to provide engineers with methods to assess or predict flicker without such extensive research as that reported above. One approach by Rogowitz utilizes a flicker-matching technique in which an observer adjusts a flickering light source to the same perceived rate of flicker as a refreshed display (31). Figure 20 portrays the measurement system. Initially, a large number of observers participated in a study in which the luminance contrast of a display was varied and matching flicker rates obtained. The resulting distribution of observer judgments permitted an estimate of what percentage of the population would perceive flicker at any given refresh rate.

To evaluate a display prototype, it is possible to have the display viewed by a small number of observers (five is recommended by Rogowitz) whose flicker sensitivity is known to represent the fifth percentile of flicker sensitivity in the normal population. This can be derived from the norms of the data collected by Rogowitz. If the prototype is judged flicker free by these fifth-percentile observers, one can assume that it will, in fact, be flicker free for 95% of the

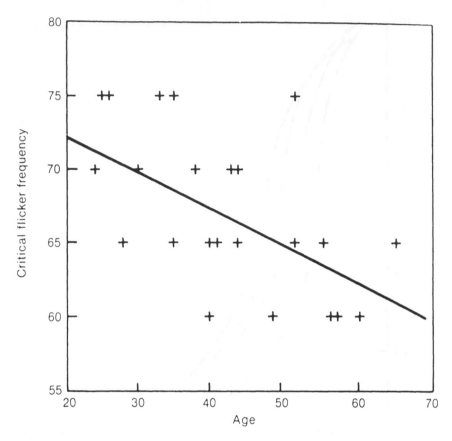

FIG. 19   Critical flicker frequency as a function of age.

viewing population. Similarly, the 90% criterion could be met by selecting test observers from the tenth percentile.

An alternative approach to flicker prediction has been developed by Farrell (32). In this case, Farrell derived a flicker metric from the extensive psychophysical research on flicker conducted by D. H. Kelly (33). Kelly, leveraging off the Ferry–Porter law, had discovered that the energy residing in the fundamental Fourier component of the temporal waveform was the critical determinant in the production of flicker perception. At some critical energy level, he noted, the flicker threshold is crossed and visually stable light appears to flicker.

The details of the metric are to be found in Farrell (32). In essence, one calculates the display luminance in trolands. Assuming the display phosphor decays exponentially, the Fourier amplitude of the critical threshold is estimated based on the parameters of display size and viewing conditions.*

---

*Note that the equation for calculation of the amplitude coefficient of the fundamental frequency of the varying luminance should be Amp $(f) = 2/[1 + (\alpha 2\pi f)^2]^{1/2}$.

If one applies the Farrell model to previously published reports on flicker, a reasonably good fit is found. In some cases, the specific viewing conditions were not specified in these publications to a degree permitting exact calculation of the metric. This requires some estimation on the exact prediction. Despite this limitation, the data, which are given in Table 2, indicate that the model does very well in postdicting the empirical results. The data reported in Table 2 are for a 90% flicker threshold rather than the typical 50% threshold. That is, a display refreshed at the specified value should be flicker free for 90% of the viewing population.

It is important to note that the model predicts a maximum refresh rate of 75 Hz for any phosphor decaying exponentially to 10% in less than about 1 ms given sufficient luminance. For most viewing conditions, that represents a luminance of about 25 fl. This suggests that the current standard of 60 Hz is inadequate and that a target of 75 Hz is reasonable for rapidly decaying phosphors used in bright displays. Obviously, longer decaying phosphors would permit a shorter refresh rate—and the associated reduction in bandwidth. Caution must be urged, however, that such long-persistence phosphors may lead to image smearing and ghosting.

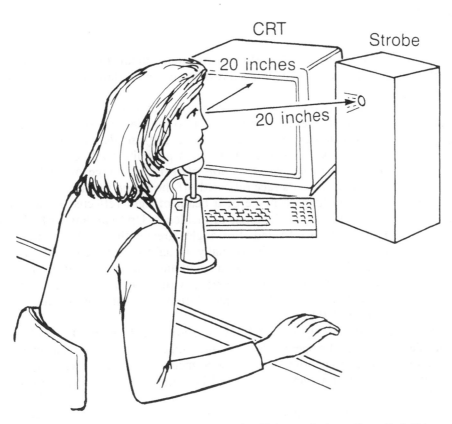

FIG. 20   A matching technique for measuring flicker on displays. (From Ref. 31.)

TABLE 2 Predicting the 90% Flicker-Free Threshold

Data Source	Phosphor	Raster Decay (10%)	Luminance (fl)	CFF Observed	CFF Predicted
Ref. 28	P-4	72 μs	4	40	41
"	"	"	5	50	52
"	"	"	10	60	64
"	"	"	13	60	65
Ref. 25	P-4	26 μs	27	82	76
"	"	"	51	84	82
"	"	"	97	91	88
Ref. 27	P-31	30 μs	7	62	62
"	"	"	15	68	68
"	"	"	29	72	72
"	"	"	58	77	79
"	"	"	117	87	87
Ref. 30	P-22	20–500 μs	9	61	54
"	"	"	22	68	66
"	"	"	30	65	69
Ref. 29	P-22	20–500 μs	24	72	73

Glare

From the operator's perspective, the most serious human-factors problem with displays is screen glare. Most user surveys have found glare to be the number one complaint.

Glare, as it affects the viewer of a display, comes in several forms. *Direct glare* refers to light sources that shine light into the viewer's eyes. An overhead light or window is often a source of direct glare. *Diffuse glare* results from light reflecting off surfaces within the room, such as white walls or other highly reflective surfaces. *Specular* or *reflection glare* is the image of such specific objects as the display viewer's face reflected off the screen surface.

Figure 21 shows the percentage of a beam of light reflected by a smooth glass surface as a function of the angle of incidence. Up to an angle of 45°, just under 5% of the light is reflected. For angles greater than 45°, an increasing amount of light is reflected, while virtually all light is reflected at an angle of incidence of 90°. While 5% seems like a small amount, recall that an overhead light of 500 lux might illuminate the surface with about 200 lux, resulting in a reflected glare of 10 cd/m^2. Light-colored clothing typically will produce a 5–10 cd/m^2 image on the CRT screen.

The effects of glare upon the CRT viewer are multifaceted. Direct glare, for example, influences the operator's operating range of adaptation. The eye seems to adapt to the average brightness of the immediate field of view, so a proportionally greater display luminance would be required. Specular glare provides an alternative viewable image on the screen at a very different optical distance than the screen from the operator. Viewing the images requires a refocusing of the eye, which can cause fatigue. Diffuse glare reduces the contrast of the

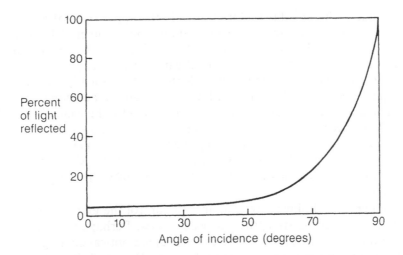

FIG. 21 The percent of light reflected from a smooth untreated glass surface as a function of the angle of incidence.

display and masks the images on the screen. In effect, both specular and diffuse glare reduce the display resolution.

The total elimination of glare is not a practical possibility. Its reduction is a combination of the proper structuring of the display environment and the construction of the display itself. Figure 22, taken from the German DIN 66234 display guideline (14), points out the conditions leading to glare as well as some

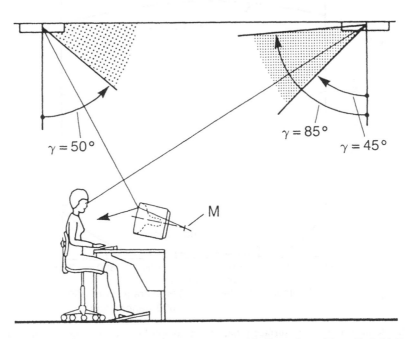

FIG. 22 Angles of light sources producing direct and indirect glare. (From Ref. 14, Part 2.)

ways of reducing it. The direct-glare source in the upper right of Fig. 22 can be controlled by restricting the angle of illumination. Louvered light covers or recessed lighting can be used. Similarly, diffuse glare can be reduced as shown in the upper right of the figure. Locating the display surface parallel to windows and not directly below overhead lights aids the control. Ideally, an environment for a display system can be structured architecturally to accommodate the display. Such an environment would contain controlled lighting, low-reflective walls and surfaces, separate source lighting, and window-light control. The utility of controlling glare by eliminating the source of the glare is made clear in a study by Beaton and Snyder (34), which found that the MTF of a display as measured in a dark room was reduced by almost 97% in the presence of an intense glare source.

A number of techniques are available to control the glare from the display itself; each has specific strengths and weaknesses. Perhaps the top of the line for controlling glare (they tend to be expensive) are optical coatings that reduce the reflection from the front surface of the glass. Figure 23 shows the capability of a coating marketed by OCLI of Santa Rosa, CA. The reflectance averages 0.3% across the spectrum. Besides the high price, the coatings show fingerprints that will present difficulties for such screen interfaces as touch panels. Specular reflections can be reduced by etching the front surface of the CRT or by a filter

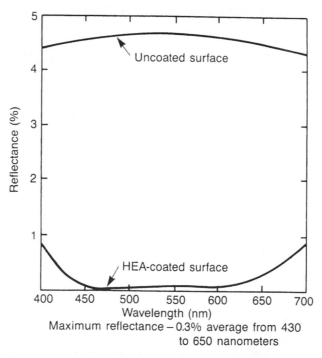

FIG. 23   Reduction in front-surface reflections accomplished by an antiglare optical coating. The maximum reflectance averages 0.3% from 430 to 650 nanometers (nm).

mounted in front of the CRT. Of course, this technique turns a specular reflection into a diffuse reflection and results in a loss of contrast; the resolution of the display also is reduced. Similar problems are associated with mesh filters. Both mesh filters and etching also produce and odd visual effect when such high-luminance saturated colors as green are viewed. Any head movement produces a scintillation pattern on the screen. Conversely, for lower-resolution displays, etching and mesh filters provide cost-effective alternatives to optical coatings. A final approach is to place a hood to shield the screen from light sources. One often encounters homemade versions of hoods. While inexpensive, the hoods may create adaptation problems by increasing the operating range of vision.

In the report by Beaton and Snyder (35), free-standing antiglare filters were evaluated on etched and nonetched (polished glass) CRTs. The filters evaluated included optically coated filters and mesh filters. Using the MTF metric, they found the etched screen to reduce the MTF by about 11% relative to an untreated screen. As one might expect, the best performance resulted from the optically coated filter on a polished CRT. While the mesh filter, compared to a bare CRT, improved the MTF by 20%, the optically coated filter improved MTF by 72%.

A recent study by Hunter et al. (35) evaluated glare control via changes in MTF as well as with subjective judgments of image quality. In this case, a special CRT was prepared in which one-half the surface was etched and one-half was left untreated. To each of these halves optical antiglare coating was applied to one-half while the other was left untreated. The result was a single CRT containing four different surfaces: etched with and without optical coating and unetched with and without optical treatment. The use of a single CRT eliminated any variation in results due to inherent differences in the quality of images between different CRTs. The MTF measurements yielded results similar to those reported by Beaton and Snyder. The ratings of perceived image quality showed a clear perceptual disadvantage for the etched surface under all viewing conditions: dark room, diffuse glare, and specular glare. The antiglare optical coating showed no improvement in perceived image quality under diffuse glare conditions but a significantly higher image quality rating under specular glare conditions. In general, the image quality ratings were lowest for text viewed with specular glare.

In light of the research findings that glare is considered a major problem by display users and that the presence of a glare source dramatically reduces the MTF of a display, it would seem prudent to encourage workplace design that eliminates glare sources. While such antiglare treatments as optical coating clearly improve the perceived quality of the image, nothing is as good as removing the glare itself.

Polarity

Traditionally, most monochrome devices display light characters on a dark background. Although the descriptive terminology often is inconsistent, this

TABLE 3   Subjective Ratings of Positive- and Negative-
Image Displays

	Positive Image	Negative Image
Legibility	2.42	2.07*
Contrast	2.33	2.11
Sharpness	2.20	1.9*
Flicker freeness	1.84	2.09*

Source: Ref. 38.
*Statistically significant difference at the 3% level.

usually is called *normal video* or *negative image*. More recently, monochrome displays, particularly those used in conjunction with paper documents, have favored *reverse video* or *positive image* (dark characters on a light background). The debate on the superiority of each polarity format in terms of viewability has been extensive. The most vehement proponents of positive image have been the German authors of the ZH 1/618 Ergonomic Standard for Visual Displays (36).

From a visual perspective, neither format is inherently superior. The visual system appears to adjust its operating range to a combination of the average luminance in the field weighted by the peak luminance (37). While this might suggest a potential superiority of positive-image display in high-luminance environments, the extensive visual operating range of over one log unit indicates that either polarity should work equally well.

Each polarity has a specific advantage with regard to glare and flicker. With a fixed frame rate, flicker is less noticeable on a negative-image display. Grimes (28) had subjects adjust the luminance of a monochrome display until flicker could just be detected. With a refresh rate of 60 Hz, the flicker threshold for a group of subjects averaged 200 cd/m^2 for a positive-format display and 260 cd/m^2 when the same information, a memo, was shown in negative image. The positive-image display reduces the level of specular glare by reducing the contrasts between the glare source and the background. The impact of these advantages for the two formats is reflected in the evaluation of displays of both formats conducted by the VDMA in Germany. Table 3 reports the average ratings on a 4-point scale (1 = excellent, 2 = very good, 3 = good, 4 = fair) for 27 viewers of positive- and negative-format displays (38).

A study reported by Bauer and Cavonius (39) found a performance difference in favor of the positive-image display. Subsequent studies such as Moeller's and the study by Kokoschka and Fleck (40) failed to find any differences. The resolution of these conflicting results points out an important design consideration for positive and negative formats. Bauer and Cavonius simply reversed the video on their displays without adjusting the font to compensate for the widening of the characters under negative and narrowing under positive contrasts. When each display is optimized for the chosen format, no performance differences occur.

Legge et al. tested the impact of polarity of a text display on a subject's reading performance (41). Using the criterion of the maximum sustainable reading speed, discussed in more detail in conjunction with a similar study on contrast, these researchers found polarity to have no influence upon reading performance. Similarly, Zwahlen measured eye movement patterns and obtained subjective estimates of visual discomfort from subjects working for two eight-hour days at positive- and negative-image displays (42). Again, neither index showed a superiority of one format over the other.

The lack of either physiological, subjective discomfort or performance data indicates that any preferences for positive- or negative-polarity displays are due to personal choice. To some extent, a preference for positive polarity might be expected due to the extensive experience most individuals have with printed paper. The one study to report a performance difference between polarities also noted a subjective preference for a positive polarity (39). Yet, as a performance difference was obtained, some differences may have existed between the displays beyond polarity.

Monochrome Color

The issue of the choice of a single color for monochrome displays has spawned a discussion equal in magnitude to the polarity issue. Despite numerous advertising claims to that effect, no performance data has demonstrated clearly the superiority or inferiority of a given color. Radl (43) compared white, green, yellow, and orange displays and found "only relatively small differences with respect to readability." Significant differences were found on ratings over a 10-point scale derived from paired comparisons (see Table 4).

Clearly, the orange filter was not liked by Radl's subjects. The orange-filtered CRT also produced poorer text-reading performance, which may indicate inherent image-quality problems with the display as opposed to a visual

TABLE 4 Preferences for Monochrome Cathode-Ray Tube Colors

Display Phosphor	Display Filter	Preference Rating*
Yellow	Amber	8.2
Yellow	–	7.5
White	Yellow	7.2
Green	–	6.9
White	–	6.5
White	Orange	2.2
Yellow	Orange	1.7

Source: Ref. 43.
*1 = low, 10 = high.

problem associated with orange. With the exception of a highly saturated blue on a dark background, which will appear fuzzy and unclear due to the inability of the lens to flatten sufficiently to focus this color on the retina, the color selections for monochrome displays are a question of personal preference (12).

Some misunderstanding seems to exist on the influence of the chromatic aberration of the human eye upon display color. When light is refracted at any surface, the degree of refraction depends on the wavelength of the light. This means that, as the eye accommodates to closer and closer targets, the wavelength in focus gradually shifts from long wavelengths toward shorter ones. At the typical viewing distance of 50 cm for most displays, the wavelength in focus is about 520 nanometers (nm), or green (44). This is of concern only to those who wear eyeglasses, as their normal correction would produce an over- or underaccommodation for other wavelengths. In general, the state of accommodation for a given color is relative, and a yellow viewed at 50 cm simply requires a different accommodative state than some other color.

Multicolor Displays

Multicolor displays clearly are dominant in today's display applications. Beyond the general principles described in the sections above, some specific aspects of multicolor displays need to be considered in order to optimize the interface between the user and the display.

Colorimetry

Colors can be produced by the additive combination of emitted light or the subtractive combination of inks, dyes, or filters. In an additive color mixture, such as occurs with a multicolor display, a set of such primary colors as red, green, and blue are combined in various amounts to form mixture colors. Each color can be specified numerically as a proportional mixture of the three primaries. One easily can develop a two-dimensional color diagram in which the proportional amount of one primary (i.e., red) is plotted on the abscissa (R) and the proportional amount of another primary (i.e., green) is indicated on the ordinate (G). The resulting triangle forms the color space of the three primaries (see Fig. 24). Pure red would be specified as $R = 1.0$, $G = 0$. Pure blue is $R = 0$, $G = 0$. (The amount of the blue primary can be found by subtraction as the total of all three proportions equals 1.0. Thus, $B = 1 - R - G$.) A mixture such as yellow is $R = 0.5$, $G = 0.5$. For a color containing some amount of each primary, a mixture such as pink is $R = 0.75$, $G = 0.20$, $B = 0.05$, or white is $R = 0.33$, $G = 0.33$, $B = 0.33$ results. For a detailed description of colorimetric principles of displays, see Ref. 45.

The problem is that such diagrams as Fig. 24 can specify only color mixtures made by the combination of a specific trio of primaries. To overcome this limitation, in 1931 the CIE decided to use a concept developed by Maxwell (46)

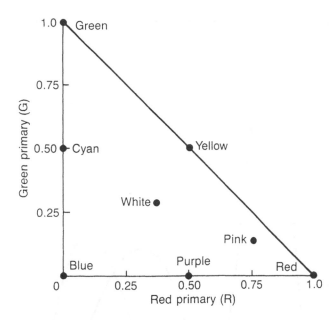

FIG. 24 A color-mixing triangle.

in which a set of imaginary primaries were determined that encompassed the entire visual spectrum. The imaginary primaries do not exist as physical colors but are mathematical abstractions, yet it is possible to indicate the position of any real primaries, and the colors that can be mixed by their additive combination, within this space. Because the imaginary primaries do not exist as real colors, they have been given the names, X, Y, and Z. Their energy distributions are shown in Fig. 25. Note that the Y curve is the curve of visual spectral sensitivity V_λ used in determining the luminance sensitivity of the eye.

To specify such a color as the green phosphor emission shown in Fig. 26, one follows the procedure shown in Fig. 27. That is, the spectral-energy distribution of the phosphor is multiplied wavelength by wavelength with the X primary and integrated to yield the quantity X. Similarly, the spectral-energy distribution of the phosphor is multiplied and integrated with Y and Z to form the quantities Y and Z. As with our previous example, the color can be specified simply as a value x, y, and z. Figure 28 shows the 1931 CIE diagram with approximate color names. The green of our example calculates to a value of $x = 0.300$, $y = 0.550$, which corresponds to a yellowish green. Note that the diagram, by using the luminosity curve V_λ as an imaginary primary, does not provide any indication of the color's brightness. The Y value for the color does, however. Thus, specifying a color in terms of Y, x, y provides an exact representation of the color.

Because the primaries used in a color display have a fixed spectral-energy distribution mixture colors are produced by varying the amount of each primary. The chromaticity coordinates (x and y) of any mixture or the proportions of each primary for a specific coordinate can be calculated using a simple nomograph (47). Figure 29 shows the 1931 chromaticity diagram with the x,y

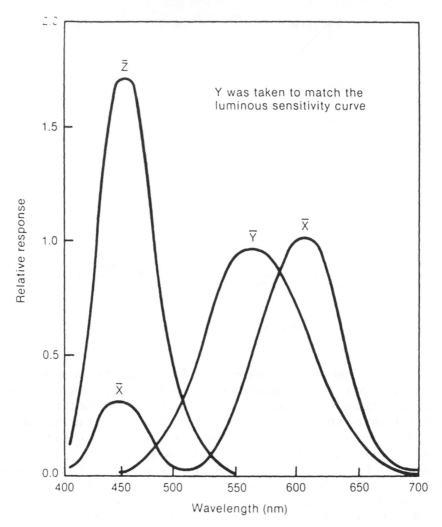

FIG. 25   Spectral tristimulus values of constant radiance stimuli.

positions of the red, green, and blue phosphors. The scales used to derive the percentage contribution of red and green are shown in the upper right of the diagram. Figure 30 shows the calculation of the percentage of red, green, and, by subtraction, blue that would produce a specific white on the display. The reverse procedure allows the calculation of the x, y coordinates resulting from a particular percentage of the primary.

The nomograph of Fig. 29 is useful in selecting the absolute level of each primary that will match a particular white when all three primaries are on full. The two whites most frequently used are D6500 K° and 9300 K°. These values refer to a white that closely matches the color temperature (expressed in degrees Kelvin, or K) of a blackbody radiator. As the color temperature match achievable by the mixture of phosphors only approximates the color temperature of a blackbody radiator, it properly is termed *correlated color temperature*. The

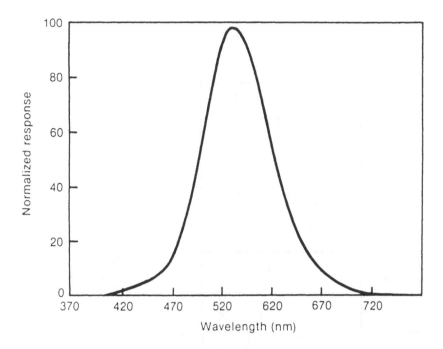

FIG. 26   The spectral radiant energy distribution of a green phosphor.

amount of deviation from the blackbody locus for a particular white is specified in minimum perceptible color difference (MPCD) units. The white mixture that approaches D6500 usually is used as the reference white for color television, while visual displays usually use the 9300 value. This latter mixture contains a higher proportion of blue and is felt by most observers to produce a better white under the fluorescent illumination found in most environments in which displays are used.

Despite some obvious strengths, the 1931 CIE system also has some drawbacks. The most serious of these is that the size of the distance between two points on the CIE diagram tells nothing about the magnitude of the perceived color difference. From a perceptual perspective, the CIE space is nonuniform.

Recently, a transformation of the CIE 1931 color space that produces a more uniform color space was adopted. This color space, the 1976 CIE Uniform Color Space, is shown in Fig. 31.

To transform x and y values to the Uniform Color Space:

$$u' = 2x/(6y - x + 1.5) \tag{9}$$

$$v' = 4.5y/(6y - x + 1.5) \tag{10}$$

Associated with the 1976 color space is a color difference metric that allows the calculation of the visual difference, or color contrast, between two colors. The metric, called ΔE, calculates the difference between two colors in relation

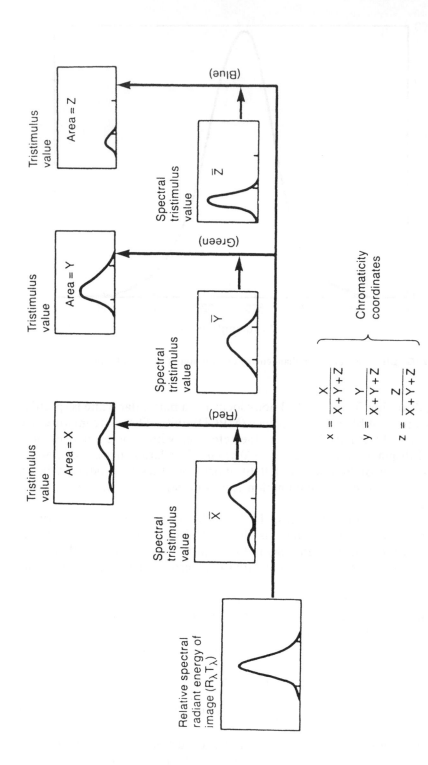

FIG. 27   The calculation of CIE chromaticity coordinates.

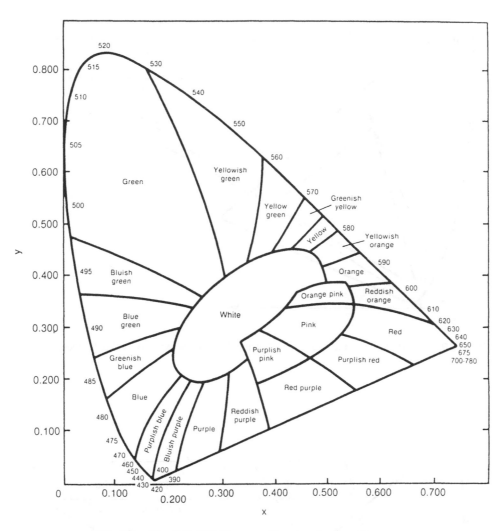

FIG. 28 The 1931 CIE diagram with approximate color means.

to each color's distance on the u', v' color diagram from some reference white (u_0', v_0').

$$\Delta E = [(\Delta L^*)^2 + (\Delta u^*)^2 + (\Delta v^*)^2]^{1/2} \tag{11}$$

in which

$$L^* = 116 \left(\frac{Y}{Y_0}\right)^{1/3} - 16 \tag{12}$$

where $Y/Y_0 > 0.01$.

The luminance component L^* is calculated for each of the two color samples to be compared. The luminance component of the comparison becomes

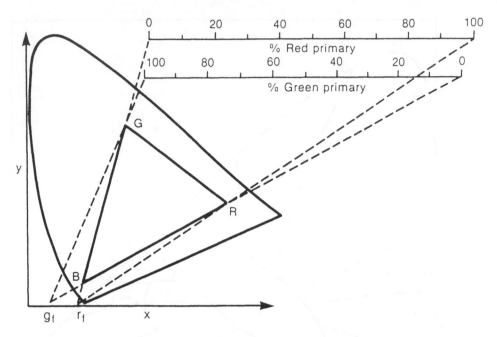

FIG. 29   The 1931 chromaticity diagram showing the color range of typical red, green, and blue phosphors.

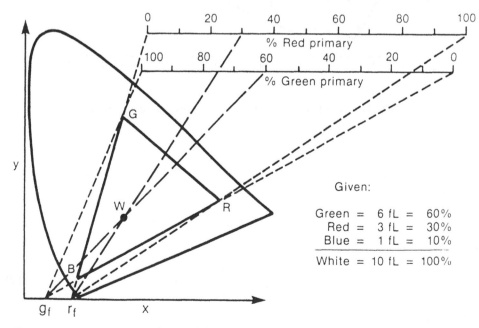

FIG. 30   The calculation of the percentages of red, green, and blue required to produce a specific white. (Adapted from Ref. 47.)

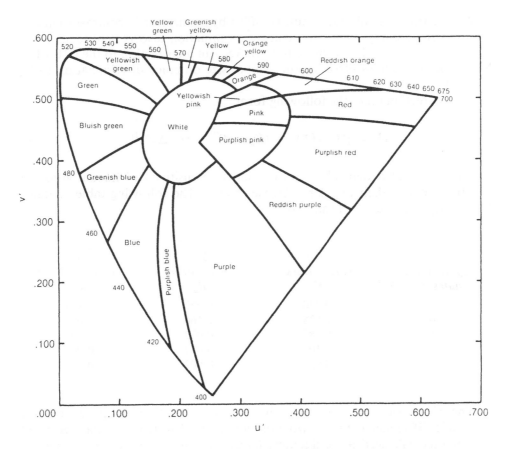

FIG. 31 The 1976 CIE Uniform Color Space with approximate color names.

$$\Delta L^* = L_1^* - L_2^* \tag{13}$$

The color component is calculated for both samples as

$$u^* = 13L^*[u' - u_0']$$
$$v^* = 13L^*[v' - v_0'] \tag{14}$$

Finally, the color difference is derived as

$$\Delta u^* = u_1^* - u_2^*$$
$$\Delta v^* = v_1^* - v_2^* \tag{15}$$

The values for the reference white (u_0', v_0') are taken as the actual white of the display. Typically, graphic displays employ a 9300°K white ($u_0' = 0.198$, $v_0' = 0.468$). Following the suggestion of Carter and Carter (48) the value of Y_0 should be taken as the maximum obtainable luminance (full drive) of the display for full-intensity white.

A reduction in symbol size must be offset by an increase in contrast to insure good discriminability. Similarly, color discrimination also is dependent upon symbol size. Judd and Yonemure (49) worked out a correction to the CIE LUV ΔE metric that allows effect of the symbol size on color discrimination to be factored in.

The correction takes the following form:

$$\Delta E = [(K_L \Delta L^*)^2 + (K_{u'} \Delta u^*)^2 + (K_{v'} \Delta v^*)^2]^{1/2} \qquad (16)$$

The values of K depend upon the relative insensitivity of human vision to specific colors as their angular subtense change. The following table contains Judd's empirically derived values of K:

Symbol Size (Minutes of Arc)	ΔL^* (Light/Dark) K_L	Δu^* (Red/Green) $K_{u'}$	Δv^* (Violet/Yellow) $K_{v'}$
32	0.850	0.270	0.133
16	0.575	0.160	0.043
8	0.285	0.072	0.003
4	0.105	0.020	0.000
2	0.032	0.003	0.000

The ΔE metric is constructed so that a value of 1 ΔE should produce a just noticeable difference between two adjacent colors (50). Colors separated from one another will need a considerably larger ΔE value to be discriminated. A color difference of 40 ΔE has been shown to permit rapid and reliable color discrimination (51).

While the ΔE metric has proven useful in assessing color discriminability, it has been less successful in predicting the legibility of one color superimposed on a background of another color. However, recent research has resulted in a proposed metric that allows the specification of minimum levels of color and luminance difference required to ensure maximum legibility of colored text on a colored background (52). The metric takes the form

$$\Delta T = [(155 \Delta Y_m / \Delta Y)^2 + (367 \Delta u')^2 + (167 \Delta v')^2]^{1/2} \qquad (17)$$

where
   Y_m = the maximum luminance of text on background
   ΔY = the difference in luminance between text and background
   $\Delta u'$ = the difference between u' coordinates of the text and background
   $\Delta v'$ = the difference between v' coordinates of the text and background.

The empirically derived weighting coefficients scale the metric such that a value of $\Delta T = 100$ indicates maximum legibility. Decreases in legibility, defined as a reduction in reading speed, occur for lower values of ΔT.

A final point on colorimetry needs to be made about ambient light. Just as the addition of ambient light reduces the contrast of the display, it also reduces the color contrast. Literally, the ambient light mixes with the emitted light to create an additive mixture on the basis of the proportional amounts of each. As most ambient light sources are broadband, the effect is one of a desaturation of all three phosphors. The amount of desaturation depends not only on the amount of the ambient light, but also on the relative proportions of spectral energy. Fluorescent light, for example, contains a greater amount of short-wavelength energy, which desaturates blue more than green or red. Sunlight, conversely, contains more long-wavelength energy. Obviously, any color difference ΔE or color legibility ΔT calculation must include the ambient light level and spectral-energy distribution in the metric.

Color Viewability

Using color effectively in any display requires exact control of the visual parameters of color. Because of the unique nature of the way in which colors are produced on visual displays, conventional color science is often of limited use in defining such visual parameters of color as brightness and, to a degree, color contrast. The control of display element brightness presents the user with the well-documented problem of the lack of correspondence between the photometric scale of luminance and the perceptual experience of brightness (53). For example, given a white light and a blue light, both with a luminance of 10 cd/m^2, the blue light will appear much brighter.

Studies have been carried out in our laboratory that compare luminance and perceived brightness for color display using the normal P$_{22}$ phosphor set (12). Results show that deviations are most extreme for colors toward the end of the visible spectrum (red, blue, and magenta). Table 5 illustrates the luminance values that render the brightness of selected video colors equal to that of a video white with luminances of 10 cd/m^2 and 20 cd/m^2. These values were obtained using the model of heterochromatic brightness matching. Fully saturated colored squares subtending 4.5° of visual angle were displayed adjacent to a similarly configured white square on a Tektronix 650HR color monitor.

TABLE 5 Palettes of Equally Bright Cathode-Ray Tube Colors

Color	Luminance (cd/m^2)	Luminance (cd/m^2)	u'	v'
White	10.0	20.0	0.2182	0.4650
Yellow	7.6	19.6	0.2353	0.5469
Cyan	7.4	17.5	0.1337	0.4259
Green	7.1	17.9	0.1151	0.5591
Red	4.7	12.7	0.4290	0.5276
Magenta	3.7	13.7	0.2922	0.3559
Blue	2.7	—	0.1674	0.1873

The 21 subjects were asked to adjust the luminance of the neutral square until it appeared equally bright as the colored square, that is, until the border between the two appeared minimally distinct (54). The testing was repeated five times, at one sitting for each subject. The perceived variation in brightness for the high-luminance test averaged 27% less than for the low-luminance condition. One exception was the magenta match, which was 68% less for the high-luminance condition. Discrepancies that result from V_λ, the luminous efficiency function of the human eye, are not as severe at higher luminances, thus ratios are smaller for the high-luminance matches. Further research will have to uncover the exact nature of the magenta discrepancy. Note that no value has been indicated for the blue at the higher luminance as this value was beyond the luminance maximum for the blue phosphor on the test monitor.

When viewing combinations of highly saturated colors from the extreme ends of the visible spectrum, two perceptual effects are relevent: a relative motion effect and a color stereoscopic depth effect (55). The relative motion effect, also known as the "fluttering hearts" phenomenon, was documented first by von Helmholtz (56). When an image consisting of highly saturated red and blue areas is moved perpendicular to the viewer's line of sight, generally the red areas appear to move into the foreground, while the blue areas are perceived as receding. The explanation most frequently advanced cites different perceptual processing latencies for the extremely different colors (57). The instability of such color combinations, which is particularly noticeable in peripheral vision, may be annoying and distracting even at moderate light levels, although the effect is most pronounced under dim illumination. For most applications, the movement of the user is restricted by the constraint of the keyboard or other input device so that images rarely are viewed as oscillating perpendicularly to the line of sight. It has been demonstrated that separating the red and blue areas with a black line causes a reduction or complete disappearance of the effect (57).

The color stereoscopic (chromosteropsis) depth effect is a phenomenon in which red and blue colors on a display seem to appear in different depth planes. The orientation of the photoreceptors within the viewer's retina dictate whether the red or blue portion of the image is seen in the foreground or whether the effect is seen at all (58). In order to avoid constant refocusing that may lead to visual fatigue, such extreme color pairs as red and blue should be avoided. Use of colors that are less saturated and less spectrally extreme reduces the effect of both phenomena and the subsequent need for refocusing (59).

While desaturating the colors solves the color stereopsis problem, it places a greater demand on the display to maintain good beam convergence. We conducted some tests on the limits of perceivable misconvergence in conjunction with the autoconvergence system for the Tektronix 4115B (60). Subjects viewed two identical images drawn in white and were asked to select the "best" image. The three beams were misconverged systematically. In a follow-up condition, the subjects were asked specifically to select the "best-converged" image. The results are shown in Fig. 32. For the "choose the best image" condition (Fig. 32-*a*), subjects chose the converged image 75% (half-way between chance of 50% and a perfect 100%) of the time up to a 0.4-millimeter (mm) misconvergence. Even when subjects are looking specifically for misconvergence (Fig. 32-*b*), the 75% point is reached at about 0.19 mm.

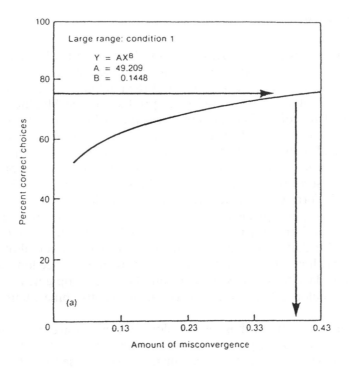

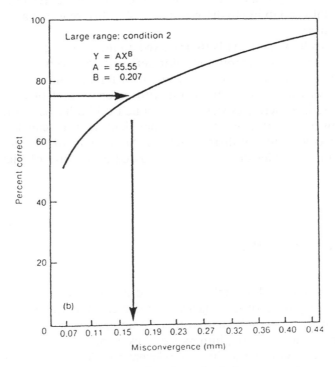

FIG. 32 Measurements of perceived misconvergence.

Color Deficiency

Some concern seems to exist regarding color displays and color-deficient users. It is unfortunate that we use the term *color blindness* to summarize the variety of color deficiencies that beset about 9% of the population. In actuality, only a tiny proportion of the color deficiencies produce a true blindness for color. Persons with true color blindness are called monochromats, and they usually experience several additional visual problems, such as foveal blindness.

Although all causes of color-deficient vision are not known, some stem directly from primary retinal receptors, the cones, and their photopigments. Perhaps the best known is the so-called red-green deficiency of dichromatism. In reality, this form of color deficiency can be produced by the lack of either the red or green photopigment. Nevertheless, these two distinct variations of the condition result in similar symptoms: affected persons have trouble discriminating any color that is dependent upon a ratio of red-to-green photopigments. They do differ, however, in the perception of brightness, since long-wavelength stimuli appear dark to the individual lacking the red photopigment. A relatively rare form of color deficiency also exists in which the blue photopigment is missing.

More common among those with color deficiencies are individuals whose response functions to the photopigments deviate significantly from normal. In one form, the peak of the red photopigment lies very close to that of the green, while in another the green is shifted toward the red. The net result is a reduction in an individual's capability to discern small color differences, particularly those of low brightness.

Verriest, Andrew, and Uvijls (61) showed an increase in color confusions on a multicolor display by dichromatic color-deficient observers. For subjects with milder forms of color deficiency, only minor differences in color-identification performance was noted in comparison with color-normal observers. In fact, older subjects, in general, experienced as much difficulty as the mildly color-deficient viewers. Confusions for color-deficient viewers can be reduced by insuring that colors do not differ solely in the amount of red or green. Both color-deficient and older display users are aided by the use of high-luminance colors.

Flat-Panel Displays

Just as with the multicolor displays described in the previous section, emerging alternative display technologies, generally referred to as *flat panels*, have both unique human-factors requirements and common requirements with CRTs. The discussions on contrast, glare, luminance, flicker, monochrome color, polarity, colorimetry, and color viewability are relevent to CRTs and flat panels. Some unique considerations must be made for flat panels in the area of resolution and addressability, range of displayable colors, ambient light, distortion, and flicker.

Resolution and Addressability

The luminance of the light spot formed by an electron beam as it hits the phosphor in a CRT has a Gaussian or near-Gaussian profile. CRT technology precludes the production of a square profile—one in which the luminance rises at the edge of the pixel and goes from zero to its full value instantly. The weakness that results from this CRT characteristic relates to the mechanism by which the human eye focuses (accommodates) upon distal objects. Research indicates that a spot or line produces in the eye an image with a Gaussian energy distribution. The eye adjusts its focal length to produce a "perfect" Gaussian spot, which is interpreted by the visual system as an indicator of image sharpness. When the target itself has a Gaussian light distribution, the ideal focus never can be found (62). The inability to accommodate properly appears to be the result of the loss in high spatial frequencies in an image constituted by Gaussian spots (63).

Thus, the square-profile luminance distribution, which maintains the higher spatial frequencies, offered by many flat-panel devices may represent a better match to the visual system than the Gaussian profile of the CRT. Unfortunately, this increase in focus and the resultant improvement in image sharpness carries a perceptual penalty. The disturbing quality of off-axis lines known as *jaggies* is accentuated by a square pixel luminance distribution because of the human eye's hyperacuity for small offsets of square luminance distributions. Offsets as low as 8–10 seconds of arc are detectable in square-wave luminance distributions while only 20–30 seconds of arc can be detected with Gaussian luminance distributions (8). This translates into a lower addressability requirement for Gaussian lines than for square profiles when the jaggies are to be rendered invisible.

Several recent research efforts have sought to establish the pixel size and spacing requirements for a square-edge pixel that would render the jaggies invisible. Beaton and Knox (64) simulated matrix-addressed alphanumeric characters on a specially designed ultra-high-resolution CRT (1000-lines-per-inch addressability with a 0.8-mil spot size). Subjects viewed the characters from a typical viewing distance (55 cm) and judged whether a given character appeared to be continuous (no visible jaggies) or digitized (visible jaggies). The size and separation of the pixels was varied. The pixel sizes of 3, 5, 7, and 9 mils could vary in either the vertical or horizontal direction. The interpixel spacings were 1, 3, or 5 mils. Figure 33 shows the averaged results for five subjects judging five alpha characters that varied in horizontal size and separation. On half of the trials, the character was a continuous character. Therefore, chance responding should produce a correct judgment on 50% of the trials. As can be seen in the figure, subjects were unable to detect the jagged edges of characters formed from pixels of a horizontal size of 3 mils with either a 2- or 3-mil spacing as well as for a 5-mil pixel separated by 1 mil. Even at a 7-mil pixel size with a 2-mil spacing, subjects experienced difficulty in detecting jaggies. When the vertical size and spacing were varied (see Fig. 34), overall performance was higher for small pixels with small separation but worse for larger separations. This finding is probably the result of the asymmetric nature of alpha characters, which are longer in the vertical than in the horizontal dimension (65). Generally, it appears that reducing the interelement spacing from 3 mils to 1 mil affords a larger gain in perceived quality than a reduction in element size from 7 mils to 3 mils.

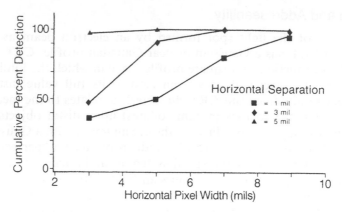

FIG. 33   Average detection scores as a function of horizontal pixel width and separation. (After Ref. 65.)

A clear advantage of the CRT over flat panels is the fact that individual pixels can be overlapped. Not only does this overlapping insure a flat field at full addressability, but pixels forming characters merge to produce smoother appearing lines. Several studies have shown that merged pixels in characters enhance display legibility (7). That is, the ratio of the RAR can be greater than one. This means that an addressability can be selected for a given CRT resolution that renders individual pixels invisible when a full-field raster is presented on the display. This is accomplished by selecting an addressability at which the luminance modulation between adjacent pixels is below the human contrast sensitivity function.

The flat panel with discrete, nonoverlapping pixels always will have an RAR of less than one. For example, a panel with a display element size of 10 mils and an interpixel spacing of 2 mils would have an RAR of 0.83 (10/10 + 2). To achieve a flat field in which the individual elements are not visible when the

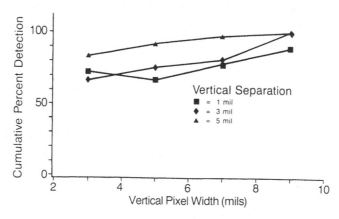

FIG. 34   Average detection scores as a function of vertical pixel width and separation. (After Ref. 65.)

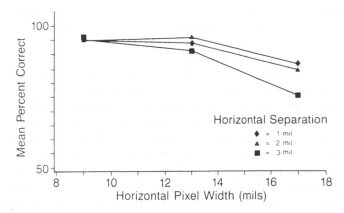

FIG. 35 Average legibility scores as a function of horizontal pixel width and separation. (After Ref. 65.)

luminance modulation between adjacent elements is 100%, the element density must be greater than the visual limit. This limit lies at about a spatial frequency of 50 cycles per degree of visual angle. If we assume a panel with a vertical height of 27.5 cm (corresponding to the value used in the preceding example) 1670 lines would be required to produce a spatial frequency equivalent to 50 cycles per degree. This example represents a worst case, however, as the spacing between active display elements was assumed to be the same as the element width. With decreasing interelement spacing, the required addressability will reduce.

The study by Weiman and Beaton (65) also assessed the legibility of flat-panel alpha characters. Figures 35 and 36 indicate the results in terms of the average correct character identifications. Clearly, very large pixels with wide separations were required before noticeable performance reductions occurred.

Flat panels will have a difficult time overcoming jaggies in hardware. Despite

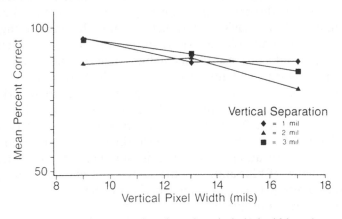

FIG. 36 Average legibility scores as a function of vertical pixel width and separation. (After Ref. 65.)

the nonoverlapping pixels, panels with gray-scale capability will benefit from antialiasing (see Ref. 66). The reported legibility results indicate that visual performance may not be as severely impaired by the discrete nature of the pixels on a flat-panel display. Perhaps this relates to the previously mentioned focusing problem with Gaussian pixels and the fact that the square edges of most flat-panel displays evoke a sharper, focused image.

The square luminance profile of most active matrix displays can produce noticeable losses in image quality when element or line failures occur. Generally, the detectability of a missing pixel is greater on these displays than on CRT displays. Laycock (67) simulated a variety of element failures on a matrix display. These images were photographed and evaluated by experienced observers. Among the findings, Laycock noted that a permanently "on" pixel was more disturbing than a permanently "off" pixel. In fact, noticeable degradation in text images were noted with 0.01% "on" pixel failures while up to 1.0% "off" failures could be tolerated.

Color

The range of colors that can be obtained with typical phosphor sets used in shadow-mask CRTs was depicted in Fig. 29. All colors located within the triangle can be produced by the device given adequate intensity levels for each gun. While this range encompasses roughly half the colors the human eye can detect, it does provide a reasonable range of hues with acceptable brightness and saturation levels as witnessed by the images we have come to accept on home television. The range is not as good as can be obtained with film dyes, as often is noted when objects with colors lying outside the color triangle of the CRT are portrayed in film. The TV portrayal of such colors as emerald green or titian red become desaturated versions of the original.

For those flat-panel technologies that do not rely upon phosphor emissions for color production, a significant opportunity to improve over the color range of the CRT exists. In consideration of the fact that dyes exist for the production of colored filters that create primary colors well beyond the boundaries of CRT phosphors, a much wider color range can be produced. Further, the three-color primary system used in today's CRT is dictated more by engineering considerations of gun design than colorimetric requirements. Four- or even five-color primary systems would increase the available range of colors and improve the saturation and brightness of mixed colors as well. Saturation improvements occur because colors produced through the mixture of pairs of primaries generally are less saturated than either of the component primaries. The loss of saturation increases as a function of the spectral difference between the primaries. By decreasing the spectral difference through an increased number of primaries, greater saturation of mixtures is obtained. As for brightness, the luminance efficiency of blue and, to some degree, red phosphors used in today's CRTs is disappointing. High luminance levels and, hence, improved brightness can be obtained with appropriately chosen filters.

Another weakness noted in the CRT is the lack of color uniformity. A full-field color often varies in luminance by 50% from center to periphery and in color by 20 to 25 ΔE units. While improvements in process control can reduce nonuniformity resulting from phosphor deposition, the cost of the tube assembly rises. Colors produced by filters placed over active elements, as is done in several flat-panel technologies, should show less variation in terms of luminance and purity due to the inherent stability of dye-coupled filters.

The number of primaries and their spatial arrangement in the CRT has been limited by the arrangement of the electron guns. Flat-panel metric displays are not subject to this limitation, which has led to some interesting research on primary pixel arrangements. For liquid crystal TV (LC-TV) Tsuruta et al. (68) obtained subjective ratings of photographs created by computer that simulated four pixel arrangements. Two arrangements simulated the typical vertical strip and diagonal mosaic found in current shadow-mask CRTs. The other two arrangements were an offset triangular layout $\left({}_R{}^G{}_B{}^R{}_G{}^B \right)$ and a quad arrangement with replication of the green pixel $\left(\begin{smallmatrix} G & B & G & B \\ R & G & R & G \end{smallmatrix} \right)$. Subjects ordered the photographs depicting a girl's face and a mandrill by image-quality preference. The offset triangular layout was preferred, followed by the quad arrangement. A similar finding has been reported by Gomer et al. (69) in a comparison of the standard vertical and diagonal red, green, blue (RGB) layouts with the quad pattern. In this case, the arrangements were simulated on a high-resolution monitor viewed through an optical imaging system that produced an image characteristic of a color matrix display (see Ref. 70). In several measures, the quad arrangement always produced higher ratings of image quality.

Ambient Light

A number of human visual interface concerns stem from the influence of the ambient light level on the CRT. Among those of greatest impact are front-surface glare and contrast reduction.

A key causal agent affecting glare and contrast reduction is the reflectivity of the CRT's phosphor layer. Typically, 30% or more of the incoming light is reflected back to the operator by the phosphor layer. This produces glare and reduces the contrast on the screen. While most flat-panel technologies utilize a glass front surface and, hence, are faced with the same reflectivity problem as the CRT, reflectivity from nonactive areas of the screen can be reduced. Obviously, nonemissive displays already hold an edge over the CRT in high ambient conditions as the contrast between written and unwritten screen areas increases with increasing ambient light. In general, the incorporation of a polarizing layer in most flat-panel technologies carries the added benefit of contrast enhancement in high-ambient-illuminance environments.

Ironically, the widespread use of CRT displays has led to a general reduction in overall office illumination. Also, task lighting is becoming popular as a means of combating glare and contrast loss on the CRT. Both these tendencies create a problem for passive flat panels that rely upon contrast as a direct function of ambient light level. Just as with a CRT, a contrast in excess of 3 : 1 and a minimum luminance of 35 cd/m^2 is required to insure good legibility and perceived image quality.

Distortion

While a piece of glass usually constitutes the first layer of most display technologies, the thickness and shape of the glass represent a further weakness in CRT technology. As the outer envelope of the CRT must provide implosion protection, the front-surface glass must be relatively thick. For off-axis viewing, the light traveling through the glass is refracted and a distortion of the image results. Obviously, as screen size increases, so does the magnitude of the distortion created by the implosion shield. This is exacerbated further by the screen curvature of the CRT, which produces numerous distortions in the image. Most flat-panel technologies eliminate distortion by the flat, thin, front-surface glass. More recently, flat CRT screens have been developed to overcome the problem of distortion.

Flicker

Also high on the list of user complaints with CRT technology is image flicker, or the inverse of image smearing (71). A delicate balance is struck between the refresh rate of a display system and the visibility of the temporal luminance modulation. If the phosphor is too fast, full-screen flicker is visible. If the phosphor is too slow, noticeable smearing and ghosting occurs. For modern systems capable of animation, the problem is even more severe as the temporal limits over which smooth motion is perceived are fairly narrow. Although information is available that allows the system designer to select the appropriate rates for given phosphors (32), the range of phosphors with the correct decay rates is limited. Many flat-panel systems are less dependent upon the physical properties of an emissive phosphor and, as a result, can tailor the temporal relations to match visual capacity.

References

1. Murch, G. M., Visual Accommodation and Convergence to Multichromatic Display Terminals, *Proc. SID*, 24:67–72 (1982).
2. Ginsburg, A., Perceptual Capabilities, Ambiguities and Artifacts in Man and Machine, *SPIE 3D Machine Perception.*, 283:78–82 (1981).
3. Murch, G. M., Visual Perception Basics, SID Seminar Lecture Notes, Vol. I, SID, Playa del Rey, CA, 1987, pp. 2.1–2.36.
4. Snyder, H. L., The Visual System: Capabilities and Limitations. In: *Flat Panel Displays and CRTs* (L. E. Tannas, ed.), Van Nostrand Reinhold, New York, 1985, pp. 54–69.
5. Rogowitz, B., The Human Visual System: A Design Guide for the Display Technologist, *Proc SID*, 24:235–252 (1983).
6. Infante, C., CRT Systems. *SID Seminar*, SID, Playa del Rey, CA, 1986, pp. 8.1, 1–126.
7. Snyder, H. L., Human Visual Performance and Flat Panel Display Image Quality,

Virginia Polytechnic Institute Technical Report HFL-80-1, Virginia Polytechnic Institute, Blacksburg, VA, 1980.

8. Murch, G. M., and Beaton, R. J., Matching Display Resolution and Addressability to Human Visual Capacity, *Displays*, 23–26 (January 1988).

9. Knox, S. T., Resolution and Addressability Requirements for Digital CRTs, *SID Digest*, 8:26–29 (1987).

10. Snyder, H., and Maddox, M., Information Transfer from Computer Generated, Dot-Matrix Displays. VPI-SU Technical Report HFL-78-31, ARO-78-1, Virginia Polytechnic Institute and State University, Blacksburg, VA (October 1987).

11. Faulkner, T. W., and Murphy, T. J., Lighting for Difficult Tasks, *Hum. Fac.*, 15: 149–162 (1973).

12. Murch, G. M., Cranford, M., and McManus, P., Perceived Brightness and Color Contrast of Visual Displays, *SID Digest*, 14:186–189 (1983).

13. Hecht, S., Vision II: The Nature of the Photoreceptor Processes. In: *Handbook of General Experimental Psychology* (C. Murchison, ed.) Clark University Press, Worcester, MA, 1934, pp. 767–789.

14. DIN 66234, VDU Workstations, Part 2, Perceptibility of Characters on Screens, Beuth Verlag, GmbH, Berlin, Germany.

15. Timmers, H., van Nes, F. L., and Blommaert, F., Visual Word Recognition as a Function of Contrast. In: *Ergonomic Aspects of Visual Display Terminals* (E. Grandjean, and E. Vigliani, eds.), Taylor and Francis, London, 1980, pp. 115–120.

16. Legge, G. E., Rubin, G. S., and Leubker, A., Psychophysics of Reading V: The Role of Contrast in Normal Vision, *Vis. Res.*, 27:1165–1178 (1987).

17. Murch, G. M., How Visible is Your Display? *Elec. Opt. Sys. Des.*, 43–49, (March 1982).

18. Schade, O., Image Reproduction by a Line Raster Process. In: *Perception of Displayed Information* (L. M. Biberman, ed.), Plenum, New York, 1973, pp. 97–164.

19. DePalma, J. J., and Lowry, G. M., Sine Wave Response of the Visual System. II, Sine Wave and Square Wave Contrast Sensitivity, *J. Opt. Soc. of Amer.*, 52:328–335 (1962).

20. Beaton, R. J., A Human Performance Based Evaluation of Quality Metrics for Hard-Copy and Soft-Copy Digital Imagery Systems, Ph.D. dissertation, Virginia Polytechnic Institute, Blacksburg, Virginia, 1984.

21. Carlson, C., and Cohen, R., A Simple Psychophysical Model for Predicting the Visibility of Displayed Information, *Proc. SID*, 21:229–246 (1980).

22. Sherr, S., *Electronic Displays*, Wiley Inter-Science, New York, 1979.

23. Keller, P., A Survey of Data-Display CRT Resolution Measurement Techniques, *SID Seminar*, Vol. I, SID, Los Angeles, CA, 1984, pp. 22a, 1–28.

24. Brown, J. L., Flicker and Intermittant Stimulation. In: *Vision and Visual Perception* (C. H. Graham, ed.) Wiley, New York, 1965, pp. 251–320.

25. Bauer, D., Bonacker, M., and Cavonius, C., Frame Repetition Rate for Flicker Free Viewing of Bright VDU Screens, *Displays*, 31–33 (January 1983).

26. Porter, T., Contributions to the Study of Flicker II, *Proc. Roy. Soc. (London)*, 70A:313–339 (1902).

27. Eriksson, S., and Bächström, L., Temporal and Spatial Stability in Visual Displays. In: *Working with Visual Displays* (B. Knave, and P. G. Wideback, eds.) North Holland, Amsterdam, 1987, pp. 461–473.

28. Grimes, J., Effects of Patterning on Flicker Frequency, *Proc. Hum. Fac. Soc.*, 27:46–50 (1983).

29. Mead, D., and Murch, G. M., The Perception of Flicker on Color Displays, Tektronix Human Factors Report Number 84-03, 1984.

30. McCulloch, B., Flicker Perception in Color CRTs, unpublished research report, Portland State University, Portland, OR, 1985.

31. Rogowitz, B., A Practical Guide to Flicker Measurement: Using the Flicker Matching Technique, *Behav. Inform. Tech.*, 5:359–373 (1986).

32. Farrell, J., Predicting Flicker Thresholds in Displays, *Proc. SID* 28(4):449–454 (1987).

33. Kelly, D. H., Diffusion Model of Flicker Responses, *J. Opt. Soc. Am.*, 59:1665–1670 (1969).

34. Beaton, R. J., and Snyder, H. L., The Display Quality of Glare Filters for CRT Terminals, *SID Digest*, 15:298–301 (1984).

35. Hunter, M. W., Pigion, R. G., Bowers, V. A., and Snyder, H. L., MTF and Perceived Image Quality for Three Glare Reduction Techniques, *SID Digest*, 18: 30–33 (1987).

36. ZH 1/618 Ergonomic Standard for Visual Displays. Trade Cooperative Assn., Hamburg, Germany.

37. Booth, J. M., and Farrell, R. J., Overview of Human Engineering Considerations for Electro-Optical Displays, *SPIE Advances in Display Technology*, 199:78–108 (1979).

38. Moeller, R. G., Study of Optimum Character Representation on Visual Display Terminals for the Office Sector, VDMA Working Group on Display Terminals, Frankfurt, 1982.

39. Bauer, D., and Cavonius, C., Improving the Legibility of Visual Display Units through Contrast Reversal. In: *Ergonomic Aspects of Visual Display Terminals* (E. Grandjean and E. Vigliani, eds.), Taylor and Francis, London, 1980, pp. 137–142.

40. Kokoschka, S., and Fleck, H. J., Experimenteller Vergleich von Negativ- und Positivdarstellung der Bildschirmzeichen, *Bericht der Lichttechnisches Institut der Universitat Karlsruhe*, 1982.

41. Legge, G. E., Pelli, D. G., Rubin, G. S., and Schleke, M. M., Psychophysics of Reading I: Normal Vision, *Vis. Res.,* 25:239–252 (1985).

42. Zwahlen, H. T., Operator Comfort with VDTs: Screen Polarity and Rest Breaks, *NCGA CAD/CAM 87*, Vol. II, Boston, June 1987.

43. Radl, G. W., Experimental Investigations for Optimal Presentation Mode of Colors of Symbols on the CRT Screen. In: *Ergonomic Aspects of Visual Display Terminals* (E. Grandjean and E. Vigliani, eds.), Taylor and Francis, London, 1980, pp. 127–136.

44. Sivak, J. G., and Woo, G. C., Color of Visual Display Terminals and the Eye: Green VDTs Provide the Optimal Stimulus to Accommodation, *Amer. J. Optom. Physio. Opt.*, 60:640–642 (1986).

45. Murch, G., *Color Displays and Color Science*, edited by H. J. Durette, Academic Press, New York, 1987.

46. Maxwell, J. C., Theory of the Perception of Colours, *Trans. Roy. Scottish Soc. Arts*, 4:394–400 (1856).

47. Merrifield, R. M. Visual Parameters for Color CRTs. In: *Color and the Computer* (H. J. Durrett, ed.), Academic Press, New York, 1987, pp. 63–82.

48. Carter, R., and Carter, E., CIE L*u*v* Color Difference Equations for Self Luminous Displays, *Col. Res. Appl.*, 8:252–253 (1983).

49. Judd, D. B., and Yonemure, G., Target Conspicuity and Its Dependence on Color and Angular Subtense for Gray and Green Foliage Surrounds, *NBS Report*, No. 10-120 (1969).

50. Wyszecki, G., and Stiles, J., *Color Science*, (2nd ed.) Wiley, New York, 1982.

51. Carter, E., and Carter, R., Color and Conspicuousness, *J. Opt. Soc. Am.*, 71: 723–729 (1981).

52. Lippert, T. M., A Standardized Color Difference Metric of Legibility, Virginia Polytechnic Institute Technical Report HFL-85-1, Virginia Polytechnic Institute, Blacksburg, VA, 1985.

53. Ware, C., and Cowen, W., Specification of Heterochromatic Brightness Matches: A Conversion Factor for Calculating Luminances of Stimuli which are Equal in Brightness, National Research Council of Canada Technical Report, 1985.

54. Kaiser, P. K., Minimally Distinct Border as a Preferred Psychophysical Criterion in Visual Heterochromatic Photometry, *J. Opt. Soc. Am.*, 61:966–971 (1971).

55. Walraven, J., Perceptual Artifacts that May Interfere with Color Coding on Visual Displays, *Proc. NATO Workshop on Color vs Monochrome Displays* (April 1984).

56. von Helmholtz, H., *Handbuch der Physiologischen Optik*, 3rd ed., Voss, Hamburg, 1911.

57. von Grunau, M. W., The Fluttering Heart and Spatiotemporal Characteristics of Color Processing, *Vis. Res.*, 15:437–440 (1975).

58. Vos, J. J., Some New Aspects of Color Stereoscopy, *J. Opt. Soc. Am.*, 50:785–791 (1960).

59. Murch, G. M., Human Factors of Color Displays. In: *Advances in Computer Graphics II* (F. Hopgood and D. Duce, eds.), Springer Verlag, London, 1986.

60. Denham, D., Meyer, W., and Murch, G., Using Autoconvergence in a Color Display, *Min. Micro Syst.*, 17–23 (November 1983).

61. Verriest, G., Andrew, I., and Uvijls, A., Visual Performance on a Multicolor VDU of Color Defective and Normal Trichromatic Subjects, IBM Hursley Technical Report, IBM Corp., Hursley, UK, 1985.

62. Westheimer, G., Spatial Interaction in Human Cone Vision, *J. Physio.*, 190:139–148 (1967).

63. Charman, W., and Tucker, J., Dependence of Accomodation Response on the Spatial Frequency Spectrum of the Observed Object, *Vis. Res.*, 17:129–139 (1976).

64. Beaton, R. J., and Knox, S. T., Flat Panel Image Quality, *SID Digest*, 18:115–118 (1987).

65. Weiman, N., and Beaton, R. J., Effects of Flat Panel Pixel Structures upon Three Human Performance Measures of Image Quality, *Proc. SAE Aerotech Conf.*, November 1987.

66. Booth, K., Bryden, M., Cowen, W., Morgan, M., and Plante, B., On the Parameters of Human Visual Performance: An Investigation into the Benefits of Anti-aliasing, *IEEE Comp. Graph. Appl.*, 34–41 (September 1987).

67. Laycock, J., The Effect of Picture Element Failure on the Legibility of a Matrix Display Image, *Displays*, 70–77 (April 1985).

68. Tsuruta, S., Mitsuhashi, K., Ichikawa, S., and Noguchi, K., Color Pixel Arrangement Evaluation for LC-TV, *IRDC*, 6:24–26 (1985).

69. Gomer, F., Silverstein, L. D., Monty, R. M., Huff, J. W., and Johnson, M. J., A Perceptual Basis for Comparing Pixel Selection Algorithms for Binary Color Matrix Display, *SID Digest*, 435–437 (1988).

70. Monty, R. M., Silverstein, L. D., Frost, K., and Boyle, L., A Color Matrix Display Image Simulation System for Human Factors Research, *SID Digest*, 18:119–122 (1987).

71. Smith, M., An Investigation of Health Complaints and Job Stress in Video Display Operators, *Hum. Fac.*, 23:387–400 (1981).

GERALD M. MURCH

Electronic Mail

Introduction and Definitions

In its simplest terms, *electronic mail* can be defined as nonsimultaneous interpersonal digital communication. This is in contrast to other sorts of communications media: traditional postal mail, the telephone, teleconferencing, electronic digital bulletin boards, facsimile (FAX), and, to a great extent, electronic messaging systems. Another popular term for electronic mail is E-mail. Figure 1 highlights the relationships among these technologies. A distinguishing characteristic of all electronic mail systems is the speed in delivery of the mail message. This singular feature permits both consumers and businesses to disseminate information in a more timely and convenient manner.

Electronic mail consists of three components: message creation, transmission, and reception. One or more of these may be oriented electrically to be considered electronic mail. This definition covers many current and historic communications services, including telegraph, telex, facsimile, and mailgrams. In fact, some experts include the telephone as a form of electronic mail, especially those telephones with voice-based electronic messaging system features. This article focuses on a form called computer-based messaging system (CBMS) or electronic mail and messaging system (EMMS). This type of electronic mail maintains its digital and electronic nature throughout the three components: message creation, transmission, and reception.

History and Evolution of Electronic Mail

Because of its similarities to traditional postal mail in form and function, electronic mail adapted quickly to computer systems. The earliest electronic mail systems evolved in the 1970s as large computing systems moved from batch-oriented processes to interactive ones. This technological breakthrough made it possible for a remote user to communicate asynchronously with a host computer and eventually with other users of that same host computer.

In the early stages of electronic mail development, all functions and capabilities were consolidated on the host computer. The user created mail messages and accessed his or her mailbox on the host computer. Virtually all mail messages were text in content. In addition, electronic mail essentially was limited to large businesses that could afford the equipment. As there were limited interconnections between host computers, electronic mail systems were independent of one another. Figure 2 represents the typical configuration of an electronic mail system during this decade.

Five trends in the early to mid-1980s drastically changed electronic mail into a viable communications medium. The first trend was the proliferation of data

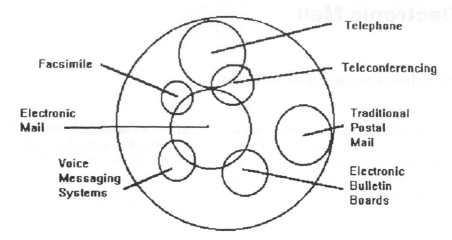

FIG. 1   Interpersonal communications technologies.

communications networks. These networks either were considered *private*, that is, dedicated to a specific business, or *public*, that is, shared by many businesses or consumers. IBMNet is an example of a private data communications network, while Telenet® and CompuServe® are examples of public data communications networks. The data communications networks permitted mail messages to travel outside of the domain of the host computer to other locations, thus making electronic mail an alternative to long-distance telephone or traditional postal mail. Data communications networks were low in cost to access since they were often excess voice-grade lines.

   A second trend was the development of dial-up access to the data communications networks. Previously, a terminal had to be linked to a single computer system. This meant that the user had to get to a terminal to communicate, and, traditionally, terminals were limited in numbers and generally concentrated in

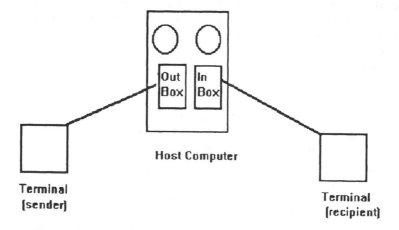

FIG. 2   Electronic mail system configuration in the 1970s.

such areas of corporations as data processing departments. The combination of dial-up access and data communications networks meant that users could access electronic mail anywhere a telephone line was available.

A third trend was the proliferation of the personal computer. The low cost of the personal computer and its computing power moved many of the electronic mail functions (e.g., message creation) from the host computer. Personal computer technology also provided a means to generate the hardcopy equivalent of the electronic mail message at the recipient's location. Previously, such peripherals as printers were expensive and, consequently, like the host computer, were centralized. Therefore, hardcopy printing of electronic mail generally required hand delivery of the hardcopy and thus defeated much of its benefit.

Personal computers' processing capability increased the functionality and ease of use of electronic mail. Previously, electronic mail programs were highly functional and limited by the capabilities of the terminals accessing the host computer. However, personal computer programs made it possible to handle mail messages in a different manner than transmitted.

The popularity of personal computers also created greater acceptance of the technology and such applications as electronic mail thus permitting a larger audience to have access to electronic mail. Marketing, accounting, purchasing, and manufacturing departments now could use electronic mail instead of just the programmers using it. Personal computers also found their way into homes and the hands of the consumer market.

A fourth trend was the increasing popularity of information services, particularly those services targeted toward the consumer (e.g., CompuServe, GEnie®, The Source, and Prodigy®). Clearly, this popularity took electronic mail out of the business domain and put it into the hands of everyone. All early consumer-based information services evolved from traditional time-sharing services and electronic mail was an application readily available. With little effort, the application was modified for the consumer marketplace. Electronic mail has become a prerequisite for all information services.

A fifth trend was the cost efficiencies of electronic mail versus other forms of communication. From the outset, electronic mail had proven its immediacy as a communications alternative, but was fairly costly. All the technological developments discussed above dramatically reduced costs of sending or receiving mail messages. At the same time, other forms of communications experienced increases in costs, thereby causing a number of companies to adopt electronic mail as a communications alternative, particularly for intracompany communications. Although cost efficient for the consumer, electronic mail did not have the same impact in that market since many of the electronic mail services did not interconnect and not everyone had access to electronic mail.

Electronic mail still was limited to text-based information. Such other sorts of information as digitized graphics and computer programs could be sent via electronic mail, but required the recipient to have the technology to handle these alternative forms of information. Figure 3 represents the typical configuration of an electronic mail system at this time.

More recently, other significant trends have influenced electronic mail's acceptance. Most notable has been the interconnectivity of communications networks, which provided for unrestricted electronic mail transmission and trans-

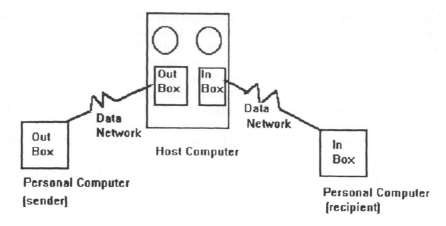

FIG. 3   Electronic mail system configuration in the 1980s.

fer. Along with this concept has been the consolidation of business- and consumer-oriented electronic mail systems. Like the traditional postal mail or the telephone, the communications networks have become impartial carriers of messages.

Acceptance of local-area network (LAN) technology also has spurred electronic mail use. It also has continued to decentralize many of the administrative electronic mail functions. Now, for instance, it is possible to think of LAN servers as local post office distribution centers to determine whether a particular electronic mail message remains within the LAN environment or if it needs to be shipped outside the environment.

Another significant trend affecting electronic mail has been the changes in electronic mail pricing, particularly within the consumer marketplace. Previously, the recipient had to pay to read an electronic mail message. More recently, many electronic mail services have adapted the traditional postal mail model and now are charging the sender. In addition, the services are charging based on the length of the message as opposed to the length of time it takes to read the message. This approach roughly approximates the cost by weight used by traditional postal mail. Such pricing changes have made it much easier for both businesses and consumers to compare costs to other forms of communications. Another benefit of this approach has been the increased motivation by the recipient to check his or her mailbox more frequently. This has increased the value of electronic mail and has eliminated the situation of charging recipients to check a mailbox only to find that there is no mail.

A recent trend has been electronic mail's ability to accommodate additional forms of information. Text no longer needs to be represented in character form. Instead, it is becoming feasible to represent multimedia information as a mail message. Figure 4 represents the more-sophisticated electronic mail configuration at this time. Currently, electronic mail systems typically are categorized into one of three configurations: mainframe large systems, integrated office systems, and LANs. Functions and capabilities are fairly similar, but the audiences and markets are different.

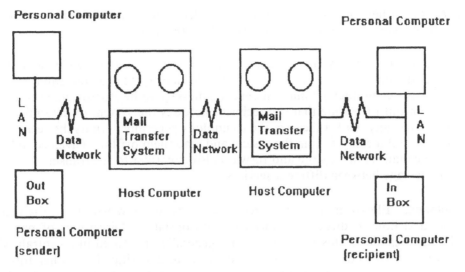

FIG. 4   Current electronic mail system configuration.

Electronic Mail Technology

Electronic mail is not a technology in and of itself. Instead, it is a tight interrelationship among hardware, software, and communications networks dedicated to an end result — communications. The following general technologies comprise electronic mail.

Workstations.   Workstations permit electronic mail users to access the system. In the past, they have been dumb terminals that connect to a host computer; more recently, intelligent terminals permit a few functions to be decentralized and handled at the user's location. Today, most workstations are personal computers running specialized electronic mail software. They permit users to create and receive mail messages and offer easy-to-use user-interface software.

Mailboxes.   Virtually all electronic mail systems maintain a notion of a place to receive a mail message. The mailbox may be local or, as in the past, on a host computer. Current electronic mail systems also may offer the notion of in boxes and out boxes as temporary places to store inbound and outbound mail.

Editors.   Editors permit message creation. As with mailboxes, this function may be local or on a host computer. Many times users will access an editor that is not part of the electronic mail application. Some electronic mail systems permit message creation as an "off-line" function that can be done without being connected to the data communications network. Editors may link with or incorporate other software to provide the sender with such additional functions as spell checking, a thesaurus, or a personal address book.

Store and Forward. A basic tenet of electronic mail is that sender and the recipient do not have to be on the electronic mail system at the same time. Instead, the technology permits the sender's message to be stored somewhere in the system until the time that the recipient accesses the system to determine if he or she has any mail. Likewise, the technology generally permits the recipient to collect mail messages from the system and read them off line. Given complex electronic mail services, store and forward also may refer to the transmission of a mail message between different services.

Directories. Electronic mail depends on its users to know who is on the mail system and how to direct mail messages appropriately. Directories of names and addresses aid in this role. Directories generally are stored in a centralized location in the electronic mail system to facilitate updating. Like other forms of communications, security may be a concern, so users are not required to be included in the directories. As discussed above, local electronic mail software may offer personal directory capabilities.

Distribution Lists. A significant advantage of electronic mail is its ability to distribute the same message to multiple individuals. Electronic mail systems tend to make the capability of creating, storing, and maintaining lists of names and addresses a standard feature. Given that each user wants to manage his or her own set of distribution lists, this capability generally is localized.

Message Headers. Electronic mail systems have developed a special header that will inform the recipient about the contents, the sender, and other summary information about the mail message. This header information is not always standard across different electronic mail systems.

Security. By its very nature, electronic mail has inherent security measures. However, most systems provide, at a minimum, for a password for a user to access his or her mailbox. Some systems permit the scrambling of messages. However, a feature of electronic mail is its accessibility. Recipients do not have to access their mailboxes from the same physical location. This inherent feature makes electronic mail less secure than other communications media.

Connectivity to Other Systems. The ultimate success of electronic mail is due to the ability of a sender to transmit a mail message to someone else regardless of the data communications network or electronic mail system the recipient is using. However, unlike the telephone and traditional postal mail, there is no single operating entity to oversee growth and development. Electronic mail systems have had to develop these connections on their own.

Communications Network. The communications network interlinks virtually all the other electronic mail components. Toward that end, the communications network becomes the electronic data highway for the mail messages. This highway may encompass a large geographic area (e.g., a wide-area network, WAN) or a smaller, well-defined area such as an office (e.g., a LAN). The network may be public, semiprivate, or private. To permit true interconnectivity between disparate network architectures, network companies have generally incorporated or adapted the International Organization for Standardization (ISO) Open System Interconnection (OSI) Reference Model. OSI consists of a seven-layer hierarchy that defines the electrical characteristics, communications standards, and software applications for computer systems. The reader may refer to Ref. 1 for a discussion of the OSI Reference Model.

Messaging Architecture. Not only do the networks need to standardize their process of handling mail messages, but the mail messages themselves require standard elements if they are to be read successfully by recipients on disparate electronic mail services. Over the years, there have been a number of standards developed, including MHS (Message Handling System), TCP/IP (Transmission Control Protocol/Internet Protocol), and X.400.

MHS, originally developed by Action Technology, has obtained near-standard status in the LAN-based messaging market. Consequently, most groupware and virtually all LAN-based electronic mail system products support MHS. MHS is designed around a hub architecture and is intended for remote access to the data communications link. MHS provides standard X.400 gateway, store-and-forward architecture, application program interfaces (APIs), standardized addressing conventions, and bridges to other messaging protocols.

TCP/IP refers to the protocol currently used on the Internet, a data communications network used by the military, research institutions, and the academic community. It is popular because it is a standard component of the UNIX® operating system and is low in cost relative to X.400.

X.400 was developed by the International Telegraph and Telephone Consultative Committee (CCITT) (2) and was specifically created to describe a standard to ensure global compatibility for electronic-message-oriented information transmission and exchange. X.400 assumes ISO compatibility between the two systems. Most systems have been successful in implementing interconnectivity with X.400. See Fig. 5 for X.400 address structure.

Electronic Mail Features

Electronic mail offers many features, few of which are standard across all systems. One of the reasons for the lack of standardization arises from the wide variety of services described as electronic mail.

Message Creation and Delivery

Prompting. Prompting refers to the ability to guide a user through a specific electronic mail task.

```
C = CC;A = AD;P = PD;S = FN;G = GN;I = MI;D = ID:DA

    where    CC = the Country Code
             AD = the Administrative Domain
             PD = Private Domain
             FN = Family Name
             GN = Given Name
             MI = Middle Initial
             DA = Domain Defined Attribute

CompuServe connection to ATTMail example

> X400:(C = US;A = ATTMAIL;S = SCALI;G = AL;D = ID:ASCALI)
```

FIG. 5   X.400 address structure.

Editor.   As discussed above, the editor refers to the ability of the sender to create or modify the mail message before it is sent. An editor also may be used by a recipient to capture all or part of a mail message for retransmission (such as a forward).

Append.   Many electronic mail packages support the ability to append already created documents or graphic files to a sender's message.

Help.   Since most electronic mail systems employ smart terminals or personal computers, assistance with electronic mail use is localized.

Carbon Copy.   Carbon copy allows users to send the same message to more than one recipient. Blind carbon copy permits the same capability but does not display the distribution list of all the recipients.

Message Length.   Most electronic mail message systems have restrictions on the size of a mail message. These restrictions usually are described in terms of number of lines or number of characters.

Distribution.   The distribution feature permits the creation, maintenance, and use of groups of recipients, including individuals not on the same system.

Scan Mailbox.   The scan mailbox feature permits the recipient to obtain summary information about pending messages, including the number of messages.

Acknowledge.   The acknowledge feature will notify the sender that the recipient has retrieved the message from his or her mailbox.

Status. The status feature, invoked by the sender, informs the sender whether or not the recipient has retrieved the message from his or her mailbox.

Priority. Electronic mail messages generally are ordered by time of receipt. This feature permits the sender to change the order and to set up priorities for a message.

Delay. The delay feature allows the sender to create a message and delay its delivery until a specified time and date.

Reply. The reply feature allows the sender more control over what the recipient does after reading a message. In this case, the recipient will be put into a create message mode.

Message Processing

Query. The query feature permits the recipient to query selected attributes of the message header such as the sender or the message title.

Storage. Like other intensively based text mechanisms, electronic mail systems have a limitation on the amount of message storage space. To handle this, limitations are placed on the size of a mail message (discussed above) or the length of retention of the storage. Some electronic mail systems permit the user to adjust the retention period for previously read stories, although most systems have a fixed time for unread mail.

Directory. The directory feature permits both senders and recipients to maintain a personal address book of names and addresses. As the technology has advanced, it now is possible for the recipient to use a single keystroke to add a name or address to a personal directory.

Tickler. The tickler feature permits the recipient to manage key activities via reminders that are triggered by certain conditions (e.g., date). Tickler files are now popular on both a centralized and local basis.

Filer. Both the sender and/or the recipient may want to keep a running history of mail messages on a certain basis (e.g., topic or sender). If the filer is maintained locally, it can become a centralized cabinet for information from such other sources as information services or spreadsheet analysis.

Given the fundamental acceptance of the application, electronic mail is often an ancillary part of a larger applications package such as groupware or information services. A product like Lotus Notes® is a cross between electronic bulletin board and electronic mail technology. In the Dialog Information Service, Dial-mail® becomes not only a way to discuss searching the information service, but also a way to order and monitor the status of article reprints. Electronic mail also has begun to appear in applications software, especially utility software, such as Central Point's PC Tools™.

Electronic mail often is linked closely with such other pure computer-based communications technologies as electronic bulletin boards or computer conferencing. This approach especially is true with consumer information services in which it is possible, for instance, to have unread electronic bulletin board messages "mailed" to the recipient's mailbox. An advantage of linking the computer-based communications technologies together is to permit the user to determine the best alternative for his or her information needs. Other information-based services also may be linked to electronic mail, thereby permitting a user to send a news story via electronic mail, for example.

Benefits and Drawbacks of Electronic Mail

When electronic mail is compared with such other asynchronous communications services as the traditional postal service, courier services, or intracompany mail, many advantages and disadvantages are evident. The most important advantage is the faster delivery of information that none of the other asynchronous communications alternatives can provide. The time it takes to complete the delivery of an item is nearly instantaneous regardless of the distance. A user of electronic mail, then, is able to focus more on the "when to deliver," not the "how to deliver." Electronic mail has a sense of urgency that makes it an extremely effective means of communication.

A second overall advantage is electronic mail's flexibility in distributing mail messages either to one recipient (i.e., narrowcasting) or to multiple recipients (i.e., broadcasting) with relatively equal amounts of ease. Many communications require a combination of both.

Electronic mail is beginning to compare favorably in cost with other communications media. Electronic mail providers are adapting their prices to make for easier comparisons. The cost of a message is constant regardless of distance. Electronic mail also has an extensive audit structure that makes it easier to know where a mail message is, and, in rare instances, why it did not get to the recipient. Finally, since electronic mail is in computer-readable form, there is more flexibility with a mail message. For instance, the sender and the recipient generally have the option of retaining the information in electronic format or having hardcopy produced. Overall, electronic mail has an extremely good cost performance ratio.

Within businesses, many experts observe that electronic mail will be a means by which office communications systems will be integrated. Electronic mail combines many of the characteristics of the telephone and conventional mail and integrates them into a single communications service. Researchers have shown that electronic mail improves productivity through its ability to capture, update, and modify information. Moreover, electronic mail tends to be handled less often. If the electronic mail service offers filing capabilities, individuals have a complete service that delivers information and permits them to digest the information at will and to handle the processed information in virtually any way they desire.

Electronic mail's most significant disadvantage is that not everyone has an electronic mailbox. This severely limits the acceptance of electronic mail as a viable alternative to traditional postal mail. Electronic mail systems are still too disparate and lack interconnectivity. It is therefore quite possible for a sender to be unable to deliver a mail message to a recipient successfully solely because the electronic mail systems do not interconnect. Another connection disadvantage is the lack of a single directory information service along the lines of the telephone directory service. A final connection disadvantage is that, with the proliferation of electronic mail systems, a recipient may have more than one mailbox. A sender, then, has to make a choice where to direct a mail message.

The recipient's mail address generally is very difficult to understand. Unlike traditional postal mail, electronic mail addresses do not require identification of a specific geographic location. Instead, the mail address generally reflects necessary system routing information, especially if the sender is on an electronic mail system different from the recipient's. Moreover, there is no standard mail address. See Fig. 6 for examples of some mail addresses.

Although actual costs to send or read electronic mail are reasonable, the technology still requires a relatively costly equipment investment, particularly for the consumer. In fact, the business market often cannot justify the one-time costs of electronic mail for all its employees despite the multifaceted capability of the equipment.

Electronic mail remains limited to text information or files of graphic or image-based material. Most systems cannot handle such standard business features as letterhead and a signature block. This limitation gives electronic mail a lower level of formality than, say, traditional postal mail in the business market. However, many businesses are accepting some of these limitations and honoring the authenticity of mail messages based on the sender information.

The text limitations for the consumer market generally are seen as a disadvantage and have limited it from being used as an advertising medium. Advertisers also tend to be limited from taking advantage of electronic mail since all systems are considered privately owned.

Electronic mail generally is secure. However, as interconnectivity increases, a mail message may traverse public as well as private data communications networks. Electronic mail systems do not enjoy the same legal protection as other communications highways like the telephone and traditional postal mail. To combat the lack of security, certain electronic mail systems provide such additional security measures as message encoding and additional password levels.

Managing mail messages is becoming an increasing problem. Senders are taking advantage of electronic mail's capabilities and increasing the amount and types of mail messages sent. This not only overwhelms recipients, but also the electronic mail system. More localization of searching, reading, and filing mail messages is facilitating the management process somewhat.

Since there is no single provider, electronic mail suffers from myriad policy

---

Mail Addresses Within Their Own Systems

Internet	GARCON.CSO.UIUC.EDU
CompuServe Mail	70007,4166
BitNet	krol@uiucvmd.bitnet
Fidonet	willie.martin 1:5/2.3
Sprintmail	John Bigboote/Yoyodyne

Mail Addresses From Internet to Other Mail Systems

CompuServe Mail	70007.4166@compuserve.com
BitNet	krol%uiucvmd@cunyvm.cuny.edu
Fidonet	wille.martin@p3.f2.n5.z1.fidonet.org
Sprintmail	/PN=John.Bigboote/O=Yoyodyne /ADMD=Telemail/C=US/@sprint.com

---

FIG. 6   Sample electronic mail addresses.

and regulation issues. For instance, will the United States benefit from a single electronic mail service similar to the United States Postal Service? Will competitive private services provide the best overall product for both businesses and the consumer? How will interconnectivity be achieved? In addition, how will electronic mail be regulated in its current form? Who will be responsible for monitoring the services to ensure that no one's privacy is violated? With what sort of regulations will electronic mail users be confronted when transmitting internationally? Who will be responsible for guaranteeing equal access to electronic mail within a country? For the most part, these issues will shape the form and success of electronic mail.

Many of the limitations cited above, such as addressing and the absence of a centralized name and address directory, limit electronic mail's ease of use. Decentralizing many of electronic mail's functions into a personal computer has helped electronic mail's image, but it will require better front-end interfaces, more powerful software, and adherence to various standards to begin to bring electronic mail more in line with the telephone or traditional postal mail.

Evaluating Electronic Mail Systems

For the business market, the evaluation of electronic mail systems takes on a number of options. First and foremost is the option to determine who needs to be part of the electronic mail service. Previously, the decision was based primarily on whether or not the service would be private or public. Today, most businesses acknowledge the necessity of quasi-public access and, therefore, need to determine what other systems clients and suppliers are using. Given the increasing acceptance of interconnectivity of disparate systems, a business no longer has to focus on a specific system.

Because systems are quasi-public, businesses are focusing on the overall security of the electronic mail system. Most important are the overall controls to other electronic mail services, message encryption capabilities, and the security of the mailboxes themselves. The directory services provided by electronic mail services also will be an important consideration.

Businesses are focusing more on ease of use for electronic mail, especially if the technology will be used by many different groups within an organization. The technology may be used by individuals with different levels of experience, so flexibility in the interface may be an important consideration. Both on-line support, in terms of context-sensitive help, and such general support as telephone assistance are necessities. Businesses need to evaluate what sort of capabilities have been localized on the user's personal computer. In general, the more that is localized, the easier the electronic mail system is to use. Businesses also need to determine how well the electronic mail system integrates with other application software and such other technologies as voice messaging systems, bulletin board systems, groupware applications, and computer conferencing.

Consumers are interested in evaluating electronic mail services along the same lines. However, consumers tend to emphasize the costs, capabilities, and ease of use of an electronic mail service above interconnectivity and security. In addition, consumers generally have interest in electronic mail applications available through such large information services as CompuServe and Prodigy. As a result, consumers do not focus solely on the electronic portion of the information service, but evaluate it along with the other applications and information sources available.

The Future of Electronic Mail

The future of electronic mail is unclear. Many experts expect electronic mail services to grow at significant rates to become a $4.9 billion industry by 1995. Large companies are expected to pump over 14 billion messages annually through data communications networks by the same year. The number of active electronic mail users is projected to increase at a 25% to 30% rate annually from the over 10 million users in 1991. Despite these predictions, electronic mail has not grown at the rate many experts originally had believed. Both businesses and consumers have embraced such other forms of electronic transmission as facsimile at the probable expense of the electronic mail market. Connectivity remains a major issue to be dealt with. The groundwork is there with various standards (X.400, X.500, among others). However, as discussed above, electronic mail systems have been reluctant to adopt the standards completely, but rather adapt portions of them to their own. Moreover, most major systems still maintain their own standards.

Bandwidth and system capabilities have limited electronic mail to either binary files of information or crude text files of information. Thanks to more sophisticated operating systems (e.g., Windows™), compound documents and multimedia presentations will be possible and, soon thereafter, electronic mail programs can be adapted to handle their transmission more easily. In the near future, electronic mail may compete with the telephone as voice technology integrates with electronic mail. Bandwidth is ever increasing with 9600-baud transmission speed becoming the accepted form today. Faster speeds will continue to appear as technology like ISDN (Integrated Services Digital Network) proliferates.

The integration of electronic mail into general applications packages will continue to further its acceptance. Many utilities packages offer rudimentary electronic mail capabilities. It is expected that such software applications as groupware, operating systems, and even such specific applications software as word processing will begin to offer more robust capabilities. As part of electronic mail's maturation, it is conceivable that applications will embrace electronic mail in a more active role. That is, the recipient will be notified as mail messages are delivered into his or her mailbox.

Continued innovation in pricing electronic mail also will influence its acceptance. Many experts thought that an early limitation to electronic mail was charging for access to one's mailbox. That limitation virtually has been eliminated today with most electronic mail systems not charging for the reading of mail originating within their service. Additional refinement to the model to make electronic mail more competitive with other forms of communication is needed, including a simple way of charging the sender and a consistent way to charge for delivery on different mail systems.

The future of microcomputer hardware could have a significant impact on the viability of electronic mail. The days of general-purpose computers have reached their zenith and current research and development are focused on such specialized machines as personal digital assistants (PDAs). This technology, coupled with cellular connectivity, essentially will guarantee that mail could be delivered to anyone, anywhere, anytime.

Acknowledgments: Telenet is a registered trademark of Telenet Communications Corporation. CompuServe is a registered trademark of CompuServe Inc. GEnie is a registered trademark of General Electric. Prodigy is a registered trademark of Prodigy Services Corporation. UNIX is a registered trademark of AT&T Bell Telephone Laboratories. Lotus Notes is a registered trademark of Lotus Development Corporation. Dialog and Dialmail are registered trademarks of Dialog Information Services, Inc. PC Tools is a trademark of Central Point Software, Inc. Windows is a trademark of Microsoft Corporation.

Bibliography

Albright, R. G., *Electronic Communications for Home and Office*, Chilton Book Co., Radnor, PA, 1989.
Connell, S., and Galbraith, I. A., *Electronic Mail: A Revolution in Business Communications*, Macmillan, New York, 1986.

Cunningham, I., Electronic Mail Standards to Get Rubber-Stamped and Go Worldwide, *Data Commun.*, 13(5):159–168 (May 1984).

Davis, S., *The Electronic Mailbox*, S. Davis Publishing, Dallas, TX, 1986.

Derfler, F. J., Jr., Building Workgroup Solutions: Voice E-Mail, *PC Magazine*, 9(13): 311–320 (July 1990).

Donnston, D., E-Mail Finds a Solid Niche in Business World; Time-Saving Organizational Tool Improves Communications, Other Corporate Functions, *PC Week*, 9(4): 123 (January 27, 1992).

Dyson, E., Mail in Modules, *Release 1.0*, 90(11):8–11 (November 30, 1990).

E-Mail Examined, *PC Sources*, 2(8):439–445 (August 1991).

Frey, D., and Adams, R., *Guide to Electronic Mail Addressing and Networks*, O'Reilly and Associates, Sebastopol, CA, 1992.

House of Representatives, *Electronic Message Service Systems: Hearings before the Subcommittee on Postal Personnel and Modernization of the Committee on Post Office and Civil Service*, 96th Congress, 2d session, U.S. Government Printing Office, Washington, DC, 1980.

Krol, E., *The Whole Internet User's Guide and Catalog*, O'Reilly and Associates, Sebastopol, CA, 1992.

Manes, S., *The Complete MCI Mail Handbook*, Bantam Books, New York, 1988.

Mayer, I., *The Electronic Mailbox*, Hayden Book Co., Hasbrouck Heights, NJ, 1985.

Miles, J. B., E-Mail Gateways, *Government Computer News*, 11(6):57–60 (March 16, 1992).

Nothhaft, H., Making a Case Using Electronic Mail, *Data Commun.*, 11(5):85–93 (May 1982).

Office of Technology Assessment, *Implications of Electronic Mail and Message Systems for the U.S. Postal Service,* Report No. OTA-CIT-183, U.S. Government Printing Office, Washington, DC, 1982.

Petrosky, M., How to Make the Electronic Mail Connection, *Infoworld*, 7(45):27–28 (November 11, 1985).

Potter, R. J., Electronic Mail, *Science*, 195(4283):1160–1164 (March 1977).

Radicati, S., *Electronic Mail: An Introduction to the X.400 Message Handling Standards*, McGraw-Hill, New York, 1992.

Raizman, M. B., *Networking: An Electronic Mail Handbook*, Scott, Foresman, Glenview, IL, 1986.

Rapaport, M., *Computer Mediated Communications: Bulletin Boards, Computer Conferencing, Electronic Mail, and Information Retrieval*, Wiley, New York, 1991.

Robinson, P., *Delivering Electronic Mail: Your Guide to Understanding E-Mail Essentials*, M&T Books, Redwood City, CA, 1992.

Rose, M. T., *The Little Black Book: Mail Bonding with OSI Directory Services*, Prentice-Hall, New York, 1991.

Schiappa, B., Electronic Mail Services that Really Work, *Bus. Comput. Sys.*, 3(11):45–54 (November 1984).

Sproull, L., *Connections: New Ways of Working in the Networked Organization*, MIT Press, Cambridge, MA, 1991.

Stone, M. D., Gateways to the World, *PC Magazine*, 10(15):303–320 (September 10, 1991).

Thompson, M. K., M-Mail Adds Voice and Graphics to E-Mail, *PC Magazine*, 11(8): 47–48 (April 28, 1992).

Townsend, C., *Electronic Mail and Beyond: A User's Handbook of Personal Computer Communications*, Wadsworth Electronic Publishing, Belmont, CA, 1984.

Trudell, L., *Options for Electronic Mail*, Macmillan, New York, 1986.

Vallee, J., *Computer Message Systems*, Data Communications, New York, 1984.

Vervest, P., *Electronic Mail and Message Handling*, Quorum Books, Westport, CT, 1985.

References

1. Znati, T. F., Communication Protocols for Computer Networks: Fundamentals. In: *The Froehlich/Kent Encyclopedia of Telecommunications*, Vol. 3 (F. E. Froehlich and A. Kent, eds.), Marcel Dekker, New York, 1992, pp. 323–393.
2. Irmer, T., CCITT. In: *The Froehlich/Kent Encyclopedia of Telecommunications*, Vol. 3 (F. E. Froehlich and A. Kent, eds.), Marcel Dekker, New York, 1992, pp. 281–288.

EBEN LEE KENT

Electronic Switching   (see Digital Switching Systems)

Encryption in Telecommunications
(see Cryptology)

Engineering for Telecommunications

Introduction

Engineering functions are required for any company substantially involved in owning or leasing telecommunications facilities. These functions must be organized and performed in a competent fashion if the company is both to meet the current needs at the lowest overall cost, and also provide for the most likely needs of the future. This article describes these functions and outlines some of the considerations one should bear in mind in organizing them to achieve such quality engineering. Associated with this article are references to some of the publications detailing the mechanics of these functions.

Definition of Terms

Approach

Most of the principles outlined herein apply whether a company is either an end-user, reseller, long-distance or local telephone company, or other user of telecommunications services. To simplify this material for the reader, the article is structured to deal with procedures within a local telephone company that buys most of the facilities and equipment to provide telecommunications services. It is expected that readers from other organizations readily will perceive the application of these principles to their own needs. *Facilities*, as used herein, include leasing telecommunications services, or buying or leasing lines, cables, and equipment as required for a company's telecommunications needs.

Multiplicity of Engineering Functions

Many different functions are associated with engineering for telecommunications. As the assignment of these functions to one or more separate groups is an endless variable of choice, I choose to enumerate some of them yet leave the reader free to organize and group them as he or she chooses. Whenever the engineering organization is small, there is little organizational flexibility and the responsibility for all functions reside with what I call the *project engineer*. On the other hand, large organizations will separate the responsibilities among several groups. For simplicity, it is suggested that the reader begin by conceiving of all of these engineering functions as resting on the shoulders of the project engineer.

The project engineer, as used herein, may be known by many names (e.g., outside plant engineer, equipment engineer, etc.). The reader should realize it is an almost impossible challenge for that one person to perform alone in a superior fashion in all but the smallest telephone environment. As the workload and available force increase, the company will seek an organizational subdivision of some of these engineering functions, either geographically or technically, or both.

The functions covered in this article are identified as engineering functions as they are all critical to the desired engineering results. Some of these functions, when viewed in isolation, could be considered by some readers as nonengineering functions.

Implied Engineering Objectives

Throughout this article is the presumed engineering objective of achieving what is herein called *quality engineering* as a perpetual objective. Quality engineering as defined herein is achieved rarely, if ever.

It is the same goal that is sought after by a building architect: to meet the near- and long-term needs, reflecting current and likely future technology, at the lowest cost, all factors considered. Most architects seek to improve with their next building and so should the engineer of telecommunications constantly seek better solutions. Quality and superior engineering might be considered interchangeable for the purposes of this article.

One is inclined to associate quality engineering with solutions that have higher future value but require additional first cost. This is not true universally. Frequently, engineering knowledge and skills can produce solutions that have the desired increase in future value, but without increasing the first or current costs.

Example. The 25-pair cable episode mentioned in the section on limitations of engineering economic functions (pp. 327–328) is an excellent example of success in seeking to achieve quality engineering solutions. The revised solution achieves the same ultimate growth capacity yet it requires fewer first-cost expenditures by avoiding the metallic partition. Actually, the growth need to require the metallic partition was speculative, so the need to add more repeaters later may not materialize. Even if additional repeaters are required, their cost is less than the partition would have been and their expenditure is delayed for years, if not completely avoided.

Quality Engineering

Competent engineering performed in a superior fashion is necessary to produce a quality engineering output. *Quality engineering*, as used herein, is defined as that which usually produces products superior to engineering solutions that merely meet current needs. These superior solutions are selected to provide the lowest overall costs and are most likely to fulfill future needs as well as current needs.

Quality engineering is simpler to describe than to achieve, nor is it easy to recognize, as much of its advantage is not realized until the future unfolds.

While one may struggle to measure the total value of quality engineering, it is only a simple matter to determine its cost. Though engineering costs may be only a small part of the project's total cost, no increment of cost should be overlooked. However, frequently quality engineering is less expensive than non-quality engineering.

Quality engineering is achieved by applying the best perception of both the immediate and future needs of a company, coupled with a thorough knowledge of the current telecommunications technology, plus the likely future trends in both capabilities and costs. Advocates of the quality engineering approach recognize that even among such solutions there often remains an even better design yet to be discovered. The perfect solution is yet to be achieved. For example, an engineer reflecting on discovering a better solution to a problem, observed, "We were the leaders in the field, but why did we not think of that solution sooner."

The value of quality engineering rests in minimizing total costs, both present and future. To achieve this, the project engineer needs to know all the available technology in such a manner that he or she not only knows what is the best current fit, but has a good idea of the likely direction of future technology developments.

Quality Engineering Solutions: Recognizing Obstacles

Quality engineering must recognize a broad interpretation of the goal of achieving the lowest overall cost, when all factors are considered. Not only must it be recognized that all obvious costs and the time value of money should be evaluated, but also the availability of resources should be evaluated. For example, it is conceivable that a simplistic engineering approach might suggest a sweeping solution to replace completely all of the company's technically obsolete facilities at once, yet, from a practical view, such a solution could be unwise.

The availability of capital, vendor supply, and available employee force often tempers one's decisions. Quality engineering solutions cannot ignore their effect on the company's near-term earnings simply because the project displays a bright future.

Forecasts and Basic Decision Inputs

Company Needs

The term *company needs* as used herein is intended to have a rather broad meaning. Of course, it must include the customers' end service needs as well as marketing plans and strategy of the company. It also includes such internal needs associated with these end products as the needs of the installer, the maintainer, and the operator of these facilities. In addition, it includes all items necessary to cause the services or products to function properly.

Growth and Other Forecasts

The project engineer, when working alone, must utilize a wide variety of forecasts required for his or her work. These forecasts involve not only an estimate of quantity, but also the type of needs.

In larger organizations, various specialists can provide the project engineer with some of these forecasts. These specialists' forecasts include those for marketing, transmission engineering, power engineering, and many others, depending on the type of service facilities in question. Many of these forecasts are for company totals, as well as forecasts subdivided to indicate local needs. Both individually and collectively, they must provide an estimate of needs at various time points in the future, depending on the requirements of the project engineer.

These forecasts are used for a variety of engineering purposes, including the basic data for engineering economic cost studies required to determine plans for courses of action. They also are used to generate cost estimates of future projects for the purpose of determining the prospective company budget of expenditures for the current year and the years ahead. They are used to aid in determining all the quantities in each project and as a surveillance tool in updating the exhaust date of embedded items of a telephone plant as actual growth is realized.

In their broadest form, these forecasts are crucial in determining the total corporate spending for construction, while at the other end of the scale they may aid in determining the size and schedule for providing augmenting cable in a local residential neighborhood.

Offsets to Forecast Accuracies

All forecasts contain an element of inaccuracy. A forecast of the number of customers to be served by a central-office dial machine is more accurate than one for a small residential neighborhood served by that machine. Similarly, a near-term forecast is more accurate than one for the distant future. The engineering reaction time varies, from a few days to almost three years, by the type of forecasted item being provided. For a central office, up to three years might be required to locate and procure land; design and construct a building; engineer, order, install and cut over the cables and equipment. At the opposite end of the time scale might be a small residential builder who could cause an unexpected need for local telecommunications facilities. A short aerial cable extension to serve the development might be within the engineer's authority and construction might be accomplished in a few days using material in field stock.

These reaction time differences between those of project engineers who provide central-office equipment and those of the providers of cables to the customer's premises constituted a smaller problem in the electromechanical dial environment, as forecasting variations in one part of the area served by a central office probably would be offset by other variations. The offsetting variations could be absorbed by the switching machine, which had terminations alike for all customer services.

In the electronic era, these variations may not offset each other.

Example. The customer termination types on the electronic central-office machines are more varied based on the technology applied to serve individual neighborhoods. Some neighborhoods still are served by copper pairs and are terminated on the machine in the electronic equivalent of a line relay. Other customers (e.g., PBX machines) may enter the central office via a 1½-megabit digital line, and still others might be from a remote digital multiplex integrated into the machine.

As the growth needs for each type of customer electronic termination are precarious to forecast separately, the conventional solution would call for inflating the provision of each type of machine termination to offset this variation, thus adding to the cost and idle investment. This is necessary to offset the typically longer reaction time required to engineer, manufacture, and install central-office equipment than that required for most customer cable facilities.

Quality engineering requires effective communication and understanding between the project engineer of facilities to the customer and the central-office machine project engineer.

Example. The machine engineer could alter his or her procedure and, on regular growth jobs, provide "universal sockets" on the customer's line termination frames, and quickly order the circuit boards as the specific need develops.

This technical option applied to offset forecast imperfections also can be applied to transmission apparatus. Traditionally, a forecast of each of the various transmission component unit types necessary for specific customer services was estimated separately for each central office. The result was that errors in forecasts led to quantities of unused and expensive transmission apparatus for one specific type of service left idle in a central office, while, at the same time, new and different apparatus was being engineered and ordered on an expedited basis to meet unforecasted service needs. The current example of quality engineering follows.

Example. The provision of inexpensive universal transmission apparatus sockets in each central office in quantity is an example. Specific service needs are met from a central stock of specialized circuitboards provided on short intervals.

First, there are occasions wherein the customer's needs are very unclear beyond the near term. Second, there are occasions whereby the technology involved is about to make significant advances. Such conditions suggest solutions that focus on the near term, giving much decision weight to the near term costs, and so on. Utilization of embedded equipment might be an ideal solution. Subsequently, as the needs become more clear and the technology advances, longer-term solutions can be applied.

Specifications, Practices, and Procedures

Project Specification Package

The culmination of the project engineering process is the project specification package. These specifications as issued by the project engineer become the authority for the manufacturer and the installer to proceed and bill the telephone company, usually through the project engineer. The specifications as a total package constitute enough specificity for the manufacturer to make the items, the installer to install, and the maintainer/operator to put them to work. Specifications may be issued as words or drawings on paper or via mechanized computer output. They also provide a continuing permanent record of items installed. This package includes job specifications as well as company standard specifications and applicable vendor/standard specifications.

Job Specifications

For the purpose of this article, job specifications are the total of all specifications required by the project except those referred to as standard. These job specifications are peculiar to, and issued for, an individual project only. They must contain all the job specificity required, but not covered by the standard specifications.

In a one person organization, the project engineer must provide the entire job specification. In larger organizations, the project engineer's specification includes both his or her work and that of others who perform associated engineering functions. Whereas the specifications issued by the project engineer include information for the manufacturer to make, the installer to install, and so on, other functional specifications might be required for the equipment or cable to function.

In a large company, the functional specifications would be performed by other than the project engineer. These cover the hardware and software activity associated with connecting a customer to the equipment, cables, and so on. These types of specifications or orders can be prepared after the equipment order is placed to the manufacturer.

Job Specifications Prepared Outside the Telephone Company (Outline Specifications Option)

In very small telecommunications environments, much of the job specification work can be contracted out to such others as the manufacturer of a switching machine. In addition, many nonmanufacturing engineering firms offer this service. In such cases, the project engineer's job specification lacks detailed specificity and is called an *outline* specification. Complete specificity in these cases is developed by the outside vendor. Job specifications containing full specificity are called *detailed* specifications whether they are prepared by the project engineer or an outside group.

Whereas the outline job specification option may be the only one available to small telephone companies, it also can be useful in larger companies, especially in periods of engineering work overload. The more computer mechanization a company has adopted to support its engineering functions, the less need will be found for the outline specification option.

The outline specification option normally will increase the project's time interval as the two engineering forces involved must work sequentially. In addition, this option passes to the vendor decisions of design and material selections that always may not be in the telephone company's best interest. Of course, the telephone company's outline specifications can contain as much specificity as desired. However, the more specificity in the outline specifications, the less advantage there is in turning to others for the detailed specification. The outline specification option may increase the engineering cost of the project as the vendor's engineering charge may not always be offset by reduced telephone company's internal engineering costs.

The telephone company's preparation of the detail specification provides the project engineer with an excellent learning opportunity, as such work requires a more thorough knowledge of the subject than does preparation of the outline specification option.

Telephone Company Standard Specifications

There is considerable repetitive content among different engineering projects that can be published in advance and become part of the job project specifications by cross-reference. These standard specifications sometimes are called *practices* or *procedures* and are issued by the telephone company, or prepared by others and adopted by the telephone company. There are several types of these standards.

Standard installation specifications are one type. They are used by telephone companies' forces installing the material. Whenever the vendor's installation force performs the work, the vendor's standards apply. However, the telephone company often has the option of electing some variations.

Maintenance/operations standards cover the work performed by operational or maintenance forces, and usually are prepared by them. The project engineer would be irresponsible to provide material not covered by these types of standards or practices.

Engineering standards prescribe the company's universal standards for engineering and may encompass matters as small as uniform drawing sizes or as large as standard floor plans for buildings. The availability of these serves both to guide the project engineer and thereby obtain necessary company uniformity and also to reduce the volume of the project engineer's job specification and effort.

Example. The standards could specify which of the several ringing voltage arrangements available in the industry that the company uses to ring its customer's telephones.

These engineering standards can become voluminous if one seeks absolute completeness. Their volume increases exponentially as specificity is included for every conceivable situation. The optimum content is to cover the majority of situations in which uniformity is necessary or desirable. The preparation and continuous updating of these specifications can be a significant engineering cost.

Specification Comprehension

Many of these standard specifications will be incorporated into the computer program used by the telecommunications engineer, resulting in their automatic inclusion in the ultimate project specification. The project engineer must posses a full understanding of both the technology and the standards involved, whether they are applied manually by the engineer or automatically via their inclusion in the computer software.

The following example depicts a situation in which a "hidden decision" was incorporated into the engineer's computer software without the engineer's knowledge, which could have resulted in the application of a suboptimum solution.

Example.   In one company, arguments continued for years concerning the design value versus cost of providing composite cables (cables with pairs that are not all of the same gauge). Sometime after applying a computer aid to the design of cables between the central office and customer premises, it was observed that all of the solutions specified only one gauge for each cable. The single gauge advocates rejoiced until it was discovered the programmer omitted composite cable solutions from the decision tables included in the computer program.

The project engineer seeking quality results must know all such hidden decisions included in manual standards or the software of any computer aid the engineer is using. The project engineer must be prepared to, and capable of, exercising an override whenever such hidden decisions are significantly suboptimum for a particular application.

Subdivision of Engineering Function

While the telephone company always requires a project engineer, traditionally this engineer has been supported by a variety of organization structures. In different times and situations, different engineering leaders will choose to subdivide the functions in different fashions to suit local conditions and the changing technology, as well as the personal preference of the leadership. As an engineering organization grows beyond the simple one-person organization, there becomes an opportunity to subdivide the effort, based both on technical work content and geographical assignment. This article is not intended to support a model engineering organization, but rather to suggest the principles for consideration whenever one is subdividing the engineering functions.

Organizing for the Engineering Function

When one reviews the plethora of engineering functions associated with the engineering process, it is easy to presume a desire for a large organization. For companies seeking to perform these functions with a small organization of only a few engineers, it usually is necessary to seek outside help. Vendors providing the equipment may provide these engineering functions as services. Independent engineering firms also are available. Small engineering organizations even may look to others to aid in establishing engineering standards.

Obviously, as one's engineering force is enlarged, it can become more self-reliant and there is a tendency to subdivide the effort and assign different engineering functions to different groups or specialists. The advantage of specialists is the realization of additional expertise in a certain area, either technical or geographical. The disadvantage is loss of internal communications. Of course, internal communication is best in a one-person organization. The example of electronic switching machine terminations for customer facilities' terminations, in the section, "Offsets to Forecast Accuracies" (pp. 318–319), is an excellent example.

Another hazard of subdividing the engineering function is for the engineer to assume that others providing him with input are without error or flexibility. Such a passive posture toward the existence of errors or an acceptance of an inability to implement desirable changes on the part of anyone performing one or more of these engineering functions is detrimental to a quality engineering process. It increases the probability that such quality engineering will not be achieved.

Geographically Divided Engineering Functions

The nature of engineering work and records is such that, as an organization grows, choice of geographical subdivision of work may be associated with the engineering of local cable facilities, whether aerial, in underground conduit, directly buried in the ground, or submarine. There are several reasons for this placement decision.

The project engineers associated with this engineering function personally must keep appraised of the local growth needs and their frequent shifts as certain items (i.e., building construction) do or do not occur on schedule. In addition, close observation via field visits not only reveals these growth happenings, they also are essential to the physical design of the cable facilities. Of course, knowledge of local conditions includes knowledge of a town's street paving plans and cable pole permit requirements, knowledge of construction by the electric company and other utilities, and so on. The necessity to maintain an awareness of these needs not only encourages the geographical subdivision of these project engineers as to their work content, but also geographically subdivides their office work location.

Centrally Located Engineering Functions

In contrast, a project engineer concerned with engineering for central offices does not need such detailed knowledge of local happenings and is able to perform much work remotely, relying on building and equipment records. This is especially true once the building, or structures housing the equipment, is designed. The equipment housing may be multistory or single-story buildings, equipment shelters above or below ground (environmental vaults), or even pole-mounted cabinets.

While geographically subdividing the work of the project engineer who handles central-office equipment projects is less urgent, it can become the decision of choice. It should be realized, however, that even if the responsibility of this work is divided geographically, it need not be coupled with the decision to separate physically the central-office project engineer's work location. There remains an advantage to locate individual central-office project engineers centrally.

Example. It allows the company to adjust the organization structure to changing technology, without a change in work location, and facilitates subdividing work from the project engineer to other specialists, yet retains the close communication realized by physical closeness.

Whereas the assignment of the project engineer dealing with cable facility work is subdivided, usually by contiguous geographical boundaries, the central-office project engineer's assignment can be subdivided geographically to non-contiguous areas.

Example. He or she may be assigned to central offices equipped with one type of switching machine. This arrangement is applicable especially to the periods of early introduction of a new switching technology, when specialization may be advantageous.

Whereas one could generate copious arguments for or against locating engineering personnel centrally, it is suggested that a logical approach is to conceive of a small organization, such as that with a single project engineer, and then enlarge the thinking to the realities of the actual organization size, assuming the forces remain co-located. The merits of physically locating various forces grouped close to their work can be persuasive, however. An engineering leader well might choose to leave his or her forces at one location unless the counter-argument is persuasive. Two of the major arguments for centrally locating forces are flexibility and specialization. It is suggested that these centrally located forces be subdivided first by specialty to achieve a higher level of expertise, and second divided by geographical work assignments, but physically centrally located. The third choice, to separate the forces physically and geographically, is viable and should be applied where the merits are conclusive.

Uniformity and economies of scale encourage centrally locating engineering forces. Many functions simply do not lend themselves to the alternatives. Developing company engineering strategy, network switching, or routing plans developing intermachine and intercity trunking quantity needs, and so on, are examples of the need to retain engineering effort centrally.

Engineering Supervision Challenge

As the reader contemplates the various engineering functions associated with project engineering coupled with the complexities of telecommunications technology and the rapid changes in customer needs, he or she will realize that to place the total engineering task with a project engineer and still achieve a quality engineering solution is highly unlikely in a large telephone environment. This produces a strong tendency to separate some of these functions organizationally to achieve a higher level of technical and/or geographical specialization.

Even after all organizational advantages are implemented, the supervision of the project engineer and/or the other specialists must remain responsible and assure that the desired quality of the engineering work product is achieved. It is not simple for anyone to recognize quality engineering, nor for the engineering supervision personnel to know whether or not their subordinates are producing such quality work products. Too often supervision, which is not highly skilled in the needs and the technology involved in the delivery of quality engineering, will tend to measure its subordinate's performance by superficial matters.

Like all supervision, engineering supervision must motivate subordinates and such, but supervisory personnel also must have a special knowledge and understanding of the subordinates' work requirements. Failure to have this could result in large sums of money being misspent by those improperly supervised subordinates doing poor quality work. Those involved in the supervision of engineering functions need not be versed expertly in all of the details of the particular function, but must have a solid understanding of the technology being used and the needs to be satisfied, plus the opportunities for better solutions.

Selective Engineering Functions

General

In large telecommunications organizations, the subdivision and grouping of engineering functions will reveal a seemingly endless list of functions and variety of groupings. However, in one form or another, this list will include:

1. Tactical, strategic, and integrated planning
2. Corporate construction budget management
3. Connecting company relations

4. Major project coordination and cutover management
5. Integrated scheduling of related projects
6. Depreciation engineering
7. Separation of joint intercompany investments
8. Investment studies to support tariffs
9. Transmission and protection engineering
10. Radio and television engineers
11. Building engineers
12. Rights-of-way engineers
13. Joint power company and other company coordination contracts and arrangements
14. Maintenance engineers
15. Power engineers
16. Switching and traffic engineers
17. Routing trunk forecasting and arrangements
18. Forecasting associated with all services
19. Product review, evaluation, and standardization
20. Equipment engineers
21. Outside cable project engineers
22. Aerial, underground, submarine, cable support design specialists
23. Engineering, personnel, training, and procedures specialists
24. Engineering methods and mechanization specialists

Engineering Economics Functions

The economic considerations associated with project engineering decisions often are complex. They can begin with such things as the realization that buying and installing a cable containing a large number of pairs of wires costs significantly less than providing two half-size cables.

Example. Providing the larger cable now may not be the choice solution after adjusting the calculation for the time value of the extra money spent now, considering inflation and other factors, versus placing the smaller cable now and another cable sometime in the future.

Such counterforces as these precipitate engineering studies that help guide project engineering decisions.

Example. This simple cable sizing alternative can be precalculated to produce decision tables with only a few variables; maybe only the forecasted speed of growth needs to be considered.

For other more involved situations, the study might be limited to produce a set of several specific solutions applicable only to the case. Such a case study could investigate alternative facility selections involving one or more locations or the like or even guide a companywide solution for facilities for a new service.

What is sometimes called *engineering economy,* when applied to telecommunications facility decisions, is the discipline of calculating the cost of alternative solutions to a given situation. It strives to accumulate all of the associated costs for both the present and future for each of the alternatives. All costs are brought back to a common point in time using the time value of money. These costs should include all capital, maintenance, taxes, and just about all related elements that can be quantified. They must reflect not only alternative solutions to the current problem, but also must include knowledge of the direction of the technology in the future and how flexible each alternative solution to the current need will be in meeting future needs in an economical and timely fashion.

Limitations

The application of engineering economics is precise for what it does. However, it can produce an unjustified aura of accuracy, although its results may be precise as far as the manipulation of numbers are concerned. Its answers are no more accurate than the data originally entered. These entered numbers may be inaccurate, the technical assumptions suboptimized, or the assumptions about future growth may be wrong; the estimate of current or future needs and the like could be in error. Sometimes items with an effect that cannot be quantified carry significant weight.

Example.   There is negative economic value in providing emergency-power-generating capability in a telephone central office; social responsibility, however, requires it.

The practice of engineering economics is a crucial engineering function, and its superior achievement rests on both the accuracy and completeness of the input and the skill of the person interpreting the output. In larger engineering organizations, the more involved studies can be performed by specialized groups, sometimes called *strategic planning engineers* or a similar term. These groups can perform studies with results that involve more than one project engineer's workload. There is a danger, however, that just as the computer output of these studies can produce a false aura of accuracy, the final output of these study groups also can appear more conclusive than the data would justify.

As the various engineering functions are spread among specialists, one specialist may presume erroneously that another's work should be accepted without question.

Example.   A person performing a facilities' study once concluded that a 25-pair cable to a rural community would be adequate to connect that office to its toll center utilizing 1½-megabit digital bit streams if the cable was equipped with a metallic internal partition. This partition increases the electrical separation between 12 pairs and the remainder of the cable. A second specialist group was asked to find a manufacturer that would produce such a cable. The engineer in the second group had the broader technical knowledge and introduced a solution that avoided the need for the metallic partition. This solution was based on the knowledge that, if the distance between the digital regenerative repeaters was cut in half, the metallic partition would not be required. Furthermore, the engineer determined that normal spacing was adequate until significant growth actually had been realized. This solution was considerably less expensive.

Universal Application Potential

While a particular organizational structure may designate some specialist groups more than other groups to perform calculations (usually via computer) that reflect the time value of money and so on, everyone performing engineering functions should understand the basics of engineering economics and be prepared to use these principles.

Depreciation Engineering Function.   One of the cost elements pertinent to an engineering economy study is the associated depreciation cost. This function requires a determination of such cost. The engineer making a decision to buy an item must have an idea how long it will be useful. Obviously, this determination is based on many factors about the future trends in growth, customer needs, and technology. These life factors will affect both the study decision at hand and also the accounting process of booking the actual depreciation cost.

The telephone plant's life expectancy usually is affected by obsolescence, technology advances, the potential of replacements with lower maintenance costs, and growth needs. Rarely is the plant used to the point it is "worn out." Accommodating growth needs is a dominant factor.

For regulated telephone companies, this depreciation engineering knowledge is paramount in the presentation to the regulating commission of arguments about the future and consequently achieving appropriate depreciation rates for the business. The depreciation engineering function requires a technical vision of the future more than any other function. There must be a knowledge of future trends in the company's needs for its customers and the associated technology expectancy. There must be a vision of what is likely to replace every element of the telephone network, including those already in place and those about to be purchased.

Engineering Project Cost Management Function.   In addition to determining the costs associated with a project for the purposes of making engineering decisions, the engineer often must determine the cost of the project in order to determine if he or she has the final authority for approving the project or must seek higher authority. This delegation of authority normally is issued by the board of directors of the company and, while it varies by companies, usually the highest-cost projects require board approval or, at least, approval by the board's executive committee.

Following this approval, in the usual process the project engineer has the authority to place the job order with the vendor and installer. The engineer remains accountable to the approving authority to see that the project is executed as designed and to account to the authority for variations in costs. In addition, the project engineer must review and approve bills. Furthermore, to assure the integrity of the company records and investments, the project engineer must see that the expenditures for his or her projects are charged to the proper accounts. Failure to do so would destroy the integrity of the corporate depreciation strategy as well as the integrity of the booked investments.

In applying company accounting standards, the project engineer must see that the proper portions of the job expenditures are capitalized or charged to current expenses. The company's capitalized expenditures are depreciated according to several dozen classes and schedules. Accordingly, the project engineer must see that each subpart of the project that is capitalized is charged to the proper capital account. This responsibility applies to every item in the project that is being added or removed.

Maintenance Engineering Function. The design selections and arrangements made by the project engineer must provide hardware and software that not only function properly but also must be maintainable and properly supplemented by efficient test, diagnostic, and surveillance technology and practices. This responsibility includes a continuing obligation to see that the items function in the current time frame and are kept viable in the future. In order to fulfill this task, the engineer must be familiar with both the maintenance arrangements of items being ordered and with the needs of the maintainer. To help with this intelligence gathering, the project engineer requires close relations and good communications with the maintenance organization.

As the engineering organization begins to expand beyond a force of one, a specialized group called *maintenance engineers* often is established to facilitate this communication need between various project engineers and the maintenance community. The maintenance engineering function can assist in verifying that the items ordered are maintainable and supported by the most effective test and analysis technology.

Seldom is any item of telephony perfect and to adjust to such reality this function can assist by offering specialized knowledge of field maintainer's problems and, when appropriate, aid in passing to the vendor an "engineering complaint," which frequently is the procedure used to obtain hardware or software problem corrections from the vendor. Maintenance engineering also assists in processing factory hardware and software updates for application to items already in service to keep them current and viable. This two-way communication aids in economically scheduling replacement of embedded facilities with excessive maintenance costs.

Example. Whenever the project engineer, to accommodate growth, needs to augment an outside cable or to add batteries, he or she can use this maintenance engineering function to ascertain the condition of the embedded items and solve both growth and maintenance needs by replacing the cable or battery with a larger one. This is less costly than meeting the two needs separately.

Product Specifications Function.   Another engineering function that requires repetitive use in order to produce economical advantages is the preselection specification of products. The project engineer, whether specifying aerial cable, pole lines, underground conduit, or central-office equipment, can produce an individual job specification more economically if some of the items have been prespecified. This preselection can be done by the project engineer or someone else beforehand.

Prespecified items usually fall into various classes.

Example.   A telephone pole used for aerial cable can be specified in advance and the project engineer merely makes reference to the particular class of pole required. Prespecification usually is applied to items used in quantity by a local telephone company so that a quantity can be ordered in advance, based on average usage, under contract with the vendor for the best price. Different vendors could be the successful bidder from time to time.

These might be considered "commodity" items. Application of this approach to commodity items usually is limited to those items so interchangeable that the engineering, installation, and maintenance practices are common to all vendors.

At the other end of the preselection spectrum are such complex and unique items as switching machine systems. For this class, prestandardization offers advantages only if there are prospects of buying additional machines. In this case, even after standard specifications are established, the machines must be evaluated for conformance to specifications and many other considerations, including price. The switching machine engineering decision includes far-reaching commitments and costs.

Example.   An order for a particular model of machine requires the provision of spare parts and the training of the engineer, the installer, and maintainer. Furthermore, the design of current switching machines is such that all future additions to that machine must be obtained from that same vendor and the buyer also must look to that same vendor for future enhancements, other feature additions, and software changes and support.

In addition, certain items in other locations often are predetermined to be supplied by this same vendor.

Example.   A remote switching machine associated with a centralized host must be procured from the same manufacturer as the host.

Summary

Engineering for telecommunications facilities requires the performance of certain engineering functions, whether the company is an end-user, reseller of telecommunications, long-distance or local telephone company. When all of these functions within the company are performed by one person, that person can be thought of as a project engineer even though he or she may seek vendor or third-party assistance for some of the engineering functions.

As the engineering force is increased, the subdivision of tasks will vary by location, time, and choice of leadership. An early subdivision of engineering tasks is the separation of project engineers associated with central-office project work from project engineers dealing with cable lines to customer premises. Subsequent subdivisions of the cable project work should be geographical, accompanied by a physical subdivision of the workplace.

There is far less advantage to separate the workplace of the central-office project engineers physically even when their work is subdivided geographically; rather, it is more likely to locate the project engineers with specialists as the workforce is increased. There are significant advantages to retaining these specialists and the central-office project engineer at the same work location. This lessens the internal communication weakness associated with subdividing work and also facilitates reorganization without a change of work location as the work content should not be allowed to lull anyone in the process to assume that the output of others is without error or beyond question and that his or her function should proceed by rote.

The goal of any telecommunications engineering organization, whether it consists of one person or many, is to produce quality engineering results. Quality results are defined as those that produce equipment and facilities at the lowest cost, all factors considered, to meet current needs, and those most likely to facilitate meeting future needs. This can be achieved best when all members of that organization seek that goal and provide optimum interplay among themselves to achieve a synergistic result. This organization not only must have a complete understanding of the current needs and the current technology, but possess the best available comprehension of future needs and likely technical and cost trends.

In time, companies practicing quality engineering will find a greater likelihood that their introduction of new services and so on can be achieved with less delay and cost.

Example. A company that has been deploying digital technology in the customer's copper-loop cable facilities can replace those cables with optical digital cables more readily and at less cost than other companies.

Bibliography

The United States Department of Agriculture Rural Electrification Telephone Division suggests their free *Index of Publications—Telephone Program* via the following address:

Publications and Directives Management Branch, ASD
Room 0180 — South Building
Rural Electrification Administration, USDA
Washington, DC 20250-1500
FAX #: 202-720-1915

Many publications covering both general and specific areas of telecommunications engineering are available from Bellcore, the research and engineering consortium serving the local telephone companies of Ameritech, Bell Atlantic, Bell South, NYNEX, Pacific Bell, Southwestern Bell, and U.S. West. A catalog can be obtained by writing or calling

Bellcore Customer Service
60 New England Avenue
Piscataway, NJ 08854-9933
1-800-521-CORE (2673)
FAX #: 908-336-2559

AT&T suggests their free *Catalog of Publications* by contacting

AT&T
Customer Information Center
2855 North Franklin Road
Indianapolis, IN 46219
1-800-432-6600

C. WILLIAM ANDERSON

Entropy and Information Theory

Introduction

Entropy is a measure of information in a random signal. It was first introduced in communications engineering and information science by Claude E. Shannon in 1948 (1). The underlying motif was inherited from statistical thermodynamics and an idea previously proposed by Hartley as a logarithmic quantity of information provided by signaling (2). Shannon not only defined the entropy of a discrete-time random process $\{X_n\}$ with discrete alphabet A, denoted by $H(X)$, but also showed a physical meaning of the entropy as a true amount of information in the random signal by proving a coding theorem. It states that, for a given random signal $\{X_n\}$ with entropy $H(X)$: (a), there exists a pair of coding and decoding schemes that can transform the signal into a sequence of binary symbols with a fixed rate above $H(X)$ bits per unit time and can reconstruct the original signal only from the once transformed binary sequence; and (b), if a coding scheme transforms the signal at the rate below $H(X)$, then there are no decoding schemes that can reproduce reliably the original signal from the transformed binary sequence. In other words, the theorem means that the data of a random signal with entropy $H(X)$ can be compressed to the rate R very close to $H(X)$ bits per unit time.

Information theory originated from Shannon's elaborate and elegant work (1), which introduced such other basic measures of information as *conditional entropy* and *mutual information*. These two notions are genuine in information theory. When a pair of two jointly random signals $\{X_n, Y_n\}$ with entropy $H(X, Y)$ is given, the conditional entropy for given $\{Y_n\}$ is defined by

$$H(X|Y) = H(X,Y) - H(Y) \tag{1}$$

and the mutual information between $\{X_n\}$ and $\{Y_n\}$ is defined by

$$I(X;Y) = H(X) + H(Y) - H(X,Y). \tag{2}$$

By substituting Eq. (1) into Eq. (2), it is seen that the mutual information also can be described in terms of conditional entropy as

$$I(X;Y) = H(X) - H(X|Y)$$
$$= H(Y) - H(Y|X). \tag{3}$$

Hence, the mutual information can be interpreted as the information quantity contained in the signal when the other signal is known. The mutual information plays an important role in the channel coding theorem describing reliable communication through a noisy channel. When the channel is memoryless and characterized by a conditional probability $P(y|x)$, then a quantity C called *channel capacity* is defined as the maximum value of the mutual information with respect to all probability distributions of input processes to the channel. The channel coding theorem assures the existence of a pair of coding and decoding schemes that realizes with an arbitrarily specified small error probability the transmission of information over the noisy channel, provided the rate of information transmission is fixed below the capacity. If the rate is above the capacity, there are no coding schemes that can achieve reliable transmission of information over that noisy channel.

Entropy

The simplest model of information sources is a memoryless source that is regarded as an independent, identically distributed, stochastic process. Given a finite set $A = \{a_1, a_2, \ldots, a_n\}$ (called *alphabet*) and a probability distribution p over A defined as $P(a_i) = p_i$ for $a_i \in A$, $i = 1, \ldots, n$, the source model denoted by $X = \{A, p\}$ can be regarded as a finite probability space, where the probability of outcome of a letter $a_i \in A$ from the source is given by $p_i = P(a_i)$. A real-valued function defined by $I(a_i) = -\log P(a_i)$ is called the *self-information* of event $a_i \in A$ (outcome of letter a_i). The mean value of the self-information described by

$$H(X) = \sum_{i=1}^{n} - P(a_i) \log P(a_i) = \sum_{i=1}^{n} - p_i \log p_i \qquad (4)$$

is called the *entropy* of the source $X(= \{A, p\})$. This can be interpreted as a measure of the average amount of information obtained upon receiving a single letter from the source, or as a measure of the average a priori uncertainty as to which letter will emanate from the source. The unit of entropy or quantity of information for base-2 logarithms is called a *bit*, while that for base-e logarithms is called a *nat*.

The entropy $H(X)$ defined in Eq. (4) can be regarded as a function of n variables $p = (p_1, \ldots, p_n)$ under the condition $p_i \geq 0$, $i = 1, \ldots, n$, and $\sum_{i=1}^{n} p_i = 1$. The following properties concerning the function

$$H(p_1, \ldots, p_n) = \sum_{i=1}^{n} - p_i \log p_i \qquad (5)$$

follow straightforwardly from this definition.

1. $H(p_1, \ldots, p_n)$ is invariant under any permutation of $(p_1, \ldots, p_n)$.

2. $H(p_1, \ldots, p_n) \geq 0$. The maximum value of $H(p_1, \ldots, p_n)$ is equal to $\log n$, which is attained when $p_1 = p_2 = \ldots = p_n = 1/n$. In other words, the entropy becomes maximum at the uniform probability distribution.

3. For two probability distributions $p = (p_1, \ldots, p_n)$ and $q = (q_1, \ldots, q_m)$, it holds that

$$H(p_1 q_1, \ldots, p_1 q_m, p_2 q_1, \ldots, p_2 q_m, \ldots, p_n q_1, \ldots, p_n q_m)$$
$$= H(p_1, \ldots, p_n) + H(q_1, \ldots, q_m). \tag{6}$$

Let $A = \{a_1, \ldots, a_n\}$ and $B = \{b_1, \ldots, b_m\}$ be two alphabets. Let $p_{ij} = P(a_i, b_j)$ be a joint probability distribution defined on the product AB and denoted by $XY = \{AB, p_{ij}\}$, the joint source model induced by product alphabet AB and joint probability measure p_{ij}. The entropy of the joint source model is defined similarly to Eq. (4) by

$$H(X,Y) = \sum_{i=1}^{n} \sum_{j=1}^{m} - P(a_i, b_j) \log P(a_i, b_j) \left(= \sum_{i=1}^{n} \sum_{j=1}^{m} - p_{ij} \log p_{ij} \right). \tag{7}$$

The joint distribution p_{ij} gives rise to marginal distributions p_i and q_j in the following way:

$$p_i = P(a_i) = \sum_{j=1}^{m} P(a_i, b_j) = \sum_{j=1}^{m} p_{ij}, \tag{8}$$

$$q_j = P(b_j) = \sum_{i=1}^{n} P(a_i, b_j) = \sum_{i=1}^{n} p_{ij} \tag{9}$$

Equations (8) and (9) induce two probability subspaces $X = \{A, p\}$ and $Y = \{B, q\}$, respectively. A real-valued function $I(a_i | b_j) = -\log P(a_i | b_j)(= -\log P(a_i, b_j)/P(b_j) = -\log p_{ij}/q_j)$ is called the conditional self-information of a_i for given b_j. The average of $I(a_i | b_j)$ over $a_i \in A$ defined by

$$H(X | b_j) = \sum_{i=1}^{m} P(a_i) I(a_i | b_j) \tag{10}$$

is called the conditional entropy of X for given b_j. The average of $H(X | b_j)$ over $b_j \in B$ defined by

$$H(X | Y) = \sum_{j=1}^{m} q_j H(X | b_j) \tag{11}$$

is called the conditional entropy of X for given Y. The conditional entropy of Y for given X, denoted by $H(Y | X)$, is defined similarly. Then, it can be shown that

$$H(X,Y) = H(Y) + H(X | Y) = H(X) + H(Y | X), \tag{12}$$

$$H(X,Y) \leq H(X) + H(Y), \tag{13}$$

$$H(X | Y) \leq H(X), \quad H(Y | X) \leq H(Y) \tag{14}$$

where equalities in Properties 2 and 3 hold if, and only if, $p_{ij} = p_i q_j$ for all a_i $\in A$ and $b_j \in B$ as shown in Eq. (6).

Mutual Information

A real-valued function $I(a_i; b_j) = \log P(a_i, b_j)/P(a_i)P(b_j)$ $(= \log p_{ij}/p_i q_j)$ is called the *mutual self-information*. The average of $I(a_i; b_j)$ over $a_i \in A$ and $b_j \in B$ defined by

$$I(X;Y) = \sum_{i=1}^{n} \sum_{j=1}^{m} p_{ij} I(a_i; b_j) \tag{15}$$

is called the *mutual information* between X and Y. It is easy to note that if X and Y are independent of each other, that is, $p_{ij} = p_i q_j$ for all $a_i \in A$ and $b_j \in B$, then $I(X;Y) = 0$, if otherwise, then $I(X;Y) > 0$. It is also easy to see that

$$I(X;Y) = H(X) + H(Y) - H(X,Y). \tag{16}$$

Therefore, the mutual information expresses a quantity of mutual dependency between two probability spaces X and Y. Again, note that substitution of Eq. (12) into Eq. (16) yields another expression for the mutual information described by Eq. (3). Hence, the mutual information expresses a measure of information contained in the probability subspace X when the other subspace Y is given or vice versa.

Information Sources

For a given finite alphabet A, consider the product of consecutive A's as $A^N = A \times \ldots \times A$ and the doubly infinite product $A^Z = \prod_{k=-\infty}^{\infty} A_k$, where $A_k = A$, $k = 0, \pm 1, \pm 2, \ldots$. Let F_A be the σ-algebra generated by all cylinder sets in A^Z. An information source is characterized by a probability space $\{\Omega(= A^Z),$ $F_A, P\}$ or a random signal $X = \{X_n\}$ if a probability measure P/F_A is defined. When the probability measure P is invariant under the shift transformation T (shift of time axis by unit time), that is, $P(E) = P(TE)$ for any $E \in F_A$, then $\{\Omega, F_A, P\}$ is said to be a stationary source. In particular, if $P(E) = 0$ or 1 whenever $TE = E \in F_A$, then the information source is said to be ergodic. If X is an independent and identically distributed random signal, then the source X is said to be memoryless.

For a given random process $X = \{\Omega, F_A, P\}$, denote the subsequence $X_1 X_2 \ldots X_N$ of X by X^N. Then, a probability measure P on $A^N = A \times A \times$

. . . $\times A$ and a finite probability space $X^N = \{A^N, P\}$ are induced naturally. The entropy of the stationary information source X is defined as $H(X) = \lim\limits_{N \to \infty} \dot{H}(X_N | X^{N-1})$ or $H(X) = \lim\limits_{N \to \infty} H(X^N)/N$ because both limits exist and are identical. If X is memoryless, the entropy of X is equivalent to that of a single space $X^1 = \{A^1, P\}$, that is, $H(X) = \sum\limits_{i=1}^{n} - p_i \log p_i$, where $p_i = P(X_1 = a_i)$. In this case, $H(X^N) = NH(X^1) = NH(X)$.

Source-Coding Theorem

Let $x_t^N = x_{t+1}, \ldots, x_{t+N}$ be a sample sequence of N consecutive letters from the information source $X = \{\Omega, F_A, P\}$. Consider now a coding scheme that can transform each block x_t^N (for $t = 0, N, 2N, \ldots$) with block length N to a fixed-length sequence $u^L = u_1 \ldots u_L$ consisting of L letters from a code alphabet U of size r. The number $R = (\log r^L)/N$ is called the code rate per source letter. A mapping $\varphi : A^N \to U^L$ is called an encoder and $\psi : U^L \to A^N$ is called a decoder. The set U^L with a pair $[\varphi, \psi]$ of specified encoder φ and decoder ψ is called a block code with rate $R(=L(\log r)/N)$. Then, the error probability of the code is defined by

$$P_e[\varphi, \psi] = P[\psi\{\varphi(x_t^N)\} \neq x_t^N]. \tag{17}$$

The source-coding theorem states that, for a stationary ergodic source $X = \{\Omega, F_A, P\}$ with entropy $H(X)$, if $R \geq H(X) + \delta$ for a positive $\delta > 0$ there exists an encoder and decoder pair $[\varphi, \psi]$ with rate R such that the error probability $P_e[\varphi, \psi]$ can be made arbitrarily small by letting block length N be sufficiently large. Conversely, if $R \leq H(X) - \varepsilon$ for a constant $\varepsilon > 0$, then, for any block code $[\varphi, \psi]$ with rate R, $P_e[\varphi, \psi]$ must be arbitrarily close to 1 as $N \to \infty$.

For memoryless sources, the source-coding theorem mentioned above follows straightforwardly from the weak law of large numbers, which implies that, for arbitrary $\varepsilon > 0$ and $\delta > 0$, there exists an integer $N(\varepsilon, \delta)$ such that for all $N > N(\varepsilon, \delta)$

$$P\left[\left| \frac{-\log P(x_t^N)}{N} - H(X) \right| > \delta \right] < \varepsilon. \tag{18}$$

This property is called the *asymptotically equi-partition property* and was first proved by B. McMillan for stationary ergodic sources (3). The source-coding theorem now is extended to cover a wider class of sources called *asymptotically mean stationary sources* (see Ref. 4).

In case of memoryless sources, the asymptotic form of the error probability for optimal codes with rate R was given by Jelinek (5) and Csiszár and Longo (6) as

$$\lim_{N \to \infty} -\frac{1}{N} \log P_e[\varphi, \psi] = \min_{q:H(q) \ge R} D(q\|p) \qquad (19)$$

where p is the source distribution, q denotes a probability distribution on A, and $D(q\|p)$ is defined as

$$D(q\|p) = \sum_{i=1}^{n} q_i \log \frac{q_i}{p_i} \qquad (20)$$

and called *Kullback–Leibler's discrimination information* or *divergence*.

Source sequences x_1^N can be encoded into variable-length codewords consisting of letters from alphabet U of size r. Such a set of codewords denoted by Ξ contains n^N elements since the encoding is considered to be a one-to-one mapping from A^N to Ξ and is called a *variable-length code*. Further, the code is called a *prefix code* if there is no codeword equivalent to the prefix of any other codeword. A prefix code has a property called *unique decodability* in the sense that any two different sequences of the source can be transformed via the encoding scheme into different sequences of U, therefore the original sequence can be reproduced uniquely from the coded sequence of U. Denote the length of the codeword corresponding to a source sequence $x^N \in A^N$ by $L(x^N)$. The average length of the codewords per source letter denoted by $\overline{L}$ is defined by

$$\overline{L} = \sum_{x^N \in A^N} \frac{1}{N} P(x^N) L(x^N). \qquad (21)$$

The variable-length source-coding theorem states that, given a memoryless source X with entropy $H(X)$, there is a prefix code Ξ such that the average length of the codewords per source letter satisfies

$$H(X) \le \overline{L} \log r \le H(X) + \frac{1}{N} \log r. \qquad (22)$$

Conversely, there is no prefix code that satisfies $\overline{L} \log r < H(X)$. The variable-length source-coding theorem also is valid for stationary ergodic sources provided the last inequality in Eq. (22) is relaxed, that is, $(\log r)/N$ is replaced by some $\varepsilon(N)$ where $\varepsilon(N) \to 0$ as $N \to \infty$.

An optimal coding scheme that minimizes the average codeword length was discovered by Huffman together with a code construction algorithm (7). A class of arithmetic codings originated from Elias' idea, first described in Abramson's book (8), achieves asymptotically the entropy rate when the source sequence becomes long. Construction of Huffman's coding and arithmetic codings is based on the probability distribution of the source (9–11). Adaptive schemes of these codings also are proposed that can be used when the distribution of source letters is changeable. In recent years, a class of universal codings were found by Ziv and Lempel that can encode the source sequence without referring to the source statistics but can attain asymptotically the entropy rate (12,13).

Source Coding with Fidelity Criterion

The source-coding theorem in conjunction with its converse implies that a data sequence from a stationary ergodic source can be compressed to the rate (bits per source letter) very close to the source entropy above, with the requirement that the original source sequence can be reproduced reliably from the encoded sequence. In contrast to this, if an approximate reproduction of the source sequence within a given fidelity criterion is admitted, the rate of data compression must be reduced further to a certain level below the source entropy.

Suppose that a distortion measure $d(a_i, a_j)$ is defined for $a_i, a_j \in A$, where it is assumed that $d(a_i, a_j) \geq 0$ and $d(a_i, a_i) = 0$. Given any two sequences $x^N, y^N \in A^N$ of block length N, the distortion between $x^N = x_1 \ldots x_N \in A^N$ and $y^N = y_1 \ldots y_N \in A^N$ per source letter is measured by

$$d(x^N, y^N) = \frac{1}{N} \sum_{k=1}^{N} d(x_k, y_k). \tag{23}$$

A mapping from $I = \{1, 2, \ldots, M\}$ to A^N is called a *code*, which is equivalent to defining a set of codewords by picking up M blocks of length N from A^N, that is, $\Xi = [y_1^N, \ldots, Y_M^N]$. Given a source sequence x_t^N from the source $X(= \{\Omega, F_A, P\})$, the encoder finds a codeword y_m^N that minimizes $d(x_t^N, y^N)$ over $y^N \in \Xi$, that is,

$$d(x_t^N, y_m^N) = \min_{y^N \in \Xi} d(x_t^N, y^N) \tag{24}$$

and transmits the suffix m as a binary sequence. Then, the coding rate per source letter is defined as $R = (\log M)/N$ and the average distortion is

$$\overline{d}_N(\Xi) = \sum_{x^N \in A^N} P(x^N) \min_{y^N \in \Xi} d(x^N, y^N). \tag{25}$$

The problem is how far it is possible to reduce the rate below the entropy when the average distortion remains within the limit of a given fidelity criterion, which is specified as a maximum tolerable value d for the average distortion (see Ref. 14).

The source-coding theorem with a fidelity criterion states that, for a given memoryless source X and any specified $d \geq 0$, any $\delta > 0$, and any $\varepsilon > 0$, there exists a code Ξ with rate $R \geq R(d) + \delta$ and with sufficiently large block length N for which the average distortion satisfies

$$\overline{d}_N(C) \leq d + \varepsilon, \tag{26}$$

where $R(d)$ is the rate distortion function defined by

$$R(d) = \min_{W \in W(d)} I(p; W), \tag{27}$$

$$I(p;W) = \sum_{i=1}^{n} \sum_{j=1}^{n} p_i W(j|i) \, \log \frac{W(j|i)}{\sum\limits_{i=1}^{n} p_i W(j|i)} , \qquad (28)$$

$$W(d) = \left\{ W: \sum_{i=1}^{n} \sum_{j=1}^{n} p_i W(j|i) d(a_i, a_j) \leq d \right\}, \qquad (29)$$

where $W(j|i)$ denotes a conditional probability distribution referred to as a test channel and $I(p;W)$ is the mutual information of the joint probability distribution $p_{ij} = p_i W(j|i)$. It is important to note that $R(0) = H(X)$ and $R(d) < H(X)$ in general for $d > 0$. The rate distortion function $R(d)$ is a convex and monotonously decreasing function in d.

In case of source codings with fidelity criterion, no general code construction schemes have been found so far. Hence, various heuristic approaches based on physical properties of the source signal are proposed for data compression of signals and images. The source-coding theorem with a fidelity criterion is also valid for a class of analogue sources, even though in general $R(d)$ tends to infinity as $d \to 0$ and therefore the entropy rate loses its meaning (see Ref. 14).

Channel Coding

In information theory, a channel over which information is transmitted is modeled in terms of the set of possible inputs, the set of outputs, and a probability measure on the output events conditioned on each input. The simplest models are a class of discrete memoryless channels described as follows. The input and output are sequences of letters from finite alphabets A and B, respectively, and the output letter at a given time depends statistically only on the corresponding input letter at that time. Hence, a discrete memoryless channel is characterized mathematically by a fixed conditional probability distribution $P(b_j|a_i)$ for $b_j \in B$ and $a_i \in A$ because the probability measure on the input and output sequences satisfies

$$P(y^N|x^N) = \prod_{k=1}^{N} P(y_k|x_k), \qquad x_k \in A, \qquad b_k \in B. \qquad (30)$$

Given a discrete memoryless channel characterized by $P(b_j|a_i)$, the capacity for the channel is defined by

$$C = \max_{p} I(p;P)$$

$$= \max_{p} \sum_{i=1}^{n} \sum_{j=1}^{m} p_i P(b_j|a_i) \, \log \frac{P(b_j|a_i)}{\sum\limits_{i=1}^{n} p_i P(b_j|a_i)} \qquad (31)$$

where $p = (p_1, \ldots, p_n)$ is a probability distribution on A and the maximum is taken over all such input distributions.

Now, consider a block code that is defined by a mapping $\varphi : I(= \{1, 2, \ldots, M\}) \rightarrow A^N$, which is equivalent to a collection $\Xi = \{x_1^N, x_2^N, \ldots, x_M^N\}$ with elements that are picked up from A^N. Then, the rate of the block code Ξ is defined as $R = (\log M)/N$. A mapping $\psi : B^N \rightarrow I$ is called the decoding. Thus, given an encoding and decoding pair $[\varphi, \psi]$, the error probability is defined as

$$P_e[\varphi, \psi] = \frac{1}{M} \sum_{m=1}^{M} \sum_{y^N : \psi(y^N) \neq m} P(y^N | \varphi(m)) \tag{32}$$

where it is assumed implicitly that each codeword x^N is equally probable.

The fundamental theorem of channel coding that was found by Shannon is described next (1). Given a discrete memoryless channel with capacity $C > 0$, there exists a block code with block length N, rate R below C, and encoding and decoding pair $[\varphi, \psi]$ such that the error probability satisfies

$$P_e[\varphi, \psi] < \exp \{-NE(R)\} \tag{33}$$

where $E(R)$ is a function that satisfies $E(R) > 0$ for $0 \leq R < C$. The discovery of this theorem later had a favorable impact on development of digital communications because the theorem assures the existence of coding schemes that realize a sufficiently reliable transmission of information without slowing down the transmission rate.

Shannon discovered the theorem but gave only a sketch for proving it. The first rigorous proof was devised by Feinstein (15), and the precise expression of error exponent $E(R)$ subsequently was given by Elias (16), Fano (17), and Gallager (18). The best expression of $E(R)$ known is presented by Gallager (19):

$$E(R) = \max \{E_r(R), E_{ex}(R + O(N^{-1}))\} \tag{34}$$

where

$$E_r(R) = \max_{0 \leq \rho \leq 1} \max_p \left[-\rho R - \log \sum_j \left\{ \sum_i p_i P(b_j | a_i)^{1/(1+\rho)} \right\}^{1+\rho} \right] \tag{35}$$

$$E_{ex}(R) = \sup_{\rho \geq 1} \max_p \left[-\rho R - \log \sum_{i,k} p_i p_k \left\{ \sum_j \sqrt{P(b_j | a_i) P(b_j | a_k)} \right\}^{1/\rho} \right] \tag{36}$$

The converse of this channel-coding theorem is that, given a discrete memoryless channel with capacity C, for any block code with block length N, rate R above C, and any pair $[\varphi, \psi]$, the error probability satisfies

$$P_e[\varphi, \psi] > 1 - \exp \{-NE(R)\} \tag{37}$$

where $E(R)$ is a function positive for $R > C$. A converse to the channel-coding theorem in a weak sense was first shown by Feinstein (15). The strong converse in the sense that the error probability tends to unity as $N \rightarrow \infty$ was proved by

Wolfowitz (20). The precise and tightest expression for $E(R)$ in Eq. (37) was given by Arimoto (21), which is described by

$$E(R) = \max_{-1 \le \rho \le 0} \min_{p} \left[-\rho R - \log \sum_j \left\{ \sum_i p_i P(b_j|a_i)^{1/(1+\rho)} \right\}^{1+\rho} \right]. \tag{38}$$

While the entropy implies a critical rate of data compression for the existence of coding schemes that can compress the source data at the rate of the condition of unique decodability, the capacity is a critical rate of information transmission for the existence of coding schemes that can transmit the sequence reliably over the noisy channel at the rate. Unfortunately, there is in general no direct method for computing the capacity as well as for computing the rate-distortion function. However, an iterative method for computing the capacity proposed by Arimoto is effective in practice (22), as well as the one proposed by Blahut for the rate-distortion function (23). The capacity for a continuous channel with additive white Gaussian noise was derived also by Shannon (1) and is given as

$$C = W \log \left(1 + \frac{P}{WN} \right) \qquad \text{bits per second (b/s)} \tag{39}$$

where W [H_z] denotes the frequency bandwidth of the channel, P (watt) the average power of the signal, and N (watt/H_z) the average power per hertz of the white noise with flat power spectrum.

Practical methods of code construction for the channel coding have been developed in both coding theory and information theory. Among systematic error-correcting codes found and studied in algebraic coding theory are the Hamming code, the BCH (Bose–Chaudhuri–Hocquenghem) code, the Reed–Solomon code, the Goppa code, and more (24). Some of them already are implemented in communications systems and also in such memory devices as magnetic tapes and photo disks. The performance on error rates for a class of tree codes, trellis codes, and convolutional codes has been studied in information theory (5,25), together with the analysis of decoding schemes. Viterbi's algorithm is a typical decoding scheme that can decode the convolutional code and trellis code efficiently (26).

Multiterminal Information Theory

In 1973, Slepian and Wolf discovered a source-coding theorem for two mutually correlated sources $\{X_n, Y_n\}$ (27). It states that, for a pair of correlated sources $\{X_n, Y_n\}$ with entropy $H(X, Y)$ and conditional entropies $H(X|Y)$ and $H(Y|X)$, there is a pair of encoding schemes that transforms sequences X and Y into binary sequences of fixed rates R_x and R_y, respectively, and independent of each other and reproduces the original sequences from their encoded sequences with arbitrarily small error probability provided the pair of rates (R_x, R_y) satisfies

$$R_x > H(X|Y), \qquad R_y > H(Y|X), \qquad R_x + R_y > H(X,Y). \qquad (40)$$

It is from this theorem that we gain a deep insight into the notion of conditional entropy $H(X|Y)$ or $H(Y|X)$. The theorem gives rise to a physical meaning of existence of coding schemes that can compress the sequence of X to the rate close to the conditional entropy $H(X|Y)$ without knowing the sequence of Y at the source and can reproduce the original sequence of X at the receiver with use of the reproduced sequence of Y.

Immediately after the work of Slepian and Wolf, Ahlswede proved a channel-coding theorem for memoryless channels with two senders and two receivers (28). These two pioneering works opened a field of multiterminal information theory (29–31).

Summary

Basic notions of entropy and other information measures such as conditional entropy, mutual information, and channel capacity are introduced and explained in this article. Two fundamental theorems of source coding and channel coding are discussed in which such information measures play a vital role. It is concluded that the source entropy is a critical rate of the coding that transforms a source sequence into a binary sequence with unique decodability for data compression, and the channel capacity is a critical rate of the coding that allows reliable transmission of a signal over the noisy channel.

References

1. Shannon, C. E., A Mathematical Theory of Communication, *Bell Sys. Tech. J.*, 27:379–423 (Part 1), 623–656 (Part II) (1948).
2. Hartley, R. V. L., Transmission of Information, *Bell Sys. Tech. J.*, 7:535–563 (1928).
3. McMillan, B., The Basic Theorems of Information Theory, *Ann. Math. Stat.*, 24: 196–219 (1953).
4. Gray, R. M., *Probability, Random Processes, and Ergodic Properties*, Springer-Verlag, New York, 1988.
5. Jelinek, F., *Probabilistic Information Theory*, McGraw-Hill, New York, 1968.
6. Csiszár, I., and Longo, G., On the Error Exponent for Source Coding, *Studia Math. Acad. Hung.*, 6:181–191 (1971).
7. Huffman, D. A., A Method for the Construction of Minimum Redundancy Codes, *Proc. IRE*, 40:1098–1101 (1952).
8. Abramson, A., *Information Theory and Coding*, McGraw-Hill, New York, 1963.
9. Rissanen, J., Generalized Kraft Inequality and Arithmetic Coding, *IBM J. Res. Develop.*, 20:198–203 (1976).
10. Guazzo, M., A General Minimum-Redundancy Source-Coding Theorem, *IEEE Trans. Inform. Theory*, IT-26:15–25 (1986).

11. Jones, C. B., An Efficient Coding System for Long Source Sequence, *IEEE Trans. Inform. Theory*, IT-27:280–291 (1981).
12. Ziv, J., and Lempel, A., A Universal Algorithm for Sequential Data Compression, *IEEE Trans. Inform. Theory*, IT-23:337–343 (1977).
13. Ziv, J., and Lempel, A., Compression of Individual Sequences via Variable-Rate Coding, *IEEE Trans. Inform. Theory*, IT-24:530–536 (1978).
14. Berger, T., *Rate Distortion Theory*, Prentice-Hall, Englewood Cliffs, NJ, 1971.
15. Feinstein, A., A New Basic Theorem of Information Theory, *IRE Trans. Inform. Theory*, PGIT-4:2–22 (1954).
16. Elias, P., Coding for Noisy Channels, *IRE Conv. Record*, Part 4, 37–46 (1955).
17. Fano, R. M., *Transmission of Information*, MIT Press, Cambridge, MA, and Wiley, New York, 1961.
18. Gallager, R. G., A Simple Derivation of the Coding Theorem and Some Applications, *IEEE Trans. Inform. Theory*, IT-11:3–18 (1965).
19. Gallager, R. G., *Information Theory and Reliable Communication*, Wiley, New York, 1968.
20. Wolfowitz, J., A Note on the Strong Converse to the Coding Theorem for the General Discrete Finite Memory Channel, *Information and Control*, 3:89–93 (1963).
21. Arimoto, S., On the Converse to the Coding Theorem for Discrete Memoryless Channels, *IEEE Trans. Inform. Theory*, IT-19:357–359 (1973).
22. Arimoto, S., An Algorithm for Computing the Capacity of Arbitrary Discrete Memoryless Channels, *IEEE Trans. Inform. Theory*, IT-18:14–20 (1972).
23. Blahut, R. E., Computation of Channel Capacity and Rate-Distortion Functions, *IEEE Trans. Inform. Theory*, IT-18:460–473 (1972).
24. Berlekamp, E. R., *Algebraic Coding Theory*, McGraw-Hill, New York, 1968.
25. Viterbi, A. J., and Omura, J. K., *Principles of Digital Communication and Coding*, McGraw-Hill, New York, 1979.
26. Viterbi, A. J., Error Bounds for Convolutional Codes and an Asymptotically Optimum Decoding Algorithm, *IEEE Trans. Inform. Theory*, IT-13:260–269 (1967).
27. Slepian, D., and Wolf, J. K., Noiseless Coding of Correlated Information Source, *IEEE Trans. Inform. Theory*, IT-19:471–480 (1973).
28. Ahlswede, R., The Capacity Region of a Channel with Two Senders and Two Receivers, *Ann. Prob.*, 2:805–814 (1974).
29. Csiszár, I., and Körner, J., *Information Theory*, Akademiai Kiado, Budapest, and Academic Press, New York, 1981.
30. Blahut, R. E., *Principles and Practice of Information Theory*, Addison-Wesley, Reading, MA, 1987.
31. Cover, T. M., and Thomas, J. A., *Elements of Information Theory*, John Wiley and Sons, New York, 1991.

SUGURU ARIMOTO

Equalization, Adaptive (see Adaptive Equalization; Electrical Filters: Fundamentals and System Applications)

Equalization of Telecommunication Channels

Introduction

In such digital communications systems as voiceband data transmission or telephony, digital microwave radio, and digital mobile radio, information symbols are transmitted from the transmitter to the receiver at a regular rate, called the *symbol rate* or the *signaling rate*. Each symbol typically represents a number of binary digits (*bits*) in order to make an efficient use of the available channel bandwidth. Bandwidth-efficient transmission has been made possible by the progress in adaptive channel equalization to compensate for signal distortions caused by the imperfect characteristics of the channel. The equalizer is a signal processing device used to counteract signal distortions and make more reliable decisions on the transmitted data.

A simplified block diagram of digital communications systems over a band-limited linear channel is depicted in Fig. 1. The data symbols, which take their values from a finite discrete alphabet, are transmitted at the symbol rate $1/T$ (where T is the symbol period or duration). The transmit filter shapes the input signal spectrum, and the (linear) modulator shifts this spectrum to center it on the carrier frequency. The channel (transmission medium) is modeled as a linear filter followed by the addition of some noise. After downconversion to baseband by the demodulator, the received signal is passed to the receive filter, which limits the additive noise power and further shapes the signal spectrum. The output signal then is sampled in synchronism with the symbol timing clock.

The overall baseband equivalent channel comprising the transmit filter, modulator, transmission medium, demodulator, and receive filter, followed by the symbol-rate sampler can be described by the discrete input/output relationship

$$x_k(\tau) = \sum_{j=-\infty}^{+\infty} h_j(\tau) a_{k-j} + n_k \tag{1}$$

where
$\quad x_k(\tau) = x(kT + \tau) =$ the kth sample of the sampler output
$\quad\quad a_k =$ the data symbol transmitted at time kT
$\quad\quad n_k =$ the contribution of the additive noise
$\quad\quad\ \tau =$ the sampling instant
$\quad\quad h_j =$ the jth sample of the overall impulse response $h(t)$.

Such a baseband model is derived easily for any linear modulation system, such as quadrature amplitude modulation (QAM) and phase-shift keying (PSK).

Equation (1) clearly shows that each transmitted symbol is corrupted not only by additive noise, but also by interference from past and future symbols.

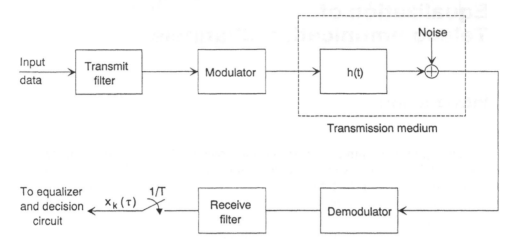

FIG. 1   Simplified block diagram of a digital communications system over a band-limited linear channel.

This phenomenon, known as *intersymbol interference* (ISI), is a major limiting factor of high-speed digital transmission. ISI-free transmission is possible only if the impulse response $h(t)$ satisfies the first Nyquist criterion, that is,

$$h_0(\tau) = 1$$

and

$$h_i(\tau) = 0, \quad i \neq 0$$

for some sampling phase τ (1, p. 532). If this condition is satisfied and the sampler phase is chosen appropriately, Eq. (1) reduces to

$$x_k(\tau) = a_k + n_k. \tag{2}$$

When the channel transfer function is known, the transmit and receive filters can be designed in such a way that the overall channel satisfies the first Nyquist criterion. Unfortunately, the response of the transmission medium usually is not known and is subject to time variations. In these situations, the transmit and receive filters are designed to form a Nyquist filter, but the unknown and possibly time-varying channel destroys this property. To combat the resulting ISI, an adaptive equalizer is required at the receiver.

Equalizer Structures

There are essentially two basic approaches to data reception over nonideal band-limited channels. The first approach is to make maximum-likelihood sequence estimation (MLSE), which provides the most likely data sequence conditional on the received noisy and distorted sequence. The second is to remove the ISI using a linear or a nonlinear equalizer prior to making decisions on a symbol-by-symbol basis. In this section, we describe both types of receivers and discuss their potentials. The adaptation algorithms required to make these devices adaptive are described in the next section.

The MLSE receiver is the optimum receiver for digital transmission over ISI channels. Its practical implementation takes the form of the well-known Viterbi algorithm because the operation of the discrete ISI channel is analogous to that of a convolutional encoder. This is apparent from Eq. (1), which shows that the channel output sequence is obtained by convolving the input data sequence with the sequence formed by the impulse response samples. The Viterbi algorithm is based on trellis representation of the channel state. In this representation, a state at time k has branches (transitions) from M states at time $k - 1$ and to M states at time $k + 1$, where M is the number of elements in the signal alphabet. At each symbol period, the receiver examines the M sequences converging to each state of the trellis and keeps only one of them as the survivor sequence and extends it to M states at the following symbol period. Selection of the survivor sequence is based on the minimization of the Euclidian distance from the received sequence. The number of survivor sequences at each time period thus is equal to the number of states of the trellis. The beauty of the Viterbi algorithm is that the number of computations per symbol period is constant and does not depend on the transmitted sequence length. The problem with Viterbi equalizers is the inherent complexity, which grows exponentially with the size of the symbol alphabet and the length of the channel impulse response. Specifically, the number of trellis states is M^{L-1}, where M is the alphabet size and L is the length of the discrete channel impulse response. The complexity involved precludes the use of the optimum receiver in many applications, and simpler linear and decision feedback equalizers (DFEs) often are preferred.

A linear equalizer usually takes the form of a nonrecursive transversal filter with adjustable tap gains (coefficients). Figure 2 shows a linear equalizer with $N = 2q + 1$ taps that have tap gains denoted by $c_{-q}, c_{-q+1}, \ldots c_0, \ldots, c_{q-1}, c_q$. The equalizer output is given by

$$y_k = \sum_{j=-q}^{+q} c_j x_{k-j} \tag{3}$$

This output is passed next to a memoryless decision circuit (threshold detector) that delivers an estimate $\hat{a}_k$ of the data symbol a_k. An essential parameter in equalizer design is the number of taps N. Increasing N improves the static performance (correction capacity) of the equalizer, but also degrades its dynamic performance (convergence and adaptation noise properties) (discussed in the next section). The equalizer is said to be synchronous when the unit delay Δ is equal to the symbol period T, and fractionally spaced when Δ is smaller than T. Fractionally spaced equalizers (FSEs) were developed to make the receiver insensitive to the sampling phase. Performance of synchronous equalizers indeed is strongly dependent on the sampling phase because the period of its periodic transfer function is $1/T$, which does not allow it to compensate inde-

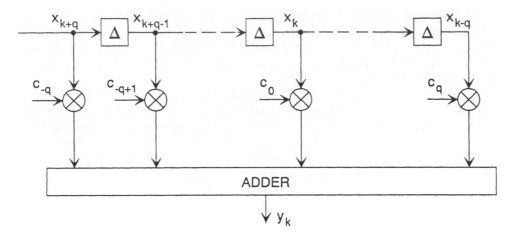

FIG. 2   Nonrecursive linear equalizer structure.

pendently for channel distortions within and outside the Nyquist bandwidth $(-1/T, +1/T)$. More specifically, the equalizer operates on the received signal spectrum folded to the Nyquist bandwidth with a shape that depends on the sampler phase. An FSE overcomes this difficulty because the period of its periodic transfer function is $2/T$, which makes it possible to correct independently the distortions in the Nyquist bandwidth and those in the excess bandwidth.

A DFE is a nonlinear device composed of two transversal filters (Fig. 3). The inputs to the feedforword filter are the received signal samples, whereas the inputs of the feedback filter are the previously detected data symbols. Using the notation of Fig. 3, the equalizer output is given by

$$y_k = \sum_{i=-N_1+1}^{0} c_i x_{k-N_1+1-i} + \sum_{j=1}^{N_2} d_j \hat{a}_{k-j} \tag{4}$$

The key idea in DFEs is that if the previous data symbols are decided correctly, their interference on the current data symbol can be removed by feeding them to the feedback filter. DFEs perform better than linear equalizers on such severely amplitude-distorted channels as radio channels that experience multipath fading. A linear equalizer can compensate for a spectral notch only at the expense of a substantial noise enhancement. In contrast, a DFE, in principle, can compensate for a spectral notch without noise enhancement, but DFEs suffer from the well-known error-propagation phenomenon. A decision error propagates in the delay line of the feedback filter and causes more errors. Despite this phenomenon, DFEs perform better than linear equalizers and turn out to be the appropriate choice for fading radio channels.

Although DFEs yield superior performance in comparison to linear equalizers, the latter often are preferred for practical implementations due to their implementation simplicity. This is true particularly in such high-speed applica-

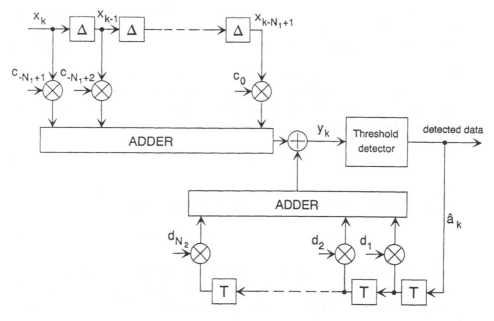

FIG. 3   Decision feedback equalizer.

tions as digital microwave radio systems that involve symbol rates of several tens of megabits per second (Mb/s).

Adaptation Algorithms

We first discuss the adaptation of linear equalizers and DFEs. The algorithms are described considering a linear equalizer, but their extention to DFEs is straightforward. Historically, the oldest algorithm for adaptive equalization is the zero forcing (ZF) algorithm developed by Lucky at the end of the 1960s (2). Referring to Fig. 2, which shows a linear equalizer with the input signal denoted x_k, the tap gains denoted $c_j (j = -q, \ldots, 0, \ldots, +q)$, and the output denoted y_k, the polarity-type ZF algorithm is

$$c_j(k + 1) = c_j(k) - \alpha . \mathrm{sgn}(a^*_{k-j}) \mathrm{sgn}(e_k) \tag{5}$$

where

α = a small positive constant denoted step size
sgn (.) = the mathematical sign function
e_k = the instantaneous output error signal given by $e_k = y_k - a_k$.

In this algorithm, the reference tap forces the main sample of the equalized impulse response to 1, and every other tap forces one sample of that response to 0. Convergence of the algorithm is guaranteed if the unequalized eye pattern is open, but, although there is no mathematical proof, simulations and experimental results show that the algorithm converges even with a closed eye pattern at the equalizer input.

The most popular algorithm in adaptive channel equalization is the stochastic gradient algorithm, which minimizes the mean-squared error (MSE) at the equalizer output. It also is known as the LMS (least-mean-square) algorithm in the literature. Referring to Fig. 2, this algorithm is given by

$$c_j(k + 1) = c_j(k) - \alpha x^*_{k-j} e_k. \tag{6}$$

In practical implementations, the polarity-type version of this algorithm often is preferred for its simplicity. This simplified algorithm is given by

$$c_j(k + 1) = c_j(k) - \alpha.\text{sgn}(x^*_{k-j})\text{sgn}(e_k) \tag{7}$$

The LMS algorithm minimizes the combined effect of ISI and additive noise, and hence leads to better performance than the ZF algorithm, especially at low signal-to-noise ratios. It is used in many applications, but the ZF algorithm is preferred in some digital microwave radios.

Unlike the ZF algorithm, the LMS algorithm is easy to analyze in terms of stability, self-noise, and convergence and tracking properties. As the step-size parameter α is increased, convergence of the algorithm quickens, but this also increases the algorithm self-noise, that is, the fluctuations of the tap gains in the steady state. The critical step size (the step size beyond which the algorithm is unstable) is given by

$$\alpha_{\text{max}} = \frac{2}{\lambda_{\text{max}}} \tag{8}$$

where α_{max} is the largest eigenvalue of the $N \times N$ dimensional autocorrelation matrix R of the equalizer input signal. Since the largest eigenvalue of a matrix is upper bounded by its trace, we have the following simple upper bound on the critical step size

$$\alpha_{\text{max}} < \frac{2}{N \cdot \sigma^2_x} \tag{9}$$

where σ^2_x is the signal power at the equalizer input and N is the number of taps. The optimum value of the step-size parameter is shown easily to be half of its critical value.

The convergence speed of the LMS algorithm is governed by the maximum-to-the-minimum eigenvalue ratio ($\lambda_{\text{max}}/\lambda_{\text{min}}$) of the signal autocorrelation matrix R (1, p. 578). The larger this ratio, the slower the convergence speed of the algorithm will be. On the other hand, $\lambda_{\text{max}}/\lambda_{\text{min}}$ is related directly to the channel amplitude distortion. Consequently, convergence of the LMS algorithm is slow on channels with strong amplitude distortion (i.e., radio channels with deep spectral nulls). Although it is difficult to demonstrate mathematically, the same problem also holds for the ZF algorithm. Therefore, neither of these algorithms is appropriate for applications in which fast convergence is required.

Rapidly Converging Algorithms

The fastest possible convergence is achieved with the least-squares algorithm that minimizes the cumulative squared error

$$J_k = \sum_{i=0}^{k} |e_i|^2 \tag{10}$$

at each instant $k > 0$. A recursive version of this algorithm is given by Ref. 3:

$$R_k = \frac{k-1}{k} R_{k-1} + \frac{1}{k} X_k^* X_k^T \tag{11-a}$$

and

$$C(k+1) = C(k) - \frac{1}{k} R_k^{-1} X_k^* e_k \tag{11-b}$$

where C is the N-dimensional vector formed by the equalizer tap gains. The R_k matrix in these expressions is the estimate at time k of the received signal correlation matrix R. For undistorted signals, $R = I$, where I is the identity matrix, and the RLS algorithm differs very little from an LMS algorithm with decreasing step size $\alpha_k = 1/k$.

The RLS algorithm as described by Eqs. (11-a) and (11-b) provides fast initial convergence but is unable to track time variations of the channel. For nonstationary channels, an appropriate cost function is an exponentially weighted sum of squared errors, that is,

$$J_k' = \sum_{i=0}^{k} \lambda^{k-1} |e_i|^2 \tag{12}$$

and the RLS algorithm takes the form

$$R_k' = \lambda R_{k-1}' + X_k^* X_k^T \tag{13-a}$$

and

$$C(k+1) = C(k) - R_k'^{-1} X_k^* e_k. \tag{13-b}$$

The RLS algorithm is computationally intensive, but fast versions have been developed that significantly reduce the computational complexity by making use of the shift property of the equalizer delay line (4–6).

Blind Algorithms

All of the algorithms described or mentioned thus far require knowledge of the transmitted data symbols to compute the error signal e_k. The symbols can be made available to the receiver during an initial training period or by periodically transmitting a known preamble. However, in many applications it is not desirable or possible to transmit a known data sequence, and the equalizer must converge in a decision-directed mode. In this mode, the equalizer uses the error signal $e_k = y_k - \hat{a}_k$, where $\hat{a}_k$ is the receiver decision, which may be different from the corresponding data symbol a_k. Convergence in decision-directed mode usually is achieved provided that the error probability is sufficiently small, but more robust algorithms are required for the equalizer to converge independently of the initial error rate. These type of algorithms are referred to as *blind* (or *self-recovering*) algorithms.

The first blind algorithm was proposed by Sato (7), and other algorithms were developed later by Godard (8) and others. The Sato algorithm for pulse amplitude modulation (PAM) signals uses $y_k - \beta.\text{sgn}(y_k)$ as error signal, where β is a constant that depends on the signal alphabet. Intuitively, the probability that $\text{sgn}(y_k) = \text{sgn}(a_k)$ in an equalized PAM system is significantly higher than the probability of $\hat{a}_k = a_k$, and this explains why the Sato algorithm can operate and converge at error rates for which the LMS algorithm fails. Later, Godard proposed a class of blind algorithms for two-dimensional (2-D) signals that minimize a cost function of the form $(|y_k|^2 - \gamma_p)^p$, where p is an integer greater than or equal to 1, and γ_p is a modulation-dependent constant. The beauty of Godard's approach is that the cost function is independent of the recovered carrier phase, which allows the equalizer to converge and track the channel variations even in the presence of carrier sync loss. Also, Benveniste and Goursat generalized the Sato algorithm to 2-D signals and analyzed its behavior (9).

Blind algorithms originally were developed for such point-to-multipoint networks as polling systems, in which a tributary must converge without interrupting the data transmission from the master terminal to the other tributaries. Today, another area in which blind equalization has gained significant interest is digital microwave radio. It is important in these systems that the equalizer remains stable and tracks the channel variations even in the presence of deep fades that cause an excessive error rate and eventually sync loss of the receiver. This significantly reduces the outage time because it suppresses the need to reconverge the equalizer after the fade events. However, due to the high data rates involved, only simple algorithms have found applications in these systems. A blind algorithm that has become popular in digital microwave radio is the maximum-level error (MLE) algorithm (10), which can be written as

$$C(k + 1) = C(k) - \alpha D_k \text{sgn}(X_k^*)\text{sgn}(e_k) \qquad (14)$$

with

$$D_k = \begin{cases} 1 & \text{if } y_k \in W \\ 0, & \text{otherwise} \end{cases}$$

where W is a predetermined set of windows of the signal constellation plane in which the sign of the output error signal is correct with probability 1. For an M^2-state rectangular QAM system in which the real and the imaginary parts of the complex data symbols take their values from the alphabet $\{\pm 1, \pm 3, \ldots, \pm(M - 1)\}$, W is defined by $|\text{Re}(y_k)| > M - 1$ and $|\text{Im}(y_k)| > M - 1$, where Re(.) and Im(.) denote real part and imaginary part, respectively. For a quaternary phase-shift keying (QPSK) (4QAM) signal, W is the dashed region in Fig. 4. It is obvious that the sign of the output error corresponding to a point in the blank region will be correct independent of the transmitted symbol. The MLE algorithm consists of enabling the original polarity-type LMS algorithm upon reception of such a point and freezing it otherwise.

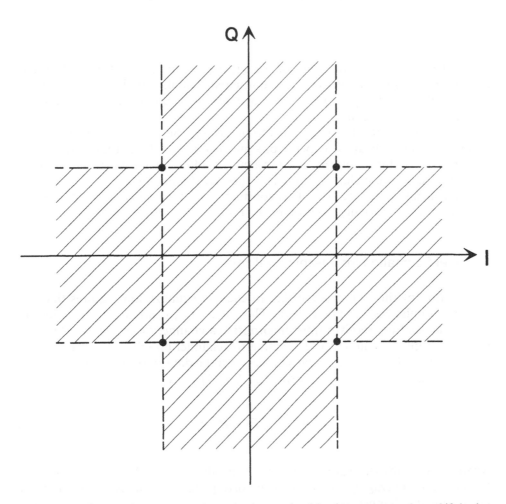

FIG. 4 Implementation of the maximum-level error algorithm in quaternary phase-shift keying systems (adaptation is frozen when the received signal point falls within the dashed region).

Adapting Fractionally Spaced Equalizers

FSE adaptation has an inherent instability that is due to the fact that the period of its periodic transfer function usually is larger than the received signal bandwidth. For example, the transfer function of a $T/2$-spaced equalizer has a period of $2/T$, which is larger than the bandwidth of a Nyquist-filtered signal for all roll-off values less than 1. In the absence of noise, the equalizer coefficients are not determined uniquely because it can synthesize an infinity of different transfer functions in the frequency intervals with vanishing signal spectrum without affecting the output MSE. The equalizer transfer function is unbounded in these regions and a consequence of this is that the tap gains may diverge after a long period of stable operation.

To stabilize the FSE operation, Gitlin, Meadors, and Weinstein developed the tap-leakage algorithm that minimizes the cost function (11)

$$J = E(|e_k|^2) + \mu\|C_k\|^2 \tag{15}$$

where the first term on the right of the equation is the conventional MSE and the second term is proportional to the squared modulus of the tap-gain vector C. The solution of Eq. (15) is the vector with the smallest modulus among all vectors that minimize the MSE.

Stabilization of the FSE operation using the tap-leakage algorithm is achieved naturally at the expense of some increase of the output MSE. It can be shown that this algorithm is equivalent to adding a virtual white noise to the signal at the equalizer input. The spectral density of this noise is given by the constant μ in Eq. (15). This constant, therefore, must be kept small to limit the performance degradation of the FSE.

Channel Estimation for Viterbi Equalizers

Application of the Viterbi algorithm to channel equalization requires that the channel impulse response be known. This, of course, is not a realistic situation in general, and the receiver must incorporate a channel estimator for this purpose. A simple technique for estimating the channel impulse response consists of correlating the received signal samples with the data symbols. Assuming that successive data symbols are uncorrelated, Eq. (1) gives

$$h_\ell = \frac{1}{\sigma_a^2} E(x_k a_{k-\ell}^*) \tag{16}$$

where σ_a^2 denotes the variance of the transmitted symbols and $E(.)$ denotes the ensemble averaging operator. The corresponding recursive algorithm for nonstationary channels is

$$h_\ell(k) = \lambda.h_\ell(k-1) + (1-\lambda)x_k a_{k-\ell}^* \tag{17}$$

where λ is the forgetting factor close to but less than 1. This algorithm is known as the *correlation estimator* and has moderate convergence rates.

As in equalizer adaptation, faster convergence is achieved using the RLS algorithm. For nonstationary channel estimation, the RLS algorithm reads

$$G_k = \lambda G_{k-1} + A_k^* A_k^T \tag{18-a}$$

and

$$H(k + 1) = H(k) - G_k^{-1} A_k^* e_k \tag{18-b}$$

where
$\quad A_k =$ the input data vector
$\quad e_k$ is the error signal given by $e_k = A_k^T H_k - x_k$
$H(k) =$ the estimate at time k of the channel impulse response vector.

Timing Synchronization

The symbol timing information at the receiver usually is extracted from the received signal through a nonlinear operation. The most well-known timing recovery techniques are the squaring synchronizer and the zero-crossing synchronizer. The squaring synchronizer for a PAM signal is shown in Fig. 5. After an eventual prefiltering operation to suppress low-frequency components, the received signal is squared and the resulting signal is passed to a phase-locked loop (PLL). The squaring operation creates a discrete spectral component at the symbol frequency, and the PLL reduces the phase noise of this signal. The phase of the clock provided by the PLL is adjusted so that the received signal is sampled at the maximum eye opening coinciding with the instant that maximizes the sampled signal energy for a Nyquist-filtered signal. As the channel distortion is varied, the squaring synchronizer tracks the clock phase that maximizes the sampled signal energy.

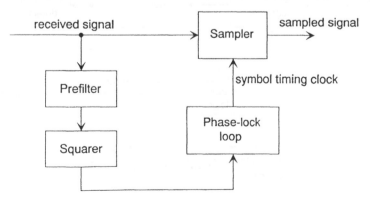

FIG. 5 Squaring-type clock synchronizer for pulse amplitude modulation systems.

Since FSEs are insensitive to timing phase, they can be used with any timing recovery scheme virtually without any performance penalty. In contrast, since synchronous equalizers are sensitive to the sampling phase, their performance is maximized using such a cost function as the minimization of the output MSE. A receiver with an adaptive equalizer and a timing recovery scheme optimized under the minimum MSE (MMSE) criterion is shown in Fig. 6. As shown, optimization of the sampling phase requires an additional equalizer with inputs that are the samples of the received signal derivative. The inputs of both equalizers are sampled in synchronism using the recovered symbol timing clock. The optimum receiver shown in Fig. 6 was derived originally by Kobayashi (12). It also incorporates a decision-directed carrier-recovery loop. This receiver obviously involves an excessive complexity due to the presence of an auxiliary equalizer; this has stimulated further search for approximate solutions with reduced complexity (13).

The major difficulty with MMSE timing recovery is that it interacts with the equalizer adaptation algorithm. The equalizer coefficients are a function of the timing phase, which in turn is a function of the equalizer coefficients. The resulting receiver tends to be difficult to stabilize, and also it is particularly sensitive to carrier phase errors. For this reason, state-of-the-art receivers with an adaptive equalizer tend to use timing recovery schemes optimized using a criterion that does not depend on the equalizer coefficient values. The conventional squaring loop is such a synchronizer that involves analog signal processing.

In digitally implemented receivers, it is of significant practical interest to extract the timing information from the sampled signal. Several criteria have been developed by Mueller and Müller (14) and others that exploit the impulse response properties of Nyquist-filtered or partial-response signals. Note that when properly sampled, the impulse response of a Nyquist filter gives a main sample $h_0 = 1$ and all other samples are zero ($h_i = 0$ for $i \neq 0$). As shown in

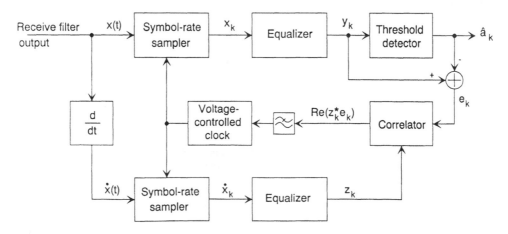

FIG. 6   Optimum timing recovery in a receiver with an adaptive equalizer.

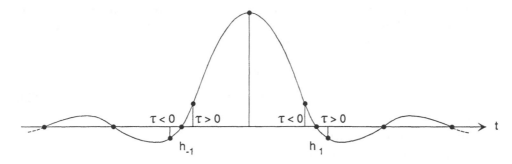

FIG. 7   Sampling of a Nyquist impulse response with a phase lead or a phase lag.

Fig. 7, a phase lead of the sampling clock gives $h_1 > 0$ and $h_{-1} < 0$, and a phase lag gives $h_1 < 0$ and $h_{-1} > 0$. Consequently, forcing to zero of either h_{-1} or h_1 can be used as a criterion to optimize the timing phase. A third possibility is to force to zero their difference, that is, $h_{-1} - h_1$. All these criteria are equivalent on a Nyquist channel, but they give different performance results on ISI channels, and $h_{-1} - h_1 = 0$ turns out to be the best choice in general. Implementation of this timing recovery scheme can be performed using

$$\varepsilon_k = x_k a_{k-1}^* - x_{k-1} a_k^* \qquad (19)$$

as control signal to the voltage-controlled clock. This signal derives from Eq. (16) in a straightforward manner.

Several other techniques are available today to derive the timing information from the digitized signal, but many of them involve sampling of the received signal at twice the symbol rate. One such technique developed by Gardner essentially maximizes the sampled signal energy and thus is equivalent to the squaring loop (15).

Carrier Synchronization

Demodulation of the received signal requires the generation of a local carrier in frequency synchronism with the carrier used to demodulate the signal at the transmitter side. Extraction of the local carrier in an N-phase PSK system usually is performed using an Nth-power loop that consists of an Nth-power device followed by a PLL. The Nth-power operation generates a discrete frequency component at the carrier frequency, and the PLL filters out the noise components around this frequency. For bandwidth-efficient systems using QAM signal formats, this technique is inappropriate and carrier recovery generally is performed from the demodulated signal using a decision feedback loop (DFL) (16) or one of its simplified versions. The carrier information in a DFL is extracted by making crossproducts of the threshold detector input and output on the in-phase (I) and quadrature (Q) channels. Using a complex notation, the control signal of this loop can be expressed as

$$\varepsilon_k = \mathrm{Im}(y_k^* \hat{a}_k)$$
$$= \mathrm{Im}(y_k^* e_k) \tag{20}$$

where

    $\mathrm{Im}(.)$ = imaginary part
    y_k = the complex equalizer output
    $\hat{a}_k$ = the detected data symbol
    e_k is the error signal given by $e_k = y_k - \hat{a}_k$.

It is verified easily that this control signal is the stochastic gradient of the output MSE with respect to carrier phase. That is, the decision feedback carrier-recovery loop initially developed on heuristic considerations turns out to minimize a well-known cost function. The equilibrium point of a DFL is the carrier phase for which $\mathrm{Im}(g_0) = 0$, where g_0 denotes the main sample of the equalized impulse response.

There are two important issues related to DFL implementation in receivers that incorporate an adaptive baseband equalizer. The first is the increased loop delay due to the presence of the equalizer in the carrier recovery loop. It is well known that in feedback control systems excessive delay leads to instability. An adaptive equalizer with $N = 2q + 1$ taps and the center tap chosen as reference, the theoretical decision delay is q times the symbol period T. This figure does not take into account the additional delay due to the multiply-and-add operations to compute the equalizer output. For N large, the equalizer delay increases the steady-state phase jitter and also decreases the acquisition range. The loop delay can be avoided by moving the equalizer from baseband to passband (carrier frequency or intermediate frequency) (17). Adaptive baseband equalizers used in digital microwave radio systems are usually short (they typically comprise from 5 to 9 taps), and the delay they introduce in the carrier-recovery loop remains acceptable. In contrast, voiceband modems employ adaptive equalizers comprising a large number of taps but the equalizers are implemented at passband, and the equalizer delay in the carrier-recovery loop is avoided.

The second important issue is the interaction of the carrier-recovery loop with the equalizer adaptation algorithm. A complex equalizer with no constraints on its coefficient values can compensate for a portion of the frequency offset between transmitter and receiver by rotating its tap gains. In other words, the equalizer can play a role that is very similar to that of the carrier-recovery loop, but its low adaptation speed does not allow compensation for the entirety of the frequency offset. The analysis shows that the interaction between the equalizer and the carrier-recovery loop is not desirable because it significantly degrades the loop steady-state jitter properties (18). It even leads to instability if the equalizer employs the ZF algorithm. For this reason, the equalizer reference tap is forced to have a fixed phase (its imaginary part is forced to zero) in many practical applications.

Bibliography

Chamberlain, J. K., Clayton, F. M., Sari, H., and Vandamme, P., Receiver Techniques for Microwave Digital Radio, *IEEE Commun. Mag.*, 24(11):43–54 (November 1986).

Proakis, J. G., Advances in Equalization for Intersymbol Interference In: *Advances in Communication Systems*, Vol. 4 (A. J. Viterbi, ed.), Academic Press, New York, 1975, pp. 123–188.

Qureshi, S. U. H., Adaptive Equalization, *IEEE Commun. Mag.*, 20(2):1–16 (March 1983).

Qureshi, S. U. H., Adaptive Equalization, *Proc. IEEE*, 53:1349–1387 (September 1985).

References

1. Proakis, J. G., *Digital Communications*, 2d ed., McGraw-Hill, New York, 1989.
2. Lucky, R. W., Techniques for Adaptive Equalization of Digital Communication, *Bell Sys. Tech. J.,* 44:547–588 (April 1965).
3. Godard, D., Channel Equalization Using a Kalman Filter for Fast Data Transmission, *IBM J. R&D*, 18:267–273 (May 1974).
4. Falconer, D. D., and Ljung, L., Application of Fast Kalman Estimation to Adaptive Equalization, *IEEE Trans. Commun.*, COM-26:1439–1446 (October 1978).
5. Proakis, J. G., and Ling, F., Recursive Least-Squares Algorithms for Adaptive Equalization of Time-Variant Multipath Channels, *ICC '84 Conf. Rec.*, 3: 1250–1254 (May 1984).
6. Cioffi, J. M., and Kailath, T., Fast Recursive Least Squares Transversal Filters for Adaptive Filtering, *IEEE Trans. ASSP*, ASSP-32:304–337 (April 1984).
7. Sato, Y., A Method of Self-Recovering Equalization for Multilevel Amplitude Modulation Systems, *IEEE Trans. Commun.*, COM-23:679–682 (June 1975).
8. Godard, D., Self-Recovering Equalization and Carrier Tracking in Two-Dimensional Data Communications Systems, *IEEE Trans. Commun.*, COM-28:1867–1875 (November 1980).
9. Benveniste, A., and Goursat, M., Blind Equalizers, *IEEE Trans. Commun.*, COM-32:871–883 (August 1984).
10. Yatsuboshi, R., Sata, N., and Aoki, K., A Convergence of Automatic Equalizers by Maximum-Level Error Control, *Nat. Conv. Rec. IECE Japan*, 2192 (1974).
11. Gitlin, R. D., Meadors, H. C., and Weinstein, S. B., The Tap-Leakage Algorithm: An Algorithm for the Stable Operation of a Digitally Implemented, Fractionally-Spaced Adaptive Equalizer, *Bell Sys. Tech. J.*, 60:275–296 (February 1981).
12. Kobayashi, H., Simultaneous Adaptive Estimation and Decision Algorithm for Carrier Modulated Data Transmission Systems, *IEEE Trans. Commun. Tech.*, COM-19:268–280 (June 1971).
13. Sari, H., Desperben, L., and Moridi, S., Minimum Mean-Square Error Timing Recovery Schemes for Digital Equalizers, *IEEE Trans. Commun.*, COM-34:694–702 (July 1986).
14. Mueller, K. H., and Miller, M., Timing Recovery in Digital Synchronous Data Receivers, *IEEE Trans. Commun.*, COM-24:516–530 (May 1976).
15. Gardner, F. M., A BPSK/QPSK Timing-Error Detector for Sampled Receivers, *IEEE Trans. Commun.*, COM-34:423–429 (May 1986).

16. Simon, M. K., and Smith, J. G., Carrier Synchronization and Detection of QASK Signal Sets, *IEEE Trans. Commun.*, COM-22:98–104 (February 1974).

17. Falconer, D. D., Jointly Adaptive Equalization and Carrier Recovery in Two-Dimensional Digital Communication Systems, *Bell Sys. Tech. J.*, 55:317–334 (March 1976).

18. Sari H., et al., Baseband Equalization and Carrier Recovery in Digital Radio Systems, *IEEE Trans. Commun.*, COM-35:319–327 (March 1987).

HIKMET SARI

Espenschied, Lloyd

Lloyd Espenschied was born on April 27, 1889 in St. Louis, Missouri, the son of Fred F. and Clara M. Espenschied. The family sent him to live with his maternal grandparents (named Espenscheid) in Brooklyn, New York, in December 1901 after his father's business failed. Espenschied was an ardent hunter and fisherman from his earliest childhood and always remembered the open spaces of Missouri with a great affection.

During his early high school years, Lloyd "became enamored of things electrical" (1) and began tinkering with many different bits of hardware. By 1904, he already was exploring wireless.

As has been documented frequently, the public high schools of New York in that era provided the background and inspiration for an extraordinary number of scientists and engineers. A student at Boy's High School of Brooklyn, Espenschied became a member of that school's after-hours Electrical Club. He quickly became an avid radio enthusiast, built himself an attic workshop, and operated an amateur station as early as 1904. By 1906, his interest in wireless had caught fire, and he was determined to make a career of the new technology.

Espenschied was a charter member of the original "Wireless Institute." In 1912, the Wireless Institute merged with the Society of Wireless Telegraph Engineers to become the IRE (Institute of Radio Engineers). Espenschied was still a member in 1962 when the IRE merged with the American Institute of Electrical Engineers (AIEE) to become the Institute of Electrical and Electronics Engineers (IEEE). He also would become an active member of various scientific societies.

Accepted at the Pratt Institute in New York City, Espenschied spent every available moment combining his love for the outdoors with his passion for wireless. He spent the summers of 1907 and 1908 as a shipboard telegraph operator for the United Wireless Telegraph Company aboard a succession of ships (2). His passion for ship-to-shore technology was spread readily to anyone who would listen via technical and popular lectures; the commitment never dimmed. Espenschied's notes record that he made one such talk at an AIEE Lehigh Valley Section Meeting in Scranton, Pennsylvania, as late as February 12, 1932. Such talks, he noted, were documented by lantern slide indices stored in the archives of Bell Labs, a technology already almost forgotten when Espenschied made his notes in April 1954 (3).

Espenschied graduated from Pratt in 1909 in applied electricity. He became an assistant engineer with the Telefunken Wireless Telegraph Company in New York City; clients of the company included both the U.S. Army and Navy.

In 1910, Espenschied joined the Bell Systems. He spent the period from 1910 to 1934 in the AT&T Engineering Department, where he worked first on loading coils and later on both high frequency electronic carrier and radio telephony.

In April 1912, he married another Pratt student, Ethel Fairfield Lovejoy, known as "Lovey." They eventually would have two children, Lloyd, Jr., and Carol.

In the years prior to World War I, Espenschied put a great deal of energy into encouraging the company's interest in radio telephony. During this period, he was part of a team concentrating on loading coil design. Competitive all his life, Espenschied took particular pride in the original work he contributed to the patents. He conceived and developed a technique for stabilizing and controlling inductance on both loading coils and transformers using air gaps and core configurations.

It was during this era that the interests Espenschied held for so long combined effectively with research at Bell; at the time, the laboratory was involved deeply in transoceanic radio telephony. It was a concept that Espenschied embraced with typical fervor. As early as 1915, he participated in the first tests of a transoceanic wireless telephone (3). He was the receiver, at Pearl Harbor, Hawaii, for those tests and assisted with the tests in various other locations as well.

At the end of World War I, Bell Laboratories made a corporate decision to apply these efforts to finding some way for the wire service to reach ships far out at sea. The determination was made that the same technology could serve both intercontinental and ship-to-shore transmissions.

From 1916 through the early 1920s, Espenschied was in charge of the Development and Research (D&R) Department for Bell, working on the first carrier telephone and telegraph systems. He specialized in "the line application end" of the research (4), and worked on early line measurements (at high frequency) with H. A. Affel and other team members.

Typically, Espenschied continued to manage this research while pursuing various other projects for the company. He also spent a great deal of time in the field, devising methods for the practical application of the hardware born from his theories. In 1918, for instance, he was involved in the installation of the first commercial carrier telephone system, which linked Baltimore, Maryland, with Pittsburgh, Pennsylvania.

Eventually Espenschied would be responsible for several dozen inventions having to do with various aspects of the line applications of carrier, "including the reduction of cross-talk on open wire lines by the grouping of all the frequencies into two separate send-receive banks" (4).

With more time available to him after the end of World War I, Espenschied applied another of his pet theories — the idea of sending out radio signals, watching what happens when they reflect, and measuring distances by the angle of reflection. He was determined that the "phenomenon of reflected electric waves" (4) might permit the engineer of a locomotive to "see" far ahead and to avoid the extremely deadly train collisions that occurred frequently during that era.

Unable to get anyone at the Bell System interested, he applied for and eventually received patents in his own name. He later would recall that "they never amounted to much" and that they "just about 'broke' him financially" (1).

That literally was true in the case of those specific patents; nonetheless, pilots and other fliers might disagree strenuously with his assessment of the value of his system. It gave him the idea for what eventually would become the radio altimeter, discussed below. This device transmitted FM (frequency modulation) signals to earth and used their reflection to measure aircraft altitude. The theory was enhanced, enlarged, and developed by Bell Laboratories

(Espenschied's notes give primary credit to Russell Newhouse) and became a significant precursor of radar.

During the early 1920s, Espenschied was part of a Bell team of scientists and engineers who conceived and developed the concept of carrier broadcasting over both telephone and power lines. Though others were working on the same concept, Espenschied was credited with this phase of the activity. At that time, he was in charge of the early testing of such a system, particularly on power lines, but also on some telephone lines in New York City.

In a moment of triumph that must seem extraordinary to students knowledgeable about Thomas Edison's often brutal treatment of competitors, Espenschied was able to create a cooperative testing arrangement that included his AT&T laboratory, the New York Telephone Company, and New York Edison. The tests revealed a great many problems with wire broadcasting and gave Bell Laboratories a head start over other organizations that later would become involved in such transmissions.

Like many inventors and scientists of the era, Espenschied devoted his personal as well as professional time to his scientific interests. As mentioned above, he, like a number of his contemporaries, also was very much intrigued with the electrical demands of the railroad. His work exploring the interference with nearby equipment generated by the nearby railroad led to a 1924 patent on a railroad warning system. It is this work that Espenschied's notes say led in 1926 to his conception of a radio altimeter for airplanes. The work was not taken seriously for some time, primarily because technological and other theorists did not understand Espenschied's concepts. As the technology (particularly new variations of vacuum tubes) developed, and as representatives of the new commercial airline began to explore their hardware options, the radio altimeter met with reluctant acceptance.

Additional personal "tinkering" during this era led him to new concepts for a crystal filter, which eventually would become a significant part of carrier wave transmission, though it would be well into the 1930s before any serious refinement would occur.

His personal tinkering also led to the invention of "pre-corrected" phonograph records (e.g., those records with distortion-correcting systems built in). The patents again were taken out by Espenschied personally, but eventually were "taken over by the company" (4).

There was strenuous resistance to many of Espenschied's ideas. He noted, for instance, that there was "a long hard row for carrier in the 1920's" (1). The idea was alien to both the theorists and the applications engineers with whom Espenschied dealt. He later would recall that his ideas about transmitting radio frequencies over wires was met with a great deal of ridicule by his colleagues in the D&R Department. Nonetheless, the 1930s saw the development of wideband coaxial cable. Espenschied credited two other members of the team, Affel and E. I. Green, with the persistence and assistance that eventually proved his theories were correct (5).

The patent for coaxial cable eventually was credited jointly to Espenschied and Affel. In his notes, Espenschied recalled that he and Affel "began in the 1930's, the age of wide-band high frequency transmission for telephony that was to prepare the way for television in the 1940's" (1).

For 13 years, beginning in 1925, Espenschied had policy-related duties in addition to his research and inventions. He represented Bell in Washington, D.C., aiding in the formulation of international radio regulations. Through this work, he became the company's logical choice as a technical contact for foreign administrations. Espenschied recalled that these roles, executed between World War I and World War II, required "technical acumen, breadth of view, important personal relationships" and that they were "discharged with credit" (4).

In 1934, Espenschied transferred to Bell Telephone Laboratories, Incorporated, in New York City, where he would remain until his retirement in 1954. During this era, one of his most important contributions, in his view, was his determination to spur on the now-famous creativity of the scientists and engineers at Bell Labs. He later would recall that he had written many internal memoranda about what he considered Bell's "lag in creativity" (1) and that he took it upon himself to stimulate the research of his colleagues at every possible chance.

During World War II, Espenschied worked at Bell Laboratories and began a collection of rare early books on electrical phenomena—an expression of his lifelong interest in the history of electricity as a technology. The collection eventually was donated to the Niels Bohr Library at the American Institute of Physics.

Espenschied was honored widely for his extraordinary creativity. He was recognized as coauthor (with Affel) of the AIEE "Best Paper" in 1935. These awards honored his creation of the radio altimeter and his significant contribution to the development of coaxial cable transceiving. In 1940, he received the Medal of Honor from the IRE. Later, there were the Television Broadcaster's Association Medal (1944), the Radio Pioneers Citation in 1954, and the Pioneer Award for the radio altimeter from the IEEE Aerospace and Electronic Systems Groups in 1967. In 1957, Espenschied received an honorary doctor of science degree from Pratt.

Although various sources indicate he had about 150 patents, Espenschied remembered the number as "a total of 130 U.S. patents" (2). Among those were several that eventually would contribute significantly to network broadcasting. In addition, Espenschied published more than 40 technical papers, many of which, along with a large collection of personal technical papers, are now in the permanent collection of the Smithsonian Institute.

Through the years, Espenschied's deep interest in the history of electricity, about which he published many papers, gradually focused on the history of radio itself. In 1972, Espenschied wrote that "the history of radio is basically one of graduating from long waves . . . to shorter and shorter waves—back to the one meter waves of Hertz and much shorter" (4).

Espenschied retired from Bell Systems in 1954 and died June 21, 1986.

Bibliography

Brittain, J. (ed.), Espenschied and Strieby Discuss a Wide-Band Coaxial Communication System. In *Turning Points in American Electrical History*, IEEE Press, John Wiley and Sons, New York, 1977.

Espenschied, L., Historic First Article on the Radio Altimeter, *Bell Laboratories Record*, January, 1948, p. 18.

Espenschied, L., Methods for Measuring Interfering Noises, *Proc. IRE*, 19:1951–1954 (November 1931); (presented in May 1931 before the International Union of Scientific Radiotelegraphy).

Espenschied, L., and Strieby, M. E., Wide Band Transmission over Coaxial Lines, *Electrical Engineering*, 53:1371–1374 (October 1934).

Mabon, Prescott C., *Mission Communications: The Story of Bell Laboratories*, Bell Telephone Laboratories, Inc., Murray Hill, NJ, 1975, 1976.

References

1. Espenschied, L., The Part of Lloyd Espenschied in the Pioneering of Radio and Carrier Telephony, notes in Espenschied's hand, August 1, 1972, from the collection of James Brittain, Georgia Institute of Technology, Atlanta, GA (original in the Smithsonian Institute Collection). Photocopy.

2. Espenschied, L., Biographical Sketch, November 15, 1969, from the collection of James Brittain, Georgia Institute of Technology, Atlanta, GA (original in the Smithsonian Institute Collection). Photocopy.

3. Espenschied, L., Talks by L. Espenschied, typewritten list of lantern slides documenting various talks, from the collection of James Brittain, Georgia Institute of Technology, Atlanta, GA (original in the Smithsonian Institute Collection). Photocopy.

4. Espenschied, L., The Creative Performance and Contributions of Lloyd Espenschied, September 25, 1944, from the collection of James Brittain, Georgia Institute of Technology, Atlanta, GA (original in the Smithsonian Institute Collection). Photocopy.

5. References to both men appear frequently in Espenschied's personal notes about the development of coaxial transmission, including Refs. 1–4.

MARYLIN K. SHEDDAN

Evolution of Cellular Switch Networking

Introduction

The cellular industry is experiencing tremendous growth throughout the world and especially in North America. In the United States alone, cellular services have expanded from a few systems in large metropolitan areas in 1983 to over 1200 systems serving more than 7.5 million subscribers by the end of 1991. Since cellular subscribers frequently travel between different cellular systems, a mechanism was needed to allow them the capability of placing and receiving calls anywhere in the United States. Such universal cellular service could be attained only through the resolution of technical and proprietary issues hindering its deployment. This article focuses mainly on the cellular switch networking developments in the United States.

Background

The widespread growth of cellular service has been quite phenomenal since its commercial introduction in 1979. Currently, more than 14 million people subscribe to cellular service on a worldwide basis. These subscribers, who generate an estimated $10 billion of service revenue annually, are provided this luxury through the use of 2000 cellular telephone switches operating across 100 countries.

With more than 700 cellular switches currently in service, the United States as a country employs the highest utilization of cellular technology in the world. Its switches, maintained and operated by a number of different cellular service providers, called *operators* or *carriers*, provide service to 7.5 million subscribers.

During the early years of cellular commercial service in the United States, however, no industrywide plan existed to provide seamless operation to subscribers on a national scale. Diversity in ownership and switching equipment presented significant obstacles to be overcome. Establishing primary cellular communication networks in the shortest amount of time became the first objective for operators to pursue. With their focus centered on providing local coverage, the cellular evolution therefore began with building solitary "islands" of coverage. Built in highly populated urban areas, the islands of operating cellular switches are known as statistical metropolitan service areas (SMSAs) or MSAs for short. They also are called *cellular markets* (see Fig. 1).

Early Roaming Service

The focus on providing local coverage, coupled with the inability of dissimilar cellular telephone switches to communicate directly with each other, prohibited

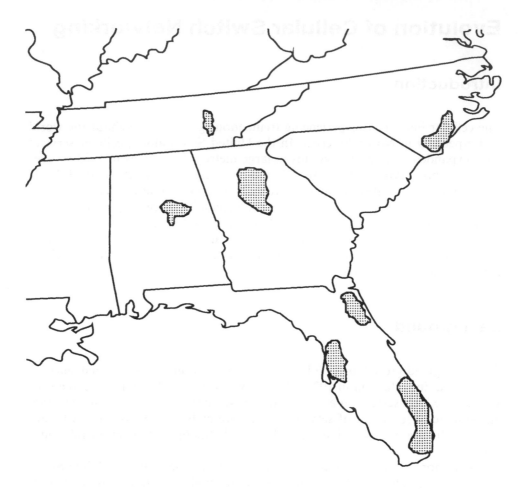

FIG. 1   Each cellular system as an island of coverage.

subscribers from using phones in all but their home cellular markets. As the number of subscribers grew, so did the demand to use their mobile phones while traveling into other cellular markets where they may not have had cellular subscription privileges. Service expansion for this type of traveling subscriber eventually was made possible through the creation of "roaming agreements" between various operators. The term *roaming* was coined to describe a subscriber's use of a mobile phone while traveling or "roaming" in cellular systems outside the home or system currently under subscription. The visited system operators classified these foreign subscribers as roamers. Agreements stipulated the operators' terms and conditions used in exchange for granting roamers access to service privileges in each other's markets. Among the terms included were usage charges, access validity, and, most important, operator liability for the service charges incurred by roaming subscribers. Providing service to roaming or foreign subscribers, however, presented operators with two issues requiring immediate attention.

The first issue was the lack of a mechanism to confirm the identity of a mobile unit requesting service in a visited system. Home markets provide such confirmation by maintaining databases linking a mobile unit's telephone or identification number (MIN, mobile identification number) to a unique electronic serial number (ESN) assigned to only one radio unit. The ESN, inserted by the manufacturer, is aimed at identifying users of stolen cellular telephones. This number in conjunction with the MIN forms a combination the home switch uses to substantiate credit worthiness and grant service access. Although the ESN/MIN combination creates an effective mechanism for local control of service access, operators did not have the capability to access the databases for making a similar evaluation of roamers who make calls in their systems.

Another problem confronting operators was the lack of an automatic means to deliver calls to subscribers who traveled away from the home market. The existing mechanism made use of a special telephone number, the *roamer access port*, assigned to each switch. This number provided callers access to a visited system's switch and supported a two-step process to support call delivery to roamers. Not only did it require the calling party to know the subscriber's location, but it also required the calling party to know the access port number of the system the subscriber currently was using. From the user's point of view, the process was laborious and complicated.

First Steps toward Automation

Third-party clearinghouses soon emerged to resolve these problems by facilitating an information interchange between the databases of the various cellular system operators. In essence, they acted as middlemen to transmit and store validation data between carriers. Clearinghouses also developed semiautomatic techniques to deliver calls to subscribers while away from home. Their emergence enhanced intersystem roaming capability within the United States.

The call-delivery process created by the clearinghouses was simple and straightforward. The subscriber only needed to dial a special code into his or her cellular telephone while in a visited system. The code alerted the clearinghouse of a roamer's desire to have calls delivered to his or her current location.

Once notified, the clearinghouse requested the visited system to supply a surrogate telephone number, the temporary local directory number (TLDN). The clearinghouse electronically forwarded the TLDN to the home system, which used it as a means to transfer incoming calls automatically to visited systems through the public switched telephone network (PSTN). No action by the caller was necessary for this to occur. However, since the system could not determine the continued presence of foreign subscribers, all roaming data was deliberately erased each midnight. This meant the code had to be reentered by the subscriber at the beginning of each day.

Validation was performed in a similar fashion but without necessitating any subscriber initiation. Confirmation of the MIN/ESN combination was relayed automatically to the visited system from the home system upon the completion of a roamer's call. Unfortunately, the employed method often took many hours and left open an opportunity for fraudulent use of the cellular system. It was, nonetheless, an improvement over not checking with the home system at all.

Finding a Common Solution

For the first time, "foreign" subscribers were validated automatically and their calls were being delivered to visited systems; a common solution was necessary as the process still had its shortcomings. First, it only validated subscribers after they completed their first call. Second, it required the subscriber to notify the clearinghouse of any change in location. This task became tedious when roamers traveled through several MSAs in a short time period. Finally, clearinghouses did nothing to address the handing off of in-progress calls when roamers cross the border from one service area into another. These shortcomings became the driving force in the development of a mechanism to allow switches from different manufacturers to share database information directly with each other. This inability to communicate was a consequence of manufacturers' use of proprietary protocols to offer features and enhancements unique to their cellular equipment. Variations in protocol were the result of different design approaches incorporated by each vendor into their switching systems.

The industry recognized that validation and call delivery could be performed in real time without third parties if dissimilar switches could communicate. Call handoffs could be treated transparently, as if under the control of a single system. In addition, direct communication between switches would enable visited systems to obtain the subscriber's full service profile, thus affording the subscriber the same level of service regardless of location.

To accomplish this goal, the industry needed to establish a forum to promote the exchange of ideas and information in a noncompetitive manner. In 1984, the Electronic Industries Association and Telecommunications Industry Association (EIA/TIA), created a working group made up of manufacturers, cellular system operators, and telecommunications experts. The purpose was to produce a protocol standard to be implemented by all manufacturers allowing direct communication between each other's switches in cellular networks based on the Advanced Mobile Phone System (AMPS). Armed with that protocol, carriers could begin to link the individual cellular islands and facilitate seamless coverage. This committee thus began its task of developing Interim Standard 41 (IS-41), a cellular networking protocol.

The IS-41 protocol defines a standard set of data messages recognizable by cellular telephone switches of different manufacturers. It is similar to the Open System Interconnection (OSI) structure for defining communications protocols between computers. IS-41 employs the physical, link, network, and application layers to define its protocol. The transport, session, and presentation levels are not used. In addition to defining the electrical characteristics of the physical links between switches, IS-41 also defines protocols for making connections, connecting entities in a network environment, and exchanging information between applications within the environment.

Introduction of Interim Standard 41

IS-41 was introduced first in the United States in September 1990 as part of a planned four-phase rollout. The initial phase of IS-41 is known as Revision 0 and outlines a protocol for addressing intersystem handoff and roamer validation. In one form or another, each switch manufacturer began to integrate IS-41 into their individual designs. It was the first step to link the databases on different switches without relying on the services of a third-party clearinghouse (see Fig. 2).

Functional operation of IS-41 relies on two types of database "registers" associated with each cellular market: the home location register (HLR) and visited location register (VLR). The HLR contains such data on each local subscriber as MIN, ESN, service feature profile, operating privileges or restrictions, and rate base. This information defines how a subscriber should be treated by his or her home switching system. The VLR holds a subset of the same information for each roamer traveling to a visited system. It is the IS-41 protocol that allows the visited systems to access the HLR of the home system and transport the data back into its own VLR. To transfer this information, IS-41 Revision 0 took advantage of existing point-to-point networks utilizing

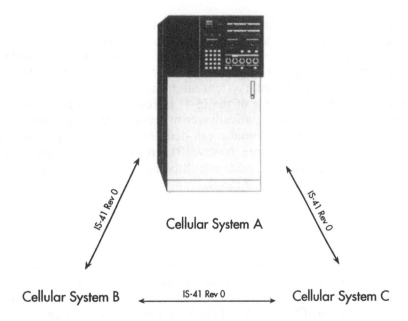

FIG. 2 Point-to-point connections used to pass Interim Standard 41 data using the X.25 packet protocol.

the X.25 protocol. X.25 provided a smooth vehicle for the introduction of IS-41 because it was an existing and acceptable carriage protocol used by the cellular industry.

Not only did direct access to the home database yield a real-time solution for validation, it performed this task before the call was initiated and connected. The new process significantly had the capability of reducing the risk associated with cellular fraud.

In addition to real-time validation, Revision 0 outlined a procedure for intersystem call handoffs. Handoffs take place when a subscriber leaves one service area and enters another while a call is in process. A handoff to an adjacent cell is initiated when a reduction in the mobile unit's signal strength is detected in the cell currently serving the subscriber. The cellular switch then chooses a cell or a group of cells to be the target cell for call handoff. A request is made of these target cells to take similar signal strength measurements to determine the best handoff candidate. The cell responding with the strongest signal strength then is requested to accept the call. The target cell must ensure an open radio channel exists before responding affirmatively to accept the handoff.

It is important to recognize that, prior to the introduction of IS-41, handoffs were limited to cells under the control of the cellular switch in the market currently serving the subscriber. Since dissimilar switches had no means to communicate with each other, handoffs to adjacent cells served by different switches in other markets could not be executed. In contrast, IS-41 facilitates handoff message transmission between switches from different manufacturers and therefore provides true seamless operation between different cellular markets.

The steps involved for doing this are complex because of the difficulty in designing coverage patterns that match service area boundaries. IS-41 offers the means for both systems to exchange information relating to the mobile unit's signal strength. Each system then decides which cell would provide the best coverage to assure a smooth intersystem call transfer.

Revision A is the next phase of the IS-41 introduction. In addition to offering cellular operators all of the advantages of Revision 0, it incorporates interswitch messages to enable automatic call delivery and feature profile transfer between systems. It furthermore removes the restriction of using only X.25 point-to-point connections and adds a multipoint networking carriage protocol known as Signaling System No. 7 (SS7).

Signaling System No. 7

SS7 is a telephony-specific real-time signaling network. Like IS-41, it has a multilayered protocol similar to the OSI structure. Its popularity in the PSTN was a natural outgrowth from the need to reduce call setup times and make more efficient use of interoffice trunks. This is accomplished through out-of-band common channel signaling techniques. In essence, SS7 is analogous to a high-speed data channel linking computers. These computers form the intelligent networking backbone controlling telephone circuit use.

SS7 offers the cellular industry numerous advantages over its X.25 counterpart. Many enhanced caller identification services (call trace, select forwarding, last number recall) currently offered to wire-line subscribers are functions supported by SS7. These services will be welcomed by a highly competitive industry as a means to increase revenue without increasing the subscriber base.

Enhanced services are made possible through the transfer of high-speed data between designated switching nodes established in the network to support SS7 traffic flow. SS7 messages are routed through a hierarchy of switching nodes and databases to provide greater data access in much less time. The average time required to process a query and response transaction is 1.5 seconds.

Another significant aspect of SS7's deployment in cellular networks will be the use of centralized databases for IS-41 message-routing information. Planned for Revision B of IS-41, a feature called *global title translation* will be implemented that makes use of SS7's intelligent networking capabilities. Global title translation will relieve operators from the cumbersome task of updating IS-41 message-routing tables within their switch every time a new switch or subscriber is added to the network. Simultaneous access to these databases by multiple switching systems is a crucial element in the SS7 network leading to a national seamless cellular roaming network.

Rural Communications

Now that cellular service is available in all MSAs, the industry is moving forward rapidly to build the smaller rural service area (RSA) markets. These areas for the most part are rural in nature and do not offer the potential of a large subscriber base as compared to their MSA counterparts. Their geographic size may be many times that of their MSA neighbors and present the RSA operators with a significant amount of roaming revenue opportunities. For these regions, IS-41 and SS7 will play a major role in providing wide-area coverage and seamless operation (see Fig. 3).

The Future

Soon U.S. cellular communications will rely on digital technology. Currently, the TIA working group is examining the unique network aspects that a digital environment presents. Their discussions have led to the development of IS-41 Revision B and coincide with the evolution of digital cellular systems. The last phase of the IS-41 rollout will be Revision C. Its purpose is to eliminate totally the proprietary protocols prohibiting cellular systems from communicating with one another.

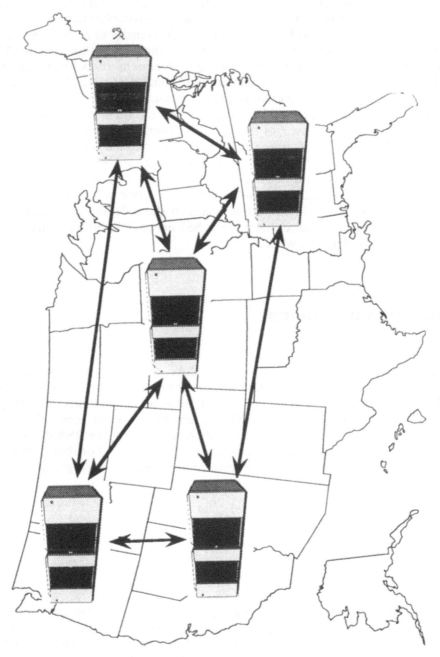

FIG. 3   Regional Signaling System No. 7 databases pass Interim Standard 41 messages between cellular switching systems.

International Solutions

The issues discussed thus far are not confined only to the United States. In Western Europe, for example, the emerging cellular telephone system is called the Global System for Mobile Communications (GSM, formerly Groupe Speciale Mobile). Fully digital in nature, it is timed for introduction in the early 1990s.

The GSM pan-European cellular radio standard is a result of an unprecedented collaboration between telecommunications administrations and industry experts from the whole of Western Europe. Many advanced features have been incorporated into the standard in an effort to maintain its predominance over the next 20 years. These advance features include data transmission, facsimile, connection to Integrated Services Digital Networks (ISDNs), short messaging service, cell broadcast, and roaming.

In order to accommodate comprehensive roaming services, GSM set out to develop a new digital signaling scheme using the mobile application part (MAP) of International Telegraph and Telephone Consultative Committee's (CCITT's) SS7 between mobile switching centers and location registers. This permits international operation between GSM networks as well as to the location registers. Additional capabilities provided by this approach include call setup, handoff, authentication, transfer of subscriber information, location updating, and supplementary services management.

Whereas IS-41 defines a standard protocol between switches, GSM extends the concept to specify standard interfaces between all network entities. This ensures a common standard for roamers and allows equipment of different manufacturers to be used anywhere in the network. GSM not only specifies the interfaces to be used by all manufacturers' switches, it also defines a standard interface between the switch and cell site equipment. Such standards allow operators greater flexibility in the selection of switch and radio equipment.

Other cellular systems also have been developed to meet the unique needs of each country or region. Some provisions for intersystem networking have been made within these systems on a regional basis. However, due to widely varying systems, operators, and regulatory restrictions, a single international standard for roaming is not expected.

Conclusion

Cellular switch networking facilitates a number of services for roaming subscribers, such as automatic access and call delivery. In addition, it provides an effective system for user validation, preventing fraudulent use of the system.

The future developments in cellular switch networking will be driven by subscriber demand and resources available to the system operators. A truly national cellular network may take years to evolve. Its emergence will be nourished by the formation of small regional networks to serve specific communities of interest. These small regional networks then will combine to form larger networks. The process will continue until all networks are connected in a cohesive manner. IS-41 and SS7 will be key tools used in this evolutionary process.

Addendum

Many things have changed since the submission of this manuscript. For example, the basic parameters of the industry have changed. Cellular systems boast more than 16 million subscribers worldwide with 11 million in the United States served by 1500 systems. (U.S. data are based on calculations by the Cellular Telecommunications Industry Association [CTIA, Washington, D.C.].)

IS-41 has also progressed as the demand for roaming and networking has increased. Revision A has been successfully deployed throughout the United States and IS-41 Revision B will be deployed in the second half of 1993.

In general, the requirement for networking cellular systems has exploded. The latter part of this decade will most likely witness the development of very large regional networks (e.g., Pan European, Latin America) that dominate in providing service. Networking products (e.g., signal transfer points [STPs], location registers) as well as the standards discussed throughout this article will be essential in the deployment of these networks.

Glossary

ELECTRONIC INDUSTRIES ASSOCIATION (EIA). An industry trade group located in Washington, D.C.

ELECTRONIC SERIAL NUMBER (ESN). The unique electronic identification number assigned to each cellular telephone unit by its manufacturer.

FEATURE PROFILE. Information related to the services accorded to a specific cellular telephone unit.

FOREIGN. *See* Roamer.

GLOBAL SYSTEM FOR MOBILE COMMUNICATIONS, formerly GROUPE SPECIALE MOBILE (GSM). A European digital cellular system standard being developed and implemented in the early 1990s.

HANDOFF. The transfer of an in-progress cellular call from one cellular switch to another.

HOME LOCATION REGISTER (HLR). The database associated with the cellular switching system in the "home" location of the cellular telephone unit. Each home subscriber's feature profile is stored in the HLR.

INTERIM STANDARD 41 (IS-41). A specification developed by members of the cellular industry under the sponsorship of the TIA and the EIA. The specification deals with interswitch communications for coordination of cellular telephone services.

INTERNATIONAL TELEGRAPH AND TELEPHONE CONSULTATIVE COMMITTEE (CCITT). A branch of the International Telecommunication Union of the United Nations. This committee creates recommendations for use by the international telephone authorities in establishing local signaling and operating standards.

MOBILE IDENTIFICATION NUMBER (MIN). The 10-digit number assigned to a cellular telephone unit used to call that telephone via the PSTN.

PACKET DATA NETWORK. A structure of computers configured to pass standardized "packets" of data from a source to another destination.

POINT-TO-POINT NETWORK. A network constructed in a manner causing messages to be passed from node to node, sometimes dealing with several nodes from the message's origin to its destination.

PUBLIC SWITCHED TELEPHONE NETWORK (PSTN). The land-line telephone network.

REGISTER. A computer database used to store information related to a specific cellular telephone unit.

ROAMER. A cellular telephone subscriber who uses the system of a cellular service provider for which the subscriber is not registered.

ROAMER ACCESS PORT. A special telephone number assigned to each cellular system and dialable via the PSTN; used to call roaming cellular telephones located in the coverage area of that system.

RURAL SERVICE AREA. Similar to the SMSA or MSA but used to define a rural or nonurban area of service.

SIGNALING SYSTEM NO. 7 (SS7). Defined by the CCITT and used to send data messages related to the establishment of telephone calls in the PSTN.

STATISTICAL METROPOLITAN SERVICE AREA (SMSA). A term used to define the geographic limits of a contiguous metropolitan area for purposes of cellular coverage.

TELECOMMUNICATIONS INDUSTRY ASSOCIATION (TIA). An industry trade group located in Washington, D.C.

TEMPORARY LOCAL DIRECTORY NUMBER. A number assigned to a roaming cellular telephone by the visited cellular system. This number is used by the home cellular system to forward calls to that telephone in the visited city.

VISITED LOCATION REGISTER (VLR). The database associated with each cellular system in which feature profile information related to roaming cellular telephones is stored.

X.25. A data protocol used in the transmission of datagrams within a packet network.

Bibliography

Beyer, Loraine, SS7 Interconnection for Cellular Mobile Carriers, presentation at Mobile Communications North America, Telocator, Toronto, Canada, April 1991.

Carlson, Kirk D., The Promise of Seamless Coverage, *Cellular Business*, 24–34 (November 1990).

Carlson, Kirk D., A Data Communications Primer for IS-41. *Cellular Business*, 78–112 (January 1991).

ElectroniCast Corporation, *Wireless Communications Market Analysis and Forecast*, ElectroniCast Corporation, San Mateo, CA, September 1991.

Electronic Industries Association, *EIA Interim Standard, Cellular Radiotelecommunications*, Intersystem Operations: Intersystem Handoff, EIA/IS-41.3, EIA Engineering Department, Washington, DC, February 1988.

Electronic Industries Association, *EIA Interim Standard, Cellular Radiotelecommunications*, Intersystem Operations: Intersystem Handoff, EIA/IS-41.2, EIA Engineering Department, Washington, DC, February 1988.

Kowalczyk, Henry M., Cellular Fraud, *Cellular Business*, 32–36 (March 1991).

Lee, William C. Y., *Mobile Cellular Telecommunications Systems,* McGraw-Hill, New York, 1989.

Steward, Shawn P. (ed.), The World Report '92, *Cellular Business*, 20–28 (May 1992).

Worthman, Ernest (ed.), Nationwide Call Delivery, *Cellular Marketing*, 35–38 (April 1991).

JACK M. SCANLON
MARTIN H. SINGER

Exchange Carriers Standards Association

The Exchange Carriers Standards Association (ECSA) was born of the Bell System divestiture with an initial charter to help develop a new environment to facilitate the development of technical network interconnection and interoperability standards. From that important, but narrowly focused beginning, the organization has grown to encompass a wide range of open industry forums that resolve a broad spectrum of access service issues, including billing, installation, testing, and maintenance. The ECSA is unique in two respects: it does not lobby in either regulatory or legislative arenas, and it is the only telecommunications industry group dedicated solely to problem solving based on consensus agreements. More than 2000 experts in technical and operational aspects of telephony, representing more than 300 companies, meet in a variety of forums open to any interested party. These forums resolve problems through creation of voluntary guidelines or develop technical standards through due process procedures sanctioned by the American National Standards Institute (ANSI).

ECSA was established in 1983 in anticipation of the break-up of the Bell System, which generally had established de facto technical interconnection standards for the United States. ECSA is a not-for-profit association of more than 130 wire-line exchange carriers ranging in size from the Regional Bell Operating Companies (RBOCs) with millions of access lines to small rural carriers with several hundred access lines. In total, the membership represents more than 95% of the nation's access lines.

The association is governed by a 21-member board of directors. It provides representation of exchange carrier interests in the spectrum of standards and related technical matters affecting the exchange carrier industry. The ECSA is located at 1200 G Street NW, Suite 500, Washington, DC 20005. ECSA's mission statement reflects the broad scope of activities its committees pursue (1):

> The Exchange Carriers Standards Association will actively promote the timely resolution of national and international issues involving telecommunications standards and the development of operational guidelines.
>
> ECSA will initiate and maintain flexible, open industry forums to address technical and operational issues affecting the nation's telecommunications facilities and services and the development of innovative technologies.
>
> ECSA will be an information source to its members, the forum participants, federal and state agencies and other interested parties.
>
> ECSA will promote industry progress and harmony with minimal regulatory or legislative intervention.

Because ECSA's forum participants represent a cross-section of the telecommunications industry, the Federal Communications Commission (FCC) frequently refers issues to ECSA for discussion and proposed resolution and acknowledges the expertise available in the forums (see Fig. 1 for an overview of ECSA's organization).

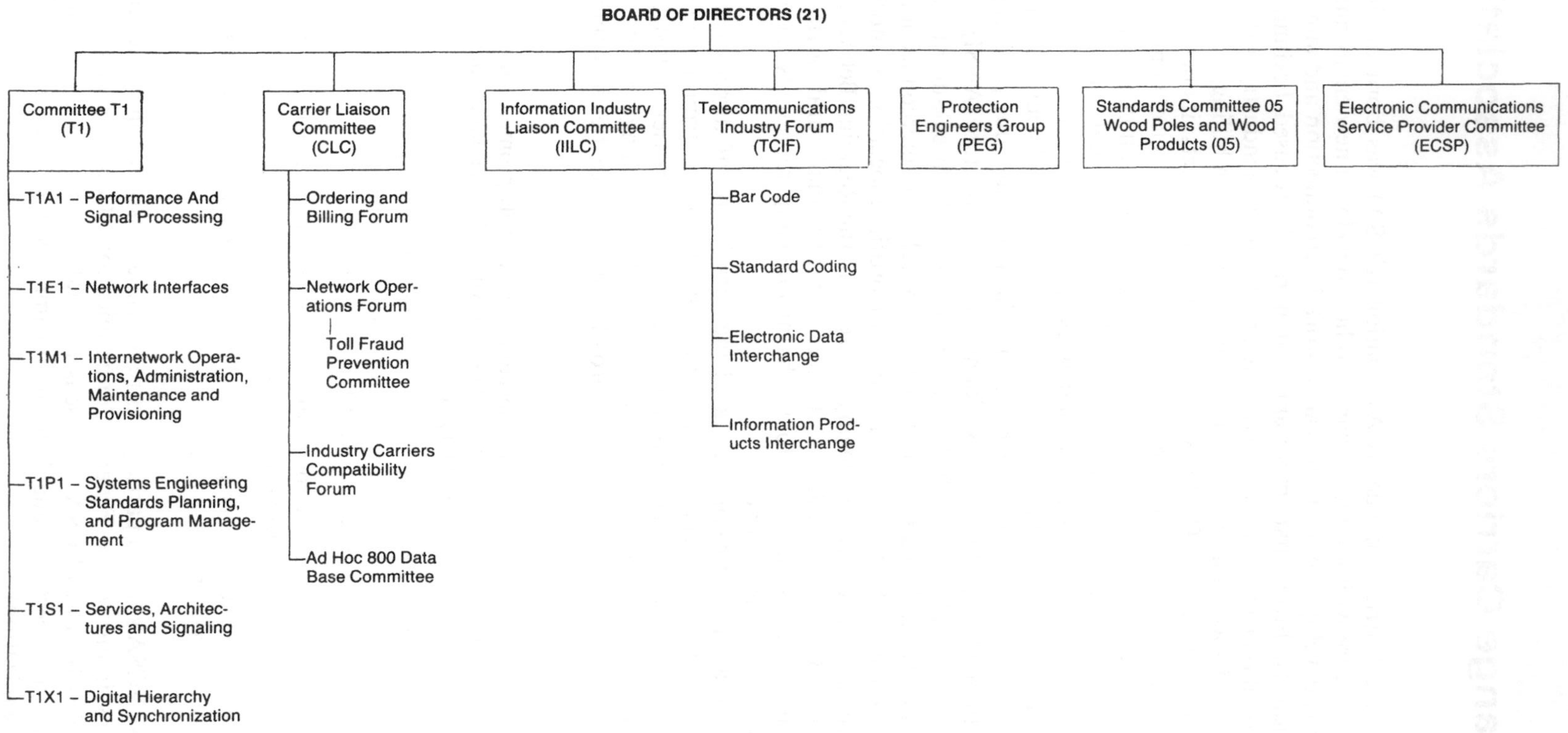

BOARD OF DIRECTORS (21)

Committee T1 (T1)

- T1A1 – Performance And Signal Processing
- T1E1 – Network Interfaces
- T1M1 – Internetwork Operations, Administration, Maintenance and Provisioning
- T1P1 – Systems Engineering Standards Planning, and Program Management
- T1S1 – Services, Architectures and Signaling
- T1X1 – Digital Hierarchy and Synchronization

Carrier Liaison Committee (CLC)

- Ordering and Billing Forum
- Network Operations Forum
  - Toll Fraud Prevention Committee
- Industry Carriers Compatibility Forum
- Ad Hoc 800 Data Base Committee

Information Industry Liaison Committee (IILC)

Telecommunications Industry Forum (TCIF)

- Bar Code
- Standard Coding
- Electronic Data Interchange
- Information Products Interchange

Protection Engineers Group (PEG)

Standards Committee 05 Wood Poles and Wood Products (05)

Electronic Communications Service Provider Committee (ECSP)

FIG. 1 Exchange Carriers Standards Association.

The ECSA sponsors six committees or forums; each sponsored group in turn may have many working subcommittees. The sponsored committees are:

1. Accredited Standards Committee T1-Telecommunications (Committee T1, or T1), which is accredited by ANSI to develop interconnection standards for United States networks
2. The Carrier Liaison Committee (CLC), which resolves nationwide problems involving the provision of exchange access services
3. The Information Industry Liaison Committee (IILC), which facilitates the exchange of information among interested parties on network capabilities and the development of Open Network Architecture (ONA) services
4. The Telecommunications Industry Forum (TCIF), which addresses issues on industry standards for bar coding and electronic data interchange in the telecommunications industry
5. The Protection Engineers Group (PEG), specialists who work on electrical protection of exchange carrier facilities
6. Accredited Standards Committee 05, which also is accredited by ANSI to develop standards for wood poles and wood products in the telecommunications industry

ECSA also represents exchange carrier interests directly through a standing committee of its board of directors. The Exchange Telephone Group Committee (ETGC) ensures that ECSA has qualified individuals representing member company interests in a wide variety of standards bodies not directly related to telecommunications. Through the ETGC, exchange carriers are represented in other ANSI-accredited organizations that develop standards for electrical, vehicular, and eyeglass safety, among others.

Membership in the ECSA itself is limited to wire-line exchange carriers. However, each of the sponsored committees or forums is open to and actively participated in by interexchange carriers, manufacturers, vendors, users, government agencies, other telecommunications trade associations, and enhanced-service providers. Following is a more detailed examination of each of the sponsored committees of the ECSA.

Accredited Standards Committee
T1-Telecommunications

Committee T1 was established in 1984. ECSA provides the sponsorship and secretariat support and T1 is accredited by ANSI, as are all of the successful standards developers in the United States. T1's mission is to develop (2)

> standards and technical reports related to interfaces for United States networks which form part of the North American telecommunications system. T1 also develops positions on related subjects under consideration in various international standards bodies.

Specifically, T1 focuses on those functions and characteristics associated with the interconnection and interoperability of telecommunications networks at interfaces with end-user systems, carriers, and information and enhanced-service providers. These include switching, signaling, transmission, performance, operation, administration and maintenance aspects.

Committee T1 is also concerned with procedural matters at points of interconnection, such as maintenance and provisioning methods and documentation, for which standardization would benefit the telecommunications industry.

Committee T1 has developed more than 100 technical standards, many of which are implemented by standards developers in other countries. In addition, Committee T1 annually prepares about 1000 technical papers for use by the United States Department of State as U.S. positions in standards discussions with the International Telecommunications Union (ITU). The structure of Committee T1 was a model for the Telecommunications Technology Council (TTC) in Japan and the European Telecommunications Standardization Institute (ETSI).

Committee T1 has six working technical subcommittees: T1A1, Performance and Signal Processing; T1E1, Network Interfaces; T1M1, Internetwork Operations, Administration, Maintenance, and Provisioning; T1P1, Systems Engineering, Standards Planning, and Program Management; T1S1, Services, Architecture and Signaling; and T1X1, Digital Hierarchy and Synchronization. Committee T1's governing body, the T1 Advisory Group, is made up of the chair, vice chair, and representatives of the various interest groups that are members of T1: exchange carriers, interexchange carriers, manufacturers, and general-interest groups (including government agencies).

Committee T1 is a leading exponent of the need for greater globalization of standards. Committee T1 convened the first Interregional Telecommunications Standards Conference in the United States in 1990. Conferences have been held annually since to promote closer cooperation and improve efficiency in the development of telecommunications interconnection standards. Committee T1 also organized the first Americas Telecommunications Standards Symposium in 1992 to encourage greater cooperation in the hemisphere in standards development to promote interoperability of telecommunications networks in the Americas, and to promote trade growth and global standards based upon users needs. The Brazilian Standards Association will be the host organization for the 1993 symposium.

Carrier Liaison Committee

The CLC was created by ECSA in 1984 and endorsed by the FCC in 1985 as an interindustry organization for discussion and voluntary resolution of nationwide issues involving the provisioning of exchange access service.

The CLC is the umbrella organization for several groups:

The Network Operations Forum (NOF), which focuses on installation, maintenance, and testing issues associated with access service for interexchange
carriers, including toll fraud prevention procedures

The Ordering and Billing Forum (OBF), which resolves issues involving exchange and interexchange carriers on ordering, provisioning, billing, and
subscription processes

The 800 Data Base (800 DB) Committee, which deals with operational issues
involving the provision of 800 service

The Industry Carriers Compatibility Forum (ICCF), which deals with national
technical (nonstandard) issues associated with numbering and network interconnection

The CLC and its subtending forums represent the broad spectrum of industry players. The voluntary guidelines and recommendations they produce represent significant industry effort and cooperation and identify technical and operational network problems, exchange information on these problems, and seek
consensus resolution of the problems in a timely manner. Thus, while a proposed resolution of a specific issue may not be a participant's first choice, an
effort is made to determine whether it is one that the participant can accept and
support.

The CLC monitors the work of the forums, but generally does not engage
in a substantive review or resolution of issues. Any interested party that feels
its position was not heard adequately in a forum may bring an issue to the
CLC. The CLC forums have resolved more than 1500 issues to date.

Network Operations Forum

A great deal of the work of the CLC's NOF on network reliability issues was
used by the FCC's Network Reliability Council to form the basis for recommendations to the FCC on how to enhance and ensure maximum reliability among
the nation's many public telecommunications networks.

The work of the NOF is codified in a series of publications. The Network
Management Guidelines and Contact Directory provides local-exchange and
interexchange carriers with alternative guidelines for traffic management in the
event of network congestion due to facility failures or abnormal calling periods,
switch or network failures, or Signaling System No. 7 (SS7) failures.

Another publication, the Installation and Maintenance Operations Reference Guide details interexchange carrier and local-exchange carrier responsibilities for the SS7 interface between carriers. It deals with the physical interconnecting trunks and link(s) as well as the software necessary to support the use
of the link(s) to transport SS7 messages between carriers.

A third publication is the Testline Guidelines and Coordinator Directory for
personnel involved in testing access services. The document also is an operations

reference for installation and maintenance of special access services, WATS (wide-area telecommunications service) access lines, switched access services feature groups A, B, C, D, and X.75 gateway services and SS7 link and trunk services provided by exchange carriers to interexchange carriers.

The Toll Fraud Prevention Committee of the NOF develops procedures and programs to deter domestic and international fraud as it affects both local-exchange carriers and interexchange carriers. The group periodically issues detailed white papers on particular types of fraud and reviews industry efforts aimed at deterrence.

Ordering and Billing Forum

The OBF has resolved nearly 1000 issues involving the ordering, provisioning, and billing of access services. The OBF regularly updates the Multiple Exchange Carrier Access Billing (MECAB) documents that provide meet-point billing procedures and distributes them to the industry and the FCC. The OBF also resolves issues associated with ordering, provisioning, and billing for services under ONA and 800 DB service.

800 Data Base Committee

The 800 DB Committee is an open forum for 800-service providers to review operational issues and develop voluntary guidelines to enhance reliable 800 DB service.

Industry Carriers Compatibility Forum

The ICCF provides open discussion and resolution of technical issues related to network interconnection. Among the issues it addresses are transmission quality, conservation of carrier identification codes and other numbering issues, SS7 interconnection, and trunk availability.

The Information Industry Liaison Committee

The IILC was chartered in 1987 as an interindustry forum for discussion and voluntary resolution of industrywide concerns about the provision of ONA services and related matters. IILC facilitates the exchange of information on network capabilities and the development of enhanced services. The IILC holds periodic public forums to discuss proposals for unbundling the public-switched network and to advise enhanced-service providers about future network technology planning and implementation schedules.

The Telecommunications Industry Forum

The TCIF was created in 1986 to respond to the growing need for voluntary guidelines to facilitate the use of new technology that offers cost savings throughout the telecommunications industry. These technologies include electronic data interchange and bar coding as well as other forms of automatic identification technology. TCIF provides a forum for purchasers, manufacturers, and suppliers of telecommunications equipment, products, and services to address issues relating to industry standards. TCIF committees research, analyze, and experiment with technologies and then issue industry-specific guidelines for use of the technology in the telecommunications industry.

The Protection Engineers Group

The PEG develops draft contributions for introduction to ANSI-accredited standards forums, provides guidance to ECSA representatives on standards organizations, and encourages the free flow of ideas among electrical protection specialists. A PEG working group develops draft technical specifications for such electrical protection apparatus as surge arrester modules and station protectors. These draft specifications provide a valuable reference to exchange carrier companies, which may adopt all or part of the document in their purchase specifications.

Accredited Standards Committee 05

Accredited Standards Committee 05 develops standards for use by the telecommunications industry in technical areas dealing with wood poles and other wood products. The committee was organized in 1924 and includes representatives of users and manufacturers of wood poles and wood products and is accredited by ANSI.

Conclusion

The Association's Board of Directors will constantly review the membership structure and initiate changes to reflect the dynamic nature of the telecommunications industry and to be responsive to its many constituencies, including the Federal Communications Commission and a variety of providers of telecommunications services.

References

1. Exchange Carriers Standards Association, *Annual Report*, ECSA, Washington, DC, 1991, p. 1.
2. Exchange Carriers Standards Association, *Committee T1 Strategic Plan* (revised), ECSA, Washington, DC, 1993.

GEORGE L. EDWARDS

Facsimile Image Coding

(see Coding of Facsimile Images)

Facsimile Standards

Although facsimile systems have existed for more than 25 years, only recently have we seen explosive growth in the use of facsimile, with a corresponding increase in facsimile standards activity. Current standards work concentrates on improving the quality and speed of image transmission and on increasing functionality. The goal is to benefit from the developing technology—higher-resolution scanners and printers, and improved processing, storage and communications technology. Topics now under discussion include (but by no means are limited to) transmission of color images, improved compression of bilevel images, facsimile routing, facsimile carried by the Integrated Services Digital Network (ISDN), and facsimile applied to teleconferencing. Personal computers (PCs) can operate as or can be used to add functionality to facsimile terminals. The result is new standards activity (e.g., binary file transfer [BFT], database storage and retrieval, computer–facsimile communication, and an applications programming interface). This article describes the facsimile and facsimile-related standards activity with particular emphasis on function and performance. Image-compression algorithms are an important factor. Communications protocols also are addressed.

Introduction

Facsimile Standards Organizations

The responsibility for international facsimile standards lies with the International Telegraph and Telephone Consultative Committee (CCITT). The CCITT is made up of a number of study groups, each working on a particular aspect of telecommunications standardization. Study Group VIII (SG VIII) develops facsimile standards (called *Recommendations* by the CCITT), while SG XVII is responsible for modems, including facsimile modems. The member bodies of the CCITT represent countries; in the United States, the Department of State is responsible for and coordinates contributions to the CCITT through its committee for the CCITT. This committee has a number of subgroups, each responsible for a part of the CCITT's work. Currently, Study Group D of the U.S. committee is responsible for facsimile. Technical support for the U.S. position is provided primarily by the Telecommunications Industry Association (TIA) TR-29

Committee, with a mission for the development of standards for facsimile terminals and facsimile systems. TR-29, organized in the early 1960s, was a major contributor to the 1980 Group 3 Recommendation. At that time, all of the facsimile-related standards activity resided in TR-29. Although TR-29 still is responsible for U.S. facsimile standards, now other U.S. organizations are responsible for related standards that might be referenced by the facsimile standards. For example, X3L3 is responsible for image compression, and X3V1 is responsible for office document architecture. Related standards organizations and their roles are described in this article as well.

Facsimile Standards Chronology

The CCITT produced the first international facsimile standard in 1968, Recommendation T.2 (Rec. T.2) for Group 1 facsimile (1). In North America, a facsimile standard was used that was similar to Rec. T.2, but with enough of a difference so that North American machines were not interoperable with the rest of the world. The Group 1 standard provided for a six-minute transmission of a page that was a nominal 210 millimeters (mm) by 297 mm at a scanning density of 3.85 lines per mm. In 1976, the first truly international facsimile standard, T.3 for Group 2 facsimile, was published (2). The U.S. manufacturers, working through TR-29, were the major contributor to the Group 2 standard. A Group 2 facsimile machine transmitted a page in half the time of a Group 1 machine with about the same quality. Both of these machines were analog, and neither used image compression.

At the same time that the Group 2 standard was being completed, the CCITT began work on Group 3. In 1980, the Group 3 standards (Recommendations T.4 and T.30) were published (3,4). Group 3 provided significantly better quality and shorter transmission time than Group 2, accomplished primarily by digital image compression. The Group 4 standard followed in 1984, defined by a number of CCITT recommendations. The intent of Group 4 was to provide higher quality (twice the resolution of Group 3), higher speed (via digital networks), and more functionality. Since 1984, work has continued on both the Group 3 and Group 4 standards, with most of the advances being made in Group 3. This article describes this work, starting with the development of the Group 3 compression algorithm, and continuing through the current activities of both Group 3 and Group 4.

Group 3 Facsimile

Group 3 Compression Algorithm Selection

In 1976, the CCITT SG XIV began work on the Group 3 recommendations. With respect to the compression algorithm, the committee quickly decided on a standard one-dimensional coding technique known as the modified Huffman code (MHC). This coding technique was mandatory for all Group 3 machines, and provided a basis for interoperability. It was relatively simple to implement even using the technology of the time, and produced acceptable results on noisy telephone lines. Work proceeded on an optional two-dimensional coding scheme with the objective of providing higher compression. The criteria for the two-dimensional algorithm selection emphasized compression and error susceptibility; of course, implementation complexity and compatibility with the one-dimensional code also were considered. Seven proposals for the optional two-dimensional code were submitted to the CCITT from Japan, IBM Europe, the British Post Office, 3M Company, AT&T, Xerox, and Germany. These algorithms are compared below.

Modified Huffman Code

A digital image to be transmitted by facsimile is formed by scanning a page from left to right and top to bottom, producing a bit map of picture elements (pels). A scanline is made up of runs of black and white pels. Instead of sending bits corresponding to black and white pels, coding efficiency can be gained by sending codes corresponding to the lengths of the black and white runs. A Huffman procedure uses variable length codes to represent the run lengths (5); the shortest codes are assigned to those run lengths that occur most frequently. Run-length frequencies are tabulated from a number of "typical" documents, and then are used to construct the code tables. True Huffman coding would require twice 1729 codewords to cover a scanline of 1728 pels. To shorten the table, the Huffman technique was modified (thus the name) to include two sets of codewords, one for lengths of 0 to 63 (terminating code table), and one for multiples of 64 (makeup code table). Run lengths in the range from 0 to 63 pels use terminating codes. Run lengths of 64 pels or greater are coded first by the appropriate makeup codeword specifying the multiple of 64 less than or equal to the run length, followed by a terminating code representing the difference. For example, a 1728-pel white line would be encoded with a makeup code of length 9 representing a run of length 1728, plus a terminating code of length 8 representing a run of length 0, resulting in a total code length of 17 bits (without the synchronizing code). Note that when the code tables for Group 3 were constructed, images containing halftones deliberately were excluded. This was done so as not to skew the code tables and degrade the compression performance for character-based documents. Although investigators were working on halftone compression, technology was not adequate for practical systems.

Compression and Error-Sensitivity Comparison

As stated above, the main criteria for the two-dimensional algorithm selection were compression and susceptibility to errors. The CCITT specified that CCITT Documents 1, 4, 5, and 7 were to be used for the evaluation at a transmission speed of 4800 bits per second (b/s) and a minimum coded line length of 96 bits. The four test images were from a set of eight images collectively known as the *CCITT Standard Images*, digitized by the French Administration at 200 pels per inch (pels/in), and recorded on magnetic tape. Although these images were never an official standard, they have been used extensively by experimenters. The error patterns were supplied on magnetic tape by Germany.

Compression and error sensitivity data for the seven candidate algorithms (6), together with the selected Modified READ (relative-address-designate) (MR) technique, are plotted in Fig. 1. The compression factor, as shown in the figure, is defined as the number of pels in the uncompressed image divided by the number of compressed bits transmitted. The error-sensitivity factor is defined as the number of incorrect pels (due to transmission errors) in the decoded image divided by the number of transmitted bits in error. Of the seven candidates' proposals for the two-dimensional code, the IBM proposal provided the best compression (not by much), with error sensitivity similar to those with comparable compression. The selected Group 3 compression technique (MR) (3) actually is closest to the British proposal, the difference being the code assignments for the two-dimensional codes.

Modified READ Code

The MR coding method makes use of the vertical correlation between black (or white) runs from one scanline to the next (called *vertical mode*). In vertical mode, transitions from white to black (or black to white) are coded relative to the line above. If the transition is directly under (zero offset), the code is only 1 bit. Only fixed offsets of zero and ±1, ±2, and ±3 are allowed. If vertical mode is not possible (e.g., when a nonwhite line follows an all-white line), then horizontal mode is used. Horizontal mode is simply an extension of MHC; that is, two consecutive runs are coded by MHC and preceded by a code indicating

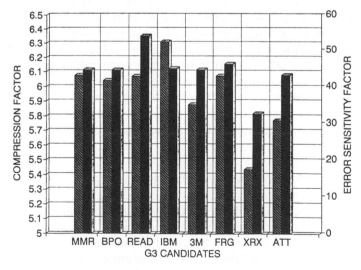

FIG. 1 Compression and error sensitivity for Group 3 candidates and Modified READ (relative-address-designate).

horizontal mode. To avoid the vertical propagation of transmission errors to the end of the page, a one-dimensional (MHC) line is sent every Kth line. The factor K typically is set to 2 or 4, depending on whether the vertical scanning density is 100 or 200 lines per inch (in). The K factor is resettable, which means that a one-dimensional line may be sent more frequently (than 2 or 4) when considered necessary by the transmitter. The synchronization code (EOL, for end of line) consists of 11 "0"s followed by a "1," followed by a tag bit to indicate whether the following line is coded one-dimensionally or two-dimensionally. In the next section, other aspects of the original Group 3 facsimile standard relevant to imaging are described briefly.

Group 3 CCITT Recommendations (1980)

The Group 3 facsimile recommendations T.4 and T.30 were first published in 1980. Recommendation T.4 covered the compression algorithm and modulation specifications, and Recommendation T.30 described the protocol. Group 3 was designed to operate on the General Switched Telephone Network (GSTN) in half-duplex mode. The protocol specifies a standard basic operation that all Group 3 terminals must provide plus a number of standard options. The various parameters and options that may be used are negotiated at 300 b/s before the compressed image is transmitted at higher speed. The protocol also accommodates the use of Group 1 and Group 2 machines, enabling a facsimile terminal that implemented all three groups to be able to communicate with whichever machine answered the call. Few Group 1 and Group 2 terminals are in service today, so this feature is not important; however, it did help Group 3 gain market acceptance with the established base of Group 2 machines.

Standard Capabilities

All Group 3 machines must provide the following minimum set of capabilities: a pel density of 204 pels/in horizontally and 98 pels/in vertically; one-dimensional compression using the modified Huffman code; a transmission speed of 4800 b/s with a fallback to 2400 b/s; a minimum time per coded scanline of 20 milliseconds (ms) (to allow real-time printer operation without extensive buffering); and a mechanism to escape the standard protocol, called *nonstandard facilities* (NSFs). The NSF has eased the addition of new features to the facsimile standards by providing an easy means of "testing the waters."

Standard Options

The standard options defined in the 1980 recommendations are the use of the Modified READ code to achieve two-dimensional compression; a vertical pel density of 196 pels/in to provide higher image quality; a higher transmission speed of 9600 b/s with a fallback to 7200 b/s; and a minimum coded scanline time of 0 to 40 ms.

Group 4 Facsimile

Group 4 CCITT Recommendations (1984)

During the time that the Group 3 recommendations were being completed, work began on Group 4. The intent for Group 4 was to transmit higher-quality images in a shorter time over digital circuits. In addition, Group 4 terminals were to be capable of sending both characters and raster graphics (facsimile). The International Organization for Standardization (ISO) seven-layer architecture was employed to provide an error-controlled communications channel, allowing the use of a two-dimensional compression algorithm. Transmission speeds of 56 or 64 kilobits per second (kb/s) allowed an image scanned at 400 pels/in to be transmitted in approximately 5 seconds (s).

At the time the Group 4 standard was first published, the application layer was specified by just two recommendations, T.5 and T.73 (7,8). Since 1984, CCITT standardization activity has resulted in a number of changes. Document transfer and manipulation (DTAM) recommendations have been expanded considerably; in particular, T.73 has been replaced with the T.400 series of recommendations. This set of recommendations, developed in cooperation with ISO, provides for transfer of processable or formatted documents containing any combination of character, raster graphics, or geometric graphics content. There are 12 new recommendations in this set, of which 7 are identical with ISO 8613 (9). Recommendations pertaining specifically to Group 4 have been recast. That is, Recommendation T.5 has been recast into three new recommendations: T.503 (G4 Document Application Profile), T.521 (Communications Application Profile), and T.563 (Terminal Characteristics) (10–13).

Group 4 Class Structure

Three classes of Group 4 terminals were defined (see Table 1):

Class 1:   A terminal able to send and receive facsimile documents
Class 2:   A terminal, in addition to having Class 1 capabilities, able to receive teletext and mixed-mode* documents
Class 3:   A terminal, in addition to having Class 1 and Class 2 capabilities, able to generate and send teletext and mixed-mode documents

These three classes provide a wide range of capability. Group 4 Class 1 provides functionality similar to Group 3, while Classes 2 and 3 permit mixed-mode operation with its promise of higher compression. All is not as well as it might be, however. The Group 4 recommendations were published at a time when the

---

*Mixed-mode documents contain a mixture of teletext and facsimile data on the same page. For example, a page consisting of line art and text could be sent as a mixed-mode document; the line art could be sent as facsimile data; and the text could be sent as teletext data.

TABLE 1 Group 4 Class Characteristics

	Class		
	1	2	3
Pel-density of scanner-printer (pels/25.4mm)	200	300	300
Pel transmission density (pels/25.4mm)	200	200/300	200/300
Pel transmission conversion capability	not required	required	required
Mixed-mode capability	not required	not required	required
Optional pel density of scanner-printer	300/400	400	400
Combined with pel transmission density (pel/25.4mm)	200/300/400	200/300/400	200/300/400
Storage	not required	not required	required

CCITT was evolving the Open Document Architecture (ODA) and aligning its application of the ISO seven-layer architecture. Since the Group 4 standards were published, both the document structure and communications protocol standards have been changed and expanded considerably by the CCITT. Unfortunately, the Group 4 recommendations have not maintained compatibility with the evolving CCITT integrated approach. As a result, Class 1 does not interoperate with Classes 2 and 3 since both the protocols and document structure are slightly different. How to rectify this situation is still an open question.

Group 4 Compression

Standard Compression Algorithm

The Group 4 bilevel image-compression algorithm, described in CCITT Recommendation T.6 (14), is very similar to that of Group 3 (Rec. T.4). The same Modified READ code is used, but only two-dimensional lines are transmitted with no EOL codes for synchronization. An error-free communications link makes this possible. A white line is assumed before the first actual line in the image. No fill is used; adequate buffering is assumed to provide a memory-to-memory transfer. The algorithm was changed slightly to accommodate the larger number of pels per line (up to approximately 4900) resulting from the 400-pels/in scanning density, without adding any new codewords. The 1984 recommendations applied to bilevel images only; gray scale and color were for further study.

Proposed Optional Pattern-Matching Algorithm

In 1983, AT&T proposed to the CCITT an optional coding scheme for Group 4 facsimile. Although their contribution was titled, "Pattern Recognition for Group 4 Facsimile" (15), the method really employs pattern matching as opposed to pattern recognition. The term *pattern recognition* usually implies some understanding of that which is recognized. For example, a specific letter of the alphabet conveys some information. With pattern matching, a pattern is not recognized, but simply is matched with already transmitted patterns and, if a correct match is detected, it is replaced by the matching pattern. It does not use the image-understanding level. The image-understanding approach has the potential advantage of a very high compression, but generally is not applicable to any document content. The matching approach yields lower compression, but keeps more of the original pictorial information. Of course, neither of the techniques is lossless since they modify the image content.

The proposed pattern-matching algorithm processes the facsimile image by extracting patterns from the image and attempting to match the extracted pattern with one in a stored library of patterns. The input image is examined line by line. When a black pel is found, an attempt to isolate the pattern to which it belongs is made. If the pattern cannot be isolated within a window (32 × 32 bits for 200 lines/in) a piece of the pattern is extracted from the image. Isolated patterns then are compared with already identified patterns stored in a library. If a match is found, the position of the pattern in the image and its location in the library are coded. If no match is found the pattern is added to the library and the bit image of the pattern and its position are coded.

Note that this technique requires standardization of only the library symbol management and communication between coder and decoder. The process of pattern extraction, matching, and the criteria for a match are open to invention. The specific matching technique described in Ref. 15 and also in a companion paper, Ref. 16, was presented only to demonstrate feasibility and not proposed as part of the standard. The algorithm described in Ref. 15 was simulated (17) on CCITT Images 1, 5, and 7 at image pel densities of 200, 300, and 400 pels/in. Figure 2 shows that the compression for the pattern-matching algorithm is approximately three times that of the T.6 encoding of Group 4. In addition to measuring the compression of the pattern-matching technique, a subjective evaluation was made of the distortions produced since the decoded image is not identical to the original. At 200 pels/in, there were a few symbol substitutions in lowercase alpha characters, punctuation, and small-size uppercase alpha characters. At 300 and 400 pels/in, no substitutions were apparent. Although the AT&T proposed pattern-matching technique showed promise of very high

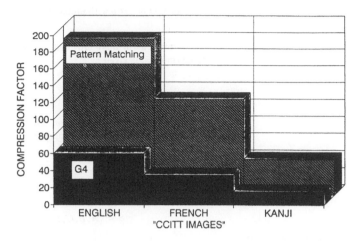

FIG. 2 Group 4 and pattern-matching compression.

compression, CCITT SG VIII showed little interest, probably because the algorithm is not image preserving.

Group 4 Compression Compared with Group 3 Compression

Many comparisons of Group 4 (T.6) compression with that of Group 3 (T.4) have been published over the years; however, very little has been reported on compression of screened halftone or electronically dithered images, probably because neither Group 3 nor Group 4 performs well on these images. Also the CCITT standard images used over the years were digitized at 200 pels/in and do not include any computer-generated images. In order to establish a baseline against which to compare the current algorithm work, it is instructive to compare Group 3 with Group 4 using a more comprehensive image set. One such image set being used currently for bilevel algorithm development is known as the *Stockholm image set*. The set contains 13 images, 11 digitized at 400 pels/in, 1 at 200 pels/in, and 1 at 800 pels/in. A sampling of this image set is illustrated in Figs. 3 and 4. Images 1 and 2 are English-text images (digitized upside down), and the Image 3 is a Japanese newspaper. The next four images (4–7) were derived from the same gray-scale source image. Each was converted to a bilevel image by an ordered dither method, but with different parameters. Image 8 contains handwriting, and Image 9 is a digitized magazine page composite. Images 10 and 11 are engineering drawings: the first is a scanned drawing and the second is a computer-generated blueprint. Image 12 was produced with an error-diffusion dithering method digitized at 200 pels/in. Image 13 is typical of computer-generated graphics, and is digitized at 800 pels/in.

Each of these images was compressed by the T.4 one-dimensional code, T.4 two-dimensional code, and T.6. The compression factor for each of these simulations is shown in Fig. 5. In general, T.6 coding outperforms T.4 two-dimensional coding, which outperforms T.4 one-dimensional coding. The differences tend to be smaller on "busy" images and greater on images containing urce alphabet of size n with probabilities $p(s_1), p(s_2), \ldots, p(s$

he information (in bits) provided by the occurrence of urce symbol s_i is given by

$$I(s_i) = \log_2 \frac{1}{p(s_i)} \quad \text{bits}$$

he average amount of information obtained per symbol om the source is called the **entropy** $H(S)$ of the source:

$$H(S) = \sum_{i=0}^{n} p(s_i) \log_2 \frac{1}{p(s_i)}$$

xample:

$$S = \{A, B, C, D\}$$

$$P(A) = \frac{1}{2}, \quad P(B) = \frac{1}{4}, \quad P(C) = \frac{1}{8}, \quad P(D) = \frac{1}{8}$$

$$H(S) = 1\frac{3}{4} \quad \text{bits}$$

8

nue
14

firm our meeting of September 9, 1988
inerary and directions for reaching the

 to meeting you next week.   If you
e feel to contact me at (555) 526-8769.

                Sincerely,

                Frank

                Frank Weiner
                Supervisor
                Image Electronic

COMPANY · U S APPARATUS DIVISION
IAD · ROCHESTER NEW YORK 14650

FIG. 3   Stockholm images 1 through 4.

FIG. 4 Stockholm images 5 through 8.

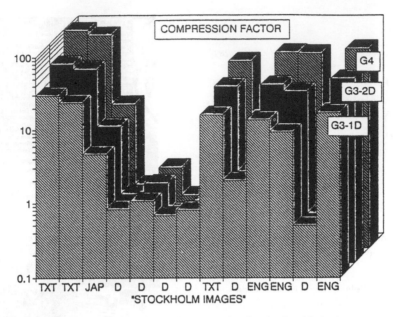

FIG. 5 Group 3 and Group 4 compression for the Stockholm images.

more white space. Note that on dithered images the compression is very poor (or even negative), and one-dimensional coding outperforms the other methods in some cases.

Group 3 1988 Options

In the CCITT study period 1985–1988, three options for Group 3 facsimile were developed and approved for inclusion in the 1988 *Blue Book*. These options included a small-page-size facsimile called *notefax*, and two optional methods of error control.

Optional A5, A6 Page Size

The A6 page size is one-fourth of an ISO A4 page, or 864 pels across a scanline of 107 mm. The A5/A6 machine was intended for home use, as opposed to business use; although the A5/A6 size is nominally "optional," the standard A4 size is not required to be implemented if the A5/A6 size is implemented. This means that the A5/A6 machine is really an alternative to, rather than an option of, the A4 machine. In any case, the A5/A6 machine does interoperate with the standard A4 machine by enlarging (or reducing) the image that it sends (or receives). Unexpected results may be obtained if one is to send a complex document to an A5/A6 machine unknowingly. That is, the document may be reduced to the point that it is unreadable. Interworking between A5/A6 and A4 facsimile machines is summarized in Table 2. So far, the A5/A6 machines have not gained market acceptance in the United States.

Error-Limiting Mode

One of the two methods for error control adopted in 1988 is the optional error-limiting mode. This technique applies only to one-dimensional coding, and is not true error correction. Each line is divided into parts (12 parts for a line composed of 1728 pels). The parts are classified as white or nonwhite and are designated as such in a line header. Each part is separately run-length coded and the number of coded bits for each part is stored following the header. When decoding, if a part is found to be corrupted it can be approximated by repeating the corresponding part from the previous line. The intent of this technique is to limit the effect of an error to a fraction of the line rather than the whole line. If an error occurs in the header or the coded bit counts, then the whole line probably is corrupted anyway. This technique, although simple to implement, has seen little use in the United States, probably because the alternative described below provides better performance.

Error-Correction Mode

The optional error-correction mode applies to one-dimensional and two-dimensional coding, and provides true error correction.

Objectives

When error correction was proposed by the United Kingdom in 1985, the idea was greeted with some skepticism. Service providers in countries with high-quality telephone networks did not want to suffer the extra overhead that error correction typically entails. Manufacturers of facsimile machines were concerned with the added complication. Therefore, the objectives included minimizing the changes to existing recommendations required for error control and minimizing the transmission overhead in channels with low error rates. The primary objective, of course, was to perform well against burst errors.

Features

The selected error-correction scheme is known as *page selective repeat ARQ* (Automatic Repeat Request). The compressed image data is embedded in HDLC (High-Level Data-Link Control) frames of length 256 octets or 64 octets, and transmitted in blocks of 256 frames. The communications link operates in a half-duplex mode; that is, the transmission of image data and the acknowledgment of the data are not sent at the same time. The technique can be thought of

TABLE 2 A5/A6 and A4 Internetworking

Terminal Capabilities Reception Side

PARAGRAPH	2.2	2.5	2.7	2.6	2.8
Horizontal Resolution	1728 pels per 215mm	864 pels per 107mm	1728 pels per 107mm	1216 pels per 151mm	1728 pels per 151mm
Vertical Resolution	3.85 1/mm / 7.7 1/mm	7.7 1/mm / 15.4 1/mm	7.7 1/mm / 15.4 1/mm	5.44 1/mm / 10.9 1/mm	5.44 1/mm / 10.9 1/mm
Pel Process	original 1728	864 (1728 * .5) note 1	original 1728	1216 (1728 * .7) note 2	original 1728
DCS DIS-DTC:	—	BIT 33 = 1 / BIT 35 = 1	BIT 33 = 1 / BIT 37 = 1	BIT 33 = 1 / BIT 34 = 1	BIT 33 = 1 / BIT 36 = 1

Situation at Transmission Side

PARA	Horizontal Resolution	Vertical Resolution	Pel Process	DCS	2.2	2.5	2.7	2.6	2.8
2.2	1728 pels per 215mm	3.85 1/mm / 7.7 1/mm	original 1728		Equal (A4)	Reduced (A4 - A6)	Reduced (A4 - A6)	Reduced (A4 - A5)	Reduced (A4 - A5)
2.5	864 pels per 107mm	7.7 1/mm / 15.4 1/mm	864 * 2 Note 1	BIT 17 = 0 / BIT 18 = 0 / BIT 33 = 0	Enlarged (A6 - A4)	Equal (A6) Note 1	Equal (A6)	Enlarged (A6 - A5)	Enlarged (A6 - A5)
2.7	1728 pels per 107mm	7.7 1/mm / 15.4 1/mm	original 1728		Enlarged (A6 - A4)	Equal (A6)	Equal (A6)	Enlarged (A6 - A5)	Enlarged (A6 - A5)
2.6	1216 pels per 151mm	5.44 1/mm / 10.9 1/mm	1216 * 1.42 Note 2	Note 1,2	Enlarged (A5 - A4)	Reduced (A5 - A6)	Reduced (A5 - A6)	Equal (A5) Note 2	Equal (A5)
2.8	1728 pels per 151mm	5.44 1/mm / 10.9 1/mm	original 1728		Enlarged (A5 - A4)	Reduced (A5 - A6)	Reduced (A5 - A6)	Equal (A5)	Equal (A5)

Note 1   Transmit Pel Process = 432(w) + 864 + 432(w)
          Receive Pel Process Extracts Central 864 Pels

Note 2   Transmit Pel Process = 256(w) + 1216 + 256(w)
          Receive Pel Process Extracts Central 1216 Pels

Bit 33 = 1
Bit 35 = 1
Bit 33 = 1
Bit 34 = 1

(w) = white pels

as an extension to the Group 3 protocol. The protocol information also is embedded in HDLC frames but does not use selective repeat for error control. Every Group 3 facsimile machine must have the mechanism to transmit and receive the basic HDLC frame structure, including flags, address, control and frame check sequence. Thus, the use of an extended HDLC scheme meets the requirement of minimizing changes to existing recommendations (and to facsimile designs).

Protocol

For the protocol, the transmitting terminal divides the compressed image data into 256-octet or 64-octet frames and sends a 256-octet block of frames to the receiving terminal. (The receiving terminal must be able to receive both frame sizes.) Each of the transmitted frames has a unique frame number. The receiver requests retransmission of bad frames by frame number. The transmitter retransmits the requested frames. After four requests for retransmission for the same block, the transmitter may stop or continue, with optional modem speed fallback.

Performance

The page selective repeat ARQ is a good performance compromise that balances complexity and throughput (18). A continuous selective repeat ARQ provides slightly higher throughput, but requires a modem back channel. Forward error correcting (FEC) schemes typically have higher overhead on good connections, can be more complex, and may break down in the presence of burst errors. In addition to providing the capability of higher throughput on noisy lines, the error-correction-mode option supports an error-free environment that has enabled many new features. Examples of these are T.6 encoding, BFT, and operation over the ISDN. These new features and other current facsimile standards activities are described in the following paragraphs.

Group 3 1992 Options

Higher-Speed Modulation Technique

Early in 1991, the CCITT approved a new modulation technique, V.17, for use with Group 3 facsimile (19). In addition to the existing data rates, 9.6 kb/s, 12.0-kb/s, and 14.4-kb/s rates were added. V.17 operates in a half-duplex mode and is intended only for facsimile. The TR-29 Committee has established an ad hoc group to concentrate on improving Group 3 performance, including data rates higher than 14.4 kb/s. Data rates to 24 kb/s theoretically are possible (20), and are highly desirable if continuous-tone color is to be added.

Operation on Digital Networks

One of the improvements that has had strong U.S. support is the modification of the Group 3 protocol to allow operation on digital networks. This modification builds on the existing Group 3 protocol and allows Group 3 equipment to take advantage of ISDN's 64 kb/s while maintaining compatibility with existing Group 3 apparatus. Two optional methods have been approved. The first method makes use of the full-duplex capability of the digital channel and provides more efficient transmission. The second method is an optional half-duplex method that is somewhat simpler to implement. The error-correction-mode option is inherent to both methods. Operation on digital networks at 64 kb/s (or 56 kb/s) not only shortens transmission time for the current bilevel facsimile imaging, but allows color and gray scale to be transmitted in a reasonable time.

T.6 Encoding

A logical extension of the optional error-correction capability is to use Group 4's T.6 compression algorithm. As shown above, T.6 requires an error-free environment to provide improved compression performance. The use of T.6 with Group 3 (as an option) also was approved by the CCITT early in 1991.

Higher Resolution

The resolution of the scanners and printers used in facsimile apparatus (and the associated transmitted pel density) has a direct effect on the resulting output's image quality. The highest resolution specified for Group 3 is 204 pels by 196 pels per 25.4 mm. Many of the scanners and printers being purchased today are capable of higher resolutions. These higher resolutions provide the quality required by desktop publishing and other applications. The Group 4 recommendations support the following standard and optional resolutions: 200×200, 240×240, 300×300, and 400×400 pels per 25.4 mm. Note that the resolutions specified for Group 3 are "unsquare," that is, not equal horizontally and vertically. Group 4 resolutions are "square." This difference causes a compatibility problem. If the Group 3 resolutions were to be extended in multiples of their current resolutions (e.g., to 408×392), then a distortion of approximately 2% horizontally and vertically occurs when communicating with a 400×400 square machine. The U.S. position has been to accept this distortion and encourage a gradual migration to the square resolutions.

The CCITT has agreed to enhance Group 3 to include higher resolutions, including both multiples of current Group 3 resolutions and square resolutions. Specifically, optional Group 3 resolutions of 408×196, 408×392, 200×200, 300×300, and 400×400 pels per 25.4 mm have been approved. The 2% distortion has been shown to be relatively unimportant.

Binary File Transfer

The Group 3 facsimile recommendations originally were geared toward facsimile terminals or paper-to-paper transmission. Today, facsimile terminals are being integrated with computers and networks. In addition, the facsimile function now is being implemented in PCs, commonly known as PC-FAX. These new products are represented by both add-in PC facsimile boards and by remote facsimile devices connected to the PC through its parallel or serial port. Along with the standard facsimile functions, these new devices offer additional capabilities inherent in network and PC environments. Examples of these new features are higher-resolution scanners and printers, color, large memories for buffering and storage, scheduling and distribution, and confidential reception. BFT was developed to enhance the use of these new features by providing a general method to transmit files error free from one computer to another. Although file transmission can be accomplished using standard data modems, there are good reasons for incorporating this feature in PC-FAX. For example, facsimile modems typically operate at higher data rates than do data modems for the same cost. In addition, facsimile transfer and file transfer can be accomplished within one call establishment.

The BFT standard was developed by the TR-29 Committee and promoted in the CCITT. After some minor revisions, it was approved in 1992 as Recommendation T.434 (21). Changes to Recommendations T.4 and T.30 required to use BFT have been documented, and specific bit assignments have been approved to signal the use of BFT in the Group 3 protocol. Three other (less important) optional file transfer modes have been approved for Group 3 facsimile. They are basic transfer mode, document transfer mode, and edifact transfer.

New Control Messages

Four new control messages have been added to the Group 3 protocol by the CCITT. They are selective polling, subaddressing, password for polling, and password for transmission. The selective polling message carries a subaddress. This subaddress may be used to indicate that a specific document is being polled or that a mailbox is to be accessed. The subaddress message may be used to provide additional routing information to a facsimile transaction. The password messages are available for additional qualification of a facsimile connection.

In addition, an optional facsimile mixed mode has been approved by the CCITT that allows pages containing both character-coded and facsimile-coded information to be transferred.

Facsimile Data Circuit Terminating Equipment– Data Terminal Equipment Interface

The data circuit terminating equipment–data terminal equipment (DCE-DTE) interface project, the primary mission of TR-29.2, is another standards activity stimulated by the PC-FAX industry. The purpose of this project is to provide standards for "splitting" a facsimile terminal into two parts, one part being a computer (DTE) and the other being a facsimile modem (DCE). The standard describes the interface between the computer, typically but not limited to a PC, and the facsimile modem, including the protocol required for the computer to control the modem.

The facsimile modem is controlled by a command set much the same way that a data modem is controlled by the Hayes command set. The combination emulates a Group 3 facsimile machine. Three classes define increasing levels of complexity for the facsimile modem. Class 1 is the simplest, performing the minimum facsimile modem function. Class 1 has been approved as Electronic Industries Association/Telecommunications Industry Association (EIA/TIA) 578 (22). Class 2 has been published as TIA 592 (23) and TIA 605 (24). TIA 592 describes the basic Class 2 commands, and TIA 605 describes the protocol that is necessary to prevent data loss between modem and PC. At this time, Class 3 is not ready for ballot. This standard can be developed independently of T.4 and T.30 since it has no effect on interoperability with standard facsimile terminals. Although it is being developed as a U.S. standard, some interest has been shown in the international community. The effect of this standard is to make it much easier for developers to create facsimile devices that can be used on a variety of hardware platforms. In effect, the software can be separated from the hardware. A new project has begun to extend the Class 2 command set to voice.

A standards activity that is closely related to the DCE–DTE interface project is the application program interface (API). The API is intended to provide a standard software interface between an application such as a word-processing program and the PC-FAX function. This approach also implies a common hardware interface and would allow developers to construct generic software packages that would operate with a variety of PC-FAX devices without special software drivers. This work is progressing in the CCITT. Recommendation T.611, Programming Communication Interfaces, was approved in 1992 (25).

Other Activities

There are a number of other facsimile-related activities that either build on facsimile technology or eventually will affect Group 3 or Group 4 facsimile. These include facsimile database access, the use of facsimile in audiographic conferencing, and facsimile access to the message-handling system (MHS). TR-29 has been working on a set of test charts, including a monochrome high-resolution chart, a monochrome gray-scale chart, and a color chart. The monochrome charts have been approved by the CCITT as Recommendation T.22 (26). The resolution of these charts will be adequate for testing advanced facsimile machines (400 to 600 pels/in). The Association for Information and Image Management (AIIM) has developed a standard, MS53, to adopt the use of the T.6 compression algorithm for binary image file exchange (27).

All of this activity and the ever-increasing popularity of facsimile is not without problems. The proliferation of different devices on the telephone network (e.g., voice, voice answering machines, facsimile machines, and voice-mail systems) can cause confusion. A telephone subscriber that wants to use a single line for facsimile and voice (probably with a voice answering machine) needs to identify the type of the calling device automatically. The problem of identifying how the telephone circuits are being used extends to the international telephone network providers. The international networks use sophisticated equipment to "compress" voice calls and maximize the throughput of the networks. Facsimile traffic at the higher data rates is not easily compressible. In fact, the networks go so far as to demodulate the facsimile signal at one end of the network and remodulate it at the other end. These problems are receiving national and international attention, but the solutions are not coming easily.

New Facsimile Compression Algorithms

Color and Gray-Scale Compression

Background

The Joint Photographic Experts Group (JPEG) has been working on a still-image color-compression standard since 1986. This work has its roots in Working Group 8 (WG8) of International Organization for Standardization/International Electrotechnical Commission (ISO/IEC) Joint Technical Committee 1/Study Committee 2 (JTC1/SC2) (Coded Representation of Picture and Audio Information), which was set up in 1982. In 1985, a CCITT special rapporteur's group was formed to investigate new forms of image communication (NIC) under Question 18 in Study Group VIII. The initial work of the NIC group concentrated on a common set of requirements for all the telematic services, including facsimile. JPEG was formed in 1986 by experts from both WG8 and the NIC group with the expressed goal of selecting a high-performance universal compression technique, working under the auspices of WG8. JPEG recently has moved to a new working group, ISO/IEC JTC1/SC2/WG10 (Photographic Image Coding). The technical work on JPEG has been completed, and the ISO Draft International Standard (DIS) is nearing approval (28).

A broad set of service requirements was established for the JPEG algorithm to meet, including both progressive and sequential build-up, softcopy and hardcopy, and a wide range of image compressions. The range included those images from very lossy highly compressed to lossless (with lower compression). Probably as a result of the diverse requirements, the JPEG compression algorithm is not a single algorithm, but a collection of techniques that often are referred to as a *toolkit*. The intent is that such applications as facsimile will have a "customized" subset of the JPEG components. The following paragraphs briefly describe JPEG and its potential application to facsimile. A more detailed description of JPEG with a comprehensive bibliography is given in Ref. 29.

Joint Photographers Experts Group Overview

JPEG specifies two classes of coding processes, lossy and lossless. The lossy processes all are based on the discrete cosine transform (DCT), and the lossless are based on a predictive technique. There are four modes of operation under which the various processes are defined: the sequential DCT-based mode, the progressive DCT-based mode, the sequential lossless mode, and the hierarchical mode.

In the sequential DCT-based mode 8×8 blocks of pixels are transformed; the resulting coefficients are quantized and then entropy coded (losslessly) by Huffman or arithmetic coding. The pixel blocks typically are formed by scanning the image (or image component) from left to right, and then block row by block row from top to bottom. The allowed sample precisions are 8 and 12 bits per component sample. Of the DCT-based methods, the sequential DCT-based mode requires the least amount of storage.

For the progressive DCT-based mode, the quantized coefficients for the complete image component are determined, stored, and processed in one or both of two complementary ways: spectral selection and successive approximation. In spectral selection, the coefficients are grouped into spatial frequency bands, and a group typically is sent for the whole image component. In successive approximation, the most significant bits of the coefficients (again for the whole image component) are sent before the least significant bits. These two techniques may be used separately or may be combined in various ways.

The sequential lossless mode is not based on DCT at all, but is a totally independent predictive coding technique. The predicted value of each pixel position is calculated from the three nearest neighbors above and to the left, and the difference between the predicted value and the actual value is entropy encoded losslessly. For the lossless mode of operation, sample precisions from 2 bits per sample to 16 bits per sample are allowed.

In the hierarchical mode, an image (or image component) is transmitted with increasing spatial resolution between progressive stages. This is accomplished by first down-sampling the image a number of times to produce a reference stage that is transmitted by one of the other three modes of operation. The output of each hierarchical stage is used as the prediction for the next stage and the difference is coded. The coding of the differences may be done using only DCT-based processes, only lossless processes, or DCT-based processes with a final lossless process for each component.

All decoders that include any DCT-based mode of operation must provide a default decoding capability, referred to as the *baseline sequential DCT process*. This is a restricted form of the sequential DCT-based mode using Huffman coding and 8-bits-per-sample precision for the source image.

Discrete Cosine Transform-Based Compression

The formal definitions of the DCT and its inverse are well documented (30). The forward DCT transforms a square block of image pixel values, typically 8×8, into a similar block of spatial frequency "coefficients." The inverse DCT transforms the coefficients back into the block of image pixels. Of all the various transforms employed for image compression, the DCT is one of the best for two important reasons. The first reason is that it has low susceptibility to the blocking artifact (30). The second reason is that the DCT comes closest to the Karhunen–Loeve (K–L) transform in energy compaction, that is, the packing of most of the energy of a block of data into a few uncorrelated coefficients. The K–L transform is picture dependent, requiring intensive computation and the transmission of the transform basis functions for each frame. The DCT is a fixed transform, known to both transmitter and receiver, and performs almost as well as the K–L transform.

To achieve data compression, one must quantize the coefficients. JPEG recommends that each coefficient be quantized linearly according to a step size assigned to that coefficient, the assigned value being just small enough so that the distortion resulting from quantizing that coefficient is barely noticeable to a human observer. The resulting quantum step numbers then are ranked into an encoding order with the object of placing those quantum numbers most likely to have values of zero last, thus reducing the data to be encoded. JPEG recommends a simple zigzag order that arranges the quantum numbers in order of increasing spatial frequency.

The positions and values of the nonzero quantum numbers then are transmitted losslessly by either Huffman coding or arithmetic coding. The receiver decodes the quantum numbers, multiplies each quantum number by the step size associated with that coefficient to obtain quantized versions of the original coefficients, and then performs the inverse DCT to obtain an approximation of the original image. The compression-versus-distortion tradeoff can be controlled by a single quantization scale factor that scales all the step sizes assigned to the coefficients by a single multiplicative constant. The larger this scale factor is, the greater the compression will be, but the distortion will be greater also.

Application to Facsimile

An ad hoc group was created early in 1990 to address color facsimile under Question 4 of CCITT Study Group VIII. This group now considers JPEG to be the leading candidate for continuous-tone color-facsimile compression, but much work remains to be done. In addition to selecting the appropriate parts of the toolkit and associated parameters, a color model must be selected.

Two types of color spaces have been discussed: "primary" color spaces such as red, green, blue (RGB), and cyan, yellow, magenta (CYM) or, with the addition of black (K), CYMK; and luminance-chromaticity color spaces such as CCIR (International Radio Consultative Committee) 601 and CIELAB (CIE = Commission Internationale de l'Eclairage). The RGB and CYM spaces are associated naturally with scanners and printers, while the luminance-chromaticity spaces offer gray-scale compatibility and probably higher DCT-based compression. Much of the JPEG research and quantization matrix optimization has been based on the YCbCr color model, where the chrominance components Cb and Cr are subsampled horizontally. The human eye is much more sensitive to luminance than chrominance, thus it is easier to optimize the quantization matrix when luminance and chrominance are separate components. Subsampling the chrominance components provides further compression. Still to be decided is whether to sample the chrominance just horizontally, or horizontally and vertically. An additional advantage of the luminance-chromaticity color space is the gray-scale/color compatibility (similar to color television).

Another parameter to be determined is the resolution or pixel density of the image. Group 3 and Group 4 combined cover the range of 200 pels/in horizontally by 100 pels/in vertically to 400 × 400 pels/in for bilevel images. At 200 × 200 pixels/in, which is the standard resolution for Group 4, a color photograph (8.5 in by 11 in) compressed to 1 bit per pixel and transmitted at 64 kb/s would require about 1 minute to send. Probably a standard (mandatory) resolution will be selected with higher resolutions as options. A related issue is the selection of 8-bit or 12-bit data precision.

Although the facsimile color-compression standards activity has been concentrated on Group 4, it is very likely that Group 3 facsimile also will add color. Compatibility between Group 3 and Group 4 certainly is desirable. To achieve compatibility, the layout structure will require some thought, since Group 4 is ODA based and Group 3 is not. Other factors to be decided are the type of entropy coding, whether to use default tables, and how to interleave the color components.

Bilevel Compression

Background

The Joint Bilevel Image Group (JBIG) was formed in 1988 and organized much the same way as JPEG, that is, as a joint group working under the auspices of WG8. In fact, JBIG can be thought of as a spin-off of JPEG because the original goals of JPEG included compression of bilevel images. However, the JPEG members were not able to produce an algorithm that worked well on both continuous-tone and bilevel images; therefore, it was decided that JPEG would concentrate on continuous-tone image compression, and JBIG would select and develop a compression technique for a general class of bilevel images. The technique would form the basis of both an ISO standard and a CCITT recommendation. In 1990, JBIG was elevated to ISO working group status (ISO/IEC/JTC1/SC2/WG9 – Joint Bilevel Image Experts Group).

It is not surprising that the goals of JBIG, given its origin, were similar to those of JPEG; that is, the compression technique was formulated for a potential range of services and applications including facsimile, audiographic teleconferencing, and image databases. To support such a range of applications, the technique should be adaptable to a wide range of image resolutions and to varying image quality. It also should be capable of providing progressive (multistage with improving quality) or sequential image build-up. In addition, it should be image preserving such that the final decoded image is identical to the original.

At a series of JBIG meetings beginning in Stockholm in July 1989, an algorithm was developed from the five remaining prior to the Stockholm meeting.

The five were progressive transmission of binary images by hierarchical coding (BIHC), progressive encoding of facsimile image using edge decomposition (PED), progressive encoding of predicted signal according to classified pel patterns (PCLAP), progressive adaptive bilevel image compression (PBIC), and progressive coding scheme using block reduction (PCSB).

Overview

The progressive bilevel coding technique consists of repeatedly reducing the resolution of a bilevel image R_0, creating images $R_1, R_2, \ldots, R_n$, image R_i having one-half the number of pels per line and one-half the number of lines of image R_{i-1}. The lowest-resolution image R_{n-1}, called the *base layer*, is transmitted losslessly (free of distortion) by binary arithmetic coding. Next, image R_{n-1} is transmitted losslessly, using pels in R_n and previously transmitted (causal) pels in R_{n-1} as predictors in an attempt to predict the next R_{n-1} pel to be transmitted. If prediction is possible (both transmitter and receiver are equipped with rules to tell whether this is the case), the predicted pel value is not transmitted. This progressive build-up is repeated until image R_0 has been transmitted losslessly (or the process stopped at the receiver's request). A sequential mode of transmission also exists. It consists of performing the entire progressive transmission on successive horizontal stripes of the original image. The algorithm performs image reduction, typical prediction, deterministic prediction, and binary arithmetic encoding and decoding, described below.

Image Reduction. Each low-resolution pel is determined by the values of several high-resolution pels and low-resolution pels that already have been determined. The objective of the reduction algorithm is to preserve as much detail as possible in the low-resolution image under the constraint that it be half as wide and high as the high-resolution image. Resolution reduction could be achieved with subsampling, which is simple but yields poor results, especially on thin lines and dithered images. JBIG recommends a reduction algorithm that has given excellent results; however, any algorithm can be used.

Prediction. When a difference layer is being encoded or decoded, much of the compression is achieved by predicting new pel values from the values of pels in a predictor template. The predictor template contains pels from the reference layer and pels already predicted or encoded from the difference layer. When the predictor state is such that the prediction is known to be correct (the receiver must know this also), the predicted pel value need not be encoded or decoded. The JBIG algorithm employs two kinds of prediction: typical prediction (TP) and deterministic prediction (DP).

TP refers to prediction in which the predicted value almost always, but not necessarily always, is correct. Since in bilevel imagery each pel carries only one bit of information, it would be wasteful for the transmitter to inform the receiver of whether the prediction is correct for each pel predicted. Instead, the transmitter looks ahead for and reports TP errors (exceptions).

DP refers to prediction in which the predicted value is always correct. DP is bound tightly to the image-reduction rules. Whether a pel is or is not deterministically predictable is determined by looking up a rule in a table indexed by the state of the predictor pels. Deterministically predictable pels are flagged and not encoded by the arithmetic coder. Provision is made for an encoder to download DP tables to a decoder if it is using a private resolution-reduction algorithm.

Binary Arithmetic Coding. The data compression achieved by a binary arithmetic coder is best when the probabilities of the two symbols are near 1 and 0, and worst when they are near 1/2. In any practical application, the probability of a 1 or 0 at any given time frequently is dependent upon the conditions under which the symbol is being encoded or decoded. Therefore, best compression is achieved by keeping separate probability estimates for those conditions under which the encoded symbol probabilities are skewed the most strongly. These conditions are called *contexts*.

Consider, for example, a bilevel image containing line drawings and text. As this image is scanned, if the previous pel was white, then there is a high probability that the current one will be white also. Therefore, if one uses the previous pel value as a predictor, there are two contexts, one for each color of the previous pel. The probabilities for each usually are much nearer 1 and 0 than is the single probability with the previous pel value ignored. In the JBIG system, there is a separate context for every possible combination of pel values in a context "template."

Adaptive-Context Templates. The purpose of adaptive-context templates (ATs) is to take advantage of horizontal periodicity, which often occurs in halftone and dithered images. Data compression is best if at least one of the pels in a context template is a good predictor of the pel being encoded. An AT contains a "floating" pel; all other pels in the template are fixed in position relative to the encoded pel. There are also other pels designated as candidate floating pels that are not currently a part of the context template. If one of the candidates becomes a much better predictor than the current floating pel, then the candidate pel and current floating pel swap status so that the better predictor becomes a part of the context template. This test is made infrequently, and the swap is made only if the candidate is a much better predictor. These restrictions are imposed because, when a swap is made, compression is degraded temporarily until the binary arithmetic coder has time to adapt to the swap.

General Description

Transmission Modes. The JBIG specification defines encodings as *compatible progressive/sequential* (31). This means that the encoded data can be created either progressively or sequentially, and then eventually decoded either progressively or sequentially. Of course, storage of the encoded data is implied if the decoding is to be done in a mode that is different from the encoding.

In *progressive transmission* (encoding and decoding), the entire starting image is reduced to half its height and width. The reduced image is reduced similarly; this process is repeated some specified number of times. The image layer produced by the last reduction is called the *base layer*. The base layer is encoded. The transmitter then performs prediction where possible, encoding those pels that cannot be predicted, to transmit the next higher resolution layer using the base layer as a reference. This higher-resolution layer then is used as a reference to predict and encode a still-higher-resolution layer. This progression is continued until the original image has been transmitted.

Sequential transmission (encoding and decoding) consists of dividing the original image into horizontal stripes and then transmitting each stripe in the progressive mode. The whole series of progressive transmissions described above is performed on each stripe before it is begun on the next. The concept of stripes and layers is illustrated in Fig. 6, in which there are three resolution layers with four stripes per layer. A progressive transmission would proceed in the order

$$1, 2, 3, 4, 5, 6, 7, 8, 9, 10, 11, 12.$$

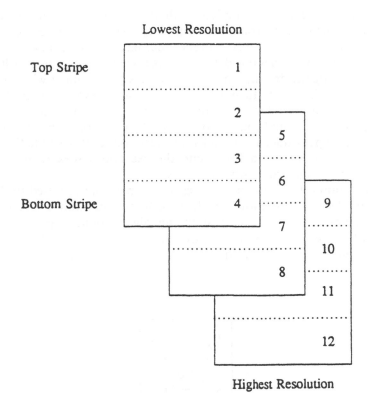

FIG. 6 Layers and stripes example.

A sequential transmission would proceed in the order

$$1, 5, 9, 2, 6, 10, 3, 7, 11, 4, 8, 12.$$

Note that the number of layers and the number of lines per stripe are free parameters with a wide range of allowable values. If the number of layers was set to one and the number of lines per stripe set large enough to permit just one stripe, a "pure" sequential transmission would result. This limiting case might be applicable to low-cost facsimile implementations in which flexibility to browse a database is lost, but simplicity is gained—resolution reduction, typical prediction, and deterministic prediction are not applicable. Figure 7 shows the relationships among and notations for pels in an image layer having a given resolution and in the layer having half this resolution. The former layer is referred to as the *high-resolution layer* and the latter as the *low-resolution layer*. High-resolution pels h_{00}, h_{01}, h_{10}, and h_{11}, represented as squares, register with the low-resolution pel l_{00}, represented by a circle. The term *phase* is used by JBIG to describe the orientation of a low-resolution pel with respect to a high-resolution pel. Refer to one of the four high-resolution pels denoted by h_{00}, h_{01}, h_{10}, and h_{11}. Phase 0 refers to h_{00}, 1 to h_{01}, 2 to h_{10}, and 3 to h_{11}.

Resolution Reduction. The simplest method of reducing an image to half its size in both dimensions is straight subsampling: keep every other pel in a given high-resolution line, and do this to every other line. In bilevel images, however, this method quickly washes out detail. In images containing a text or line drawings, lines forming the drawings or text characters grow thinner with each reduction, and soon disappear. In halftone images, gray levels become badly distorted. The resolution-reduction algorithm was formed from a number of those algorithms submitted, and consists of two parts: (1) a formula for determining a low-resolution pel value, and (2) a list of exceptions that override the formula. The formula is, in effect, a filter, and the exceptions seek to preserve such features as lines, edges, and dither patterns.

Figure 8 shows the high- and low-resolution pels that are used to determine the color of the low-resolution pel, denoted by the question mark. Its color is determined not only by the corresponding high-resolution pels, but by five peripheral high-resolution pels and three previously determined low-resolution pels. The reduction formula, in the absence of an exception, is

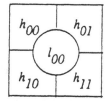

FIG. 7   High- and low-resolution pels.

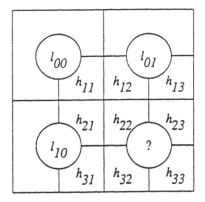

FIG. 8 Pels used for resolution reduction.

$$4h_{22} + 2(h_{12} + h_{21} + h_{23} + h_{32}) + (h_{11} + h_{13} + h_{31} + h_{33})$$
$$- 3(l_{01} + l_{10}) - l_{00} \tag{1}$$

If Eq. (1) is greater than or equal to 5, the low-resolution pel is set equal to 1. Otherwise it is set to 0. This formula is, in effect, a filter, with previously determined low-resolution pels included. The exceptions reside in a table containing exception states and exception pel values. In an actual reduction implementation, the whole algorithm, including the formula and the exceptions, is contained in a lookup table to save processing time. For each low-resolution pel to be generated, a state is assembled by concatenating pel values from the 12 variables of Eq. (1) forming a 12-bit integer. This state is used as an index to the table, which has 2^{12} or 4096 entries; the required value of the low-resolution pel is retrieved from the table.

Typical Prediction. In typical prediction, a TP cluster is defined as a low-resolution pel surrounded by a neighborhood of eight pels of the same color. The neighborhood consists of pels horizontally, vertically, and diagonally adjacent to the pel in question. If low-resolution pel l_{00} is centered in such a cluster, then high-resolution pels h_{00}, h_{01}, h_{10}, and h_{11} almost always have the same color as l_{00}. This is particularly true in images having large areas of solid color, as in line drawings and text.

Prior to encoding a pair of high-resolution image lines, the encoder performs a TP test consisting of examining the corresponding low-resolution line. (No prediction of any kind is employed for the base layer, where there is no lower-resolution layer.) For each low-resolution pel belonging to a TP cluster, the corresponding high-resolution pels are examined. If they all have the same color as the low-resolution pel, then the four high-resolution pels are predictable. If not, then a TP exception has been encountered. If a TP exception is found anywhere in the low-resolution line, then that whole line is declared an exception. After the TP exception check has been completed, an exception bit is

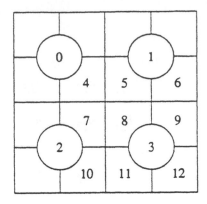

FIG. 9   Pels used to determine deterministic prediction.

encoded in a reused context to tell the decoder whether to employ TP while decoding the pair of high-resolution lines. If no exception is found, then, whenever a low-resolution pel belonging to a TP cluster is encountered, the corresponding high-resolution pels are not encoded. The decoder, having been told that there is no TP exception, merely inserts pels having the same color as that of the low-resolution pel.

Deterministic Prediction.   DP is tied tightly to the resolution-reduction rules. Some combinations of high-resolution and low-resolution pel values together with a knowledge of the reduction rules allow some of the low-resolution pels to be predicted with certainty. The encoder and decoder determine whether a low-resolution pel is deterministically predictable by a table search. Four tables, corresponding to the four phases, are created from a knowledge of the reduction rules. Figure 9 shows the labeling of the high-resolution and low-resolution pels that are used to determine DP. Figure 9 is similar to Fig. 8, but with a more convenient notation that defines the bit significance for indexing into the DP tables. Table 3 shows the reference pels that are used for each of the spatial phases to construct the DP tables. These reference pels include the low-resolution pels as well as causal pels from the higher-resolution difference layer being encoded (decoded). The appropriate DP table then is entered using the value of the reference pels as an index. There is always a table entry, whether or not the target pel is DP. If it is, the table delivers the predicted value, 1 or 0; otherwise, it delivers a 2, which means that it is not possible to make a DP. Also shown in Table 3 is the number of combinations of reference pels for each spatial phase that actually result in a DP.

Adaptive-Context Templates.   Figures 10 and 11 show the context templates for encoding (decoding) base-layer pels and difference-layer pels, respectively. Squares represent high-resolution pels for the difference layers, and all base layer pels. Circles represent low-resolution (reference-layer) pels for difference-layer processing. All pel positions are shown relative to the pel being encoded (decoded), which is labeled with a question mark.

TABLE 3   Deterministic Prediction Pels

Phase	Target Pels	Reference Pels	Number of Hits
0	8	0,1,2,3,4,5,6,7	20
1	9	0,1,2,3,4,5,6,7,8	108
2	11	0,1,2,3,4,5,6,7,8,9,10	526
3	12	0,1,2,3,4,5,6,7,8,9,10,11	1044

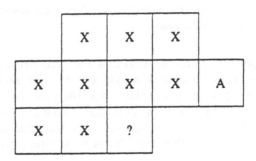

FIG. 10   Base-layer context template.

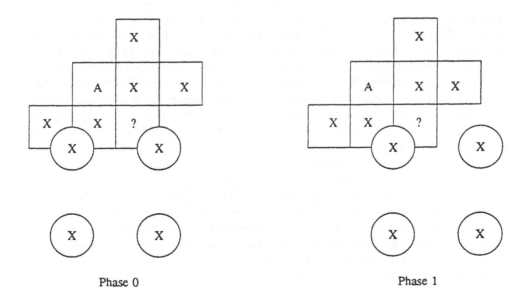

Phase 0   Phase 1

FIG. 11   Difference-layer context templates.

Squares containing an X represent pel positions that are always in the template. In each diagram, there is one square labeled with an A. This represents the initial position of the AT floating pel; the position of the AT pel may change during the process of encoding an image. The allowable new positions for the AT pel lie on the bottom row of the template to the left of the pels marked X. The number of allowable positions and the technique for deciding when and to which position the AT pel is to move (and whether or not movement is allowed) are not specified by the standard; they are decided by the application.

The low-resolution layer has 10 pels in its template. Thus there are 1024 contexts for the low-resolution layer (i.e., 1024 possible combinations of black and white pel values in the template). For each difference-layer template, there are also 10 pels, some high-resolution and some low-resolution. Thus, there are 1024 contexts for each phase, or a total of 4096 contexts for difference-layer processing. Note that it is perfectly legitimate for low-resolution pels to lie below or to the right of the encoded (decoded) pel; all low-resolution pels are known to both encoder and decoder.

Binary Arithmetic Coder.   In the general case for each stripe of each resolution layer, the arithmetic coder produces a byte stream. For each pel in the stripe being coded, there are four inputs to the coder: the value of the current pel, the context associated with that pel, a TP indicator, and a DP indicator. The coder is defined in JBIG by a set of flow charts. It is intended that the arithmetic coding operations defined by the flow charts be identical to the coding in the JPEG standard.

Performance and Application to Facsimile

Although work remains to be done on the JBIG specification, some conclusions can be drawn. A recent study of the JBIG algorithm as it existed when the study was made shows that TP, DP, ATs, and arithmetic coding all contribute to compression in varying amounts, depending on image content and resolution (32). TP contributes the most compression in line drawings and text images, including handwriting, but is of negligible importance in halftone images. DP and AT both contribute significantly in halftone images, but negligibly in drawings and text. Note that these compression results are based on the application of all parts (TP, DP, AT) of the algorithm. This does not mean that similar overall compression could not be achieved by omitting part (i.e., TP) and making up the compression with another part (i.e., the arithmetic coder). However, to achieve the highest compression in a progressive system designed to be independent of image type, all components should be applied. Typically, the highest compression was achieved when two layers were used. That is, two layers gave better compression than either a single layer (without resolution reduction, TP, or AT) or more than two layers. The JBIG algorithm is computationally complex. A proper subset could be defined to forego the progressive functionality and eliminate resolution reduction, TP, DP, and possibly even AT. Conventional Group 3 facsimile terminals, which cannot profit from progressive transmission, could take advantage of this defined subset.

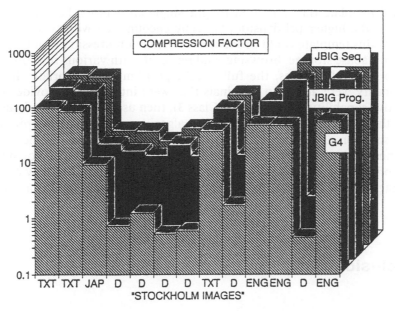

FIG. 12 T.6 compression compared to Joint Bilevel Image Group compression.

Figure 12 and Table 4 compare JBIG compression with that of Group 4 facsimile (CCITT Recommendation T.6) for the Stockholm Images. For JBIG, two sets of data are presented, one for the full algorithm (four reductions with TP, DP, and AT), and one for a single-layer subset (without TP and DP and with a fixed template). Note that both of the JBIG variations outperform T.6, especially on dithered images. The progressive JBIG mode performs better than the single-layer mode on text and line drawings, while the single-layer mode (labeled "sequential" in Fig. 12) performs better on dithered images. Note that

TABLE 4 T.6 versus Joint Bilevel Image Group Compression

				Compression Factor		
Type	Rows	Columns	Uncompressed Bytes	G4 MMR	JBIG Progressive	JBIG Sequential
TXT	4352	3072	1671168	95.96	125.35	117.22
TXT	4352	3072	1671168	81.56	107.40	105.73
JAP	4352	3072	1671168	8.98	11.41	11.10
D	2048	3072	786432	0.72	8.58	10.62
D	2048	3072	786432	1.24	7.02	7.13
D	2048	3072	786432	0.52	11.02	12.19
D	2048	3072	786432	0.59	6.99	11.83
TXT	4352	3072	1671168	35.80	51.13	49.90
D	4352	3072	1671168	1.68	7.35	8.46
ENG	4352	3072	1671168	46.70	68.80	65.05
ENG	3040	3072	1167360	45.69	170.27	194.72
D	1024	1024	131072	0.46	1.32	1.62
ENG	5856	4096	2998272	53.33	176.46	204.10

the fixed template that was used for the single-layer mode has not been optimized for the higher pel densities typically encountered with this mode. Fine-tuning of this template could produce even better compression results.

For database storage, browsing, and retrieval, with various resolutions from icons to full-scale images, the full JBIG algorithm is appropriate. If the full functionality were added to terminals that were intended to provide softcopy interactive capability (e.g., Group 4 Class 3), then users of such terminals would realize the full benefits of the progressive algorithm. For point-to-point facsimile transmission as typified by Group 3, a single-layer subset of the full JBIG algorithm might be more applicable. Group 3 terminals could not make use of the progressive functionality of the full JBIG algorithm, but they could take advantage of the excellent compression performance that the algorithm provides.

Conclusions

It is clear that, although Group 4 was intended to be the next-generation facsimile terminal, Group 3 has caught up with Group 4, at least in terms of performance. Group 3 now can use the same compression algorithm as Group 4, with the same resolutions. Group 3 soon will operate on digital networks, thereby matching the short transmission times of Group 4. Actually, Group 3 may be a little faster because its protocol overhead is lower. When color is added to Group 4, it probably will be added to Group 3 as well. The remaining difference between Group 3 and Group 4 is the method of implementation, which currently provides no advantage to Group 4. That is, Group 4 is based on ODA and the ISO seven-layer architecture, while Group 3 is not. When the use of ODA and the ISO model reaches "critical mass" at some future time, Group 4 might find its place. Meanwhile, the trend toward the integration of computers and facsimile terminals will continue. The marriage of computers and facsimile will lead to such additional functionality as audiographic conferencing and database access, all from a single desktop terminal.

References

1. CCITT, T.2, Standardization of Group 1 Facsimile Apparatus. In: *Blue Book*, Volume VII, Fascicle VII.3 (XIth Plenary Assembly. Melbourne, November 14–25, 1988), ITU, Geneva, 1989, p. 12.
2. CCITT, T.3, Standardization of Group 2 Facsimile Apparatus. In: *Blue Book*, Volume VII, Fascicle VII.3 (XIth Plenary Assembly, Melbourne, November 14–25, 1988), ITU, Geneva, 1989, p. 14.

3. CCITT, T.4, Standardization of Group 3 Facsimile Apparatus for Document Transmission. In: *Yellow Book*, Volume 7, Fascicle 7.3, ITU, Geneva, 1980, pp. 21–47.

4. CCITT, T.30, Procedures for Document Facsimile Transmission in the General Switched Telephone Network. In: *Yellow Book*, Volume 7, Fascicle 7.3, ITU, Geneva, 1980, pp. 77–174.

5. Huffman, D. A., A Method for the Construction of Minimum Redundancy Codes, *Proc. IRE*, 40:1098–1101 (September 1952).

6. Bodson, D., and Schaphorst, R., Compression and Error Sensitivity of Two-Dimensional Facsimile Coding Techniques, *IEEE Trans. Commun.*, COM-31(1): 69–81 (January 1983).

7. CCITT, T.5, General Aspects of Group 4 Facsimile Apparatus. In: *Red Book*, Volume VII, Fascicle VII.3 (VIIIth Plenary Assembly, Malaga-Torremolinos, October 8–19, 1984), ITU, Geneva, 1985, p. 509.

8. CCITT, T.73, Document Interchange Protocol for the Telematic Services. In: *Red Book*, Volume VII, Fascicle VII.3 (VIIIth Plenary Assembly, Malaga-Torremolinos, October 8–19, 1984), ITU, Geneva, 1985, p. 509.

9. ANSI, ISO 8613, Information Processing Systems, Text and Office Systems, Office Documentation Architecture (ODA), and Interchange Format, Secretariat ISO/IEC JTC 1/SC 18, American National Standards Institute, 1430 Broadway, New York, NY 10018.

10. CCITT, T.5, General Aspects of Group 4 Facsimile Apparatus. In: *Red Book*, Volume VII, Fascicle VII.3 (VIIIth Plenary Assembly, Malaga-Torremolinos, October 8–19, 1984), ITU, Geneva, 1985, p. 32.

11. CCITT, T.503, A Document Application Profile for the Interchange of Group 4 Facsimile Documents. In: *Blue Book*, Volume VII, Fascicle VII.7 (XIth Plenary Assembly, Melbourne, November 14–25, 1988), ITU, Geneva, 1989, pp. 185–193.

12. CCITT, T.521, Communication Application Profile BTO for Document Bulk Transfer Based on the Session Service. In: *Blue Book*, Volume VII, Fascicle VII.7 (XIth Plenary Assembly, Melbourne, November 14–25, 1988), ITU, Geneva, 1989, pp. 200–212.

13. CCITT, T.563, Terminal Characteristics for Group 4 Facsimile Apparatus. In: *Blue Book*, Volume VII, Fascicle VII.7 (XIth Plenary Assembly, Melbourne, November 14–25, 1988), ITU, Geneva, 1989, pp. 275–293.

14. CCITT, Facsimile Coding Schemes and Coding Control Functions for Group 4 Facsimile Apparatus. In: *Yellow Book*, Volume 7, Fascicle 7.3, ITU, Geneva, 1980, pp. 48–57.

15. AT&T, *Pattern Recognition Coding for Group 4 Facsimile*, CCITT Subgroup 8 Delayed Contribution D252, May 1983.

16. Johnsen, O., Segen, J., Cash, G. L., Coding of Two-Level Pictures by Pattern Matching and Substitution, *Bell Sys. Tech. J.*, 62(8):2513–2545 (October 1983).

17. National Communications System, *Simulation and Evaluation of the AT&T Proposed Pattern Recognition Algorithm for Group 4 Facsimile*, Technical Information Bulletin 87-4, National Communications System, January 1987.

18. National Communications System, Error Control Option for Group 3 Facsimile Equipment, Technical Information Bulletin 87-4, National Communications System, January 1987.

19. CCITT, *V.17, A 2-Wire Modem for Facsimile Application with Rates Up to 14400 Bit/s*, CCITT Recommendation, February 1991.

20. United States, *Request for a Study of High Performance Group 3 Facsimile*, CCITT Subgroup 8 Delayed Contribution D157, September 1990.

21. CCITT, *T.434*, CCITT Recommendation (new, as yet not published).
22. EIA, *ANSI/EIA/TIA 578, Asynchronous Facsimile DCE Control Standard* (October 22, 1990), Electronics Industry Association, 2001 Pennsylvania Avenue, NW, Washington, DC, 20006-1813.
23. TIA, *ANSI/TIA/EIA 592, Asynchronous Facsimile DCE Control Standard* (May 1993) Telecommunications Industry Association, 2001 Pennsylvania Avenue, NW, Suite 800, Washington, DC, 20006-1813.
24. TIA, *ANSI/TIA/EIA 605, Facsimile DCE-DTE Packet Protocol Standard, TSSC Review version* (August 26, 1992), Telecommunications Industry Association, 2001 Pennsylvania Avenue, NW, Suite 800, Washington, DC, 20006-1813.
25. CCITT, *T.611*, CCITT Recommendation (new, as yet not published).
26. CCITT, *T.22*, CCITT Recommendation (new, as yet not published).
27. AIIM, *ANSI/AIIM MS53-1993, Standard Recommended Practice—File Format for Storage and Exchange of Images—Bi-Level Image File Format: Part 1*, Association for Information and Image Management, 1100 Wayne Avenue, Suite 1100, Silver Spring, MD 20910-5699.
28. *ISO/IEC Draft International Standard 10918-1*. Available for review and comment from the X3 Secretariat.
29. Mitchell, J. L., Pennebaker, W. B., Evolving JPEG Color Data Compression Standard, paper presented at SPIE/IS&T Symposium on Electronic Imaging: Science and Technology, San Jose, CA, February 1991.
30. Rosenfeld, A., and Kak, A., *Digital Picture Processing*, 2nd ed., Vol. 1, Academic Press, New York.
31. CCITT, *Draft Recommendation T.82, IS/IEC Draft International Standard 11544, Coded Representation of Picture and Audio Information—Progressive Bi-Level Image Compression*, WG9-S1R6.1 (February 10, 1993).
32. National Communications System, *Investigation of the Progressive Bi-Level Coding Technique for the High Resolution Bi-Level Data Compression Standard*, Technical Information Bulletin 90–12, National Communications System, July 1990.

STEPHEN J. URBAN

Fading Radio Channels
(see Communication over Fading Radio Channels)

Fast, High-Performance Local-Area Networks

Introduction

Point-to-point or multipoint communications within a geographical coverage of a few kilometers (km) is referred to as *local communications*. The underlying network that supports local communications is referred to as a *local-area network* (LAN). The extension to serve a geographical area of a large city is referred to as a *metropolitan-area network* (MAN). Data communications within a local area became a hot research topic in the 1970s. The non–real-time nature of the data traffic permits store-and-forward operation. With data-only communications, it is not necessary to dedicate network (channel) bandwidth to the individual users. To make efficient use of the available network bandwidth, the messages to be transferred are partitioned into blocks called *packets*, each of a finite length and with the necessary control information for transporting the packet from the source to the destination. This mode of information transfer is called *packet switching*. Since each packet contains sufficient control information to traverse the network and does not depend on adjacent packets from the same message for reproduction at the destination, the mode of packet transport is asynchronous and is referred to as an *asynchronous transfer mode* (ATM). The term ATM has been coined by the International Telegraph and Telephone Consultative Committee Study Group XVIII (CCITT SG XVIII) standards committee (1). In the CCITT recommendation, the speed of ATM-based networks is 150 megabits per second (Mb/s). Packets, each of 53 bytes in length (48 information bytes and 5 header bytes), are called *cells*.

The popular LAN topologies are bus, ring, and star. Depending on the supporting transmission medium, signal propagation in a bus can be bidirectional or unidirectional, but must be unidirectional in a ring. A star topology supports multidrop operation. The discussions in the following sections focus on LANs with bus and ring structures and with Media Access Control (MAC) protocols suitable for supporting integrated services. Sources (users) are connected to the bus or ring through stations that normally are distributed evenly around the network. The transmission medium is thus in the public domain; every station in the network can hear every other station in the same network. In this sense, the LAN is a broadcast network; simultaneous transmissions by the individual stations constitute a multiple access situation. Efficient use of the LAN, in terms of delay-throughput performance, requires the use of flexible and efficient MAC protocols.

By increasing the network size and transmission speed, a LAN can be used to support services over a geographical area of a large city. The resultant extended LAN is referred to as a MAN. The discussions in this article on high-performance LANs also may be applied to MANs with the normalized propagation delay α, defined below as the single most important parameter that affects the efficiency of a MAC protocol.

The amount of traffic that can be transported through the network in a finite time duration is called the *throughput*. The time taken for a packet to transit through the network is referred to as the *delay*. The goodness of a MAC protocol thus is measured in terms of the delay-throughput characteristic. A MAC protocol is said to be efficient if it delivers a high throughput while sustaining a low delay.

Known Media Access Control Protocols

The simplest MAC protocol is random access, in which a station transmits whenever a packet is available for transmission, without regard to the status of other stations. Thus, random access is completely uncoordinated. Stations can transmit packets at any time instant. This mode of operation is referred to as *pure random access*. When two or more stations transmit packets simultaneously, overlap takes place, resulting in packet collision or destructive interference. The interval over which two transmitted packets can overlap is referred to as the *vulnerable interval*. Since packet transmission can take place at any time instant, partial overlap can occur. For fixed-length packets, the vulnerable interval is twice the packet length. With Poisson arrivals, the throughput of pure random access is limited to $1/2e$, where e is the natural base. If the channel time is divided into contiguous slots, each of a duration long enough to transmit one packet, transmissions can be forced to occur at the slot boundary. The resultant protocol is referred to as *slotted random access*. In this case, no partial overlap can occur and the vulnerable interval equals one slot time. With Poisson arrivals, the maximum throughput of slotted random access is $1/e$, which is twice the throughput range of pure random access. The constraint imposed on the stations to transmit packets only at specific time instants represents a form of coordination among the participating stations. Intuitively, the inclusion of better forms of coordination among the transmitting stations should improve the throughput range further. The simplest form of coordination is to have the stations monitor, or sense, the transmission channel. An active station transmits its packet only if the channel is sensed idle. This action of sensing the transmission channel is termed *carrier sensing*, and the MAC protocol in which stations sense the channel prior to transmission is referred to as Carrier Sense Multiple Access (CSMA). Since stations transmit only when the channel is sensed idle, the probability of packet collision tends to decrease, resulting in an improvement in throughput over the random-access method. CSMA is basically a probabilistic access protocol, so that collisions still exist. By detecting the presence of collisions and introducing techniques to resolve collisions efficiently, it is possible to reduce the probability of collision further. A MAC protocol that combines carrier sense and collision detection is referred to as Carrier Sense Multiple Access with Collision Detection (CSMA/CD). Ethernet (2), the earliest LAN available, uses CSMA/CD for MAC with bidirectional transmission on a single bus structure. CSMA/CD is a reasonable MAC for non–real-time service. Because of contention possibilities, media access protocols of the CSMA/CD variety are not suitable for handling time-critical services (e.g., process control, voice or video signals).

The Institute of Electrical and Electronics Engineers (IEEE) 802.5 token ring (3) is better suited for the integration of real-time and non–real-time services. Other LANs with scheduled MAC protocols that are suitable for service integration include Fasnet (4), Expressnet (5), and the IEEE 802.4 token bus (6). All of these MACs use in-band scheduling access. Welnet (7,8) uses an out-of-band scheduling MAC scheme in which the control signal is transmitted in an auxiliary channel, either by means of space-division or wavelength-division methods.

Earlier LANs considered by the international standards committees have been low speed with a transmission rate of 20 Mb/s or less. At these transmission speeds, there may be only one packet propagating in the transmission medium at any one time. In a ring network, the transmitting station sees the front end of its transmitted packet before it finishes transmitting the packet. By defining α to be the normalized propagation delay given by the ratio of the end-to-end propagation delay to the packet transmission time, then low speed usually implies a value of $\alpha < 1$. The value of α is determined by a combination of the transmission speed and the network size; high-speed LANs have values of $\alpha > 1$. With high-speed LANS, there can be more than one packet propagating in the ring or bus simultaneously.

High-Speed Networks

The advent of fiber-optic and optoelectronic technologies is pushing telecommunications toward the broadband era. The evolution toward a single unified network as a transport mechanism for all traffic types, including narrowband and wideband traffic with real-time and non–real-time requirements, has introduced new dimensions into the design and development of high-speed LANs and MANs. MAC protocols of the CSMA/CD variety cannot readily be extended to handle high-speed operations.

A high-performance LAN is a network with a value of $\alpha \geqslant 1$, which exhibits good delay-throughput characteristics and is capable of transporting real-time and non–real-time traffic in a flexible and efficient manner. To be able to handle multimedia traffic, the network is expected to have a high transmission speed, a flexible network architecture, a MAC protocol that allows for easy expansion to accommodate new services, and a coverage to handle a large user population. As mentioned above, the transmission speed and the size of the network together determine the value of α and hence the number of packets that the network can carry simultaneously. We focus attention on LANs with $\alpha \geqslant 1$, and refer to these as large networks. For high-speed networks, the propagation delay is a major factor that affects the system performance. With optical fiber, the propagation delay is approximately 5 microseconds per kilometer (μs/km). The transmission speed of high-speed LANs is in the range of 100 Mb/s and above, and the network coverage is in the range of several km. For example, the end-to-end (or round-trip) propagation delay of a 6-km LAN with a transmission speed of 150 Mb/s is about 30 μs. With a packet size of 53 bytes, the packet transmission time is then 2.827 μs. This yields a normalized propagation delay of $\alpha = 10.6$. Thus it is theoretically possible to pack 10 packets in the ring (or bus) during any single use of the network. A high-speed LAN with a large geographical coverage, say the size of a metropolitan area, often is referred to as an *extended LAN* or a MAN. An extended LAN or MAN may have a network diameter of 50 to 100 km. At a length of 60 km, the above example yields a value of $\alpha = 106$. The most efficient MAC protocol would be one that permits the packing of 106 cells in the network under a heavy traffic condition, that is, when every station has a packet or packets for transmission, and transmissions by all users can take place concurrently. For a large network, that is, one with a value of $\alpha \gg 1$, the key element to an efficient MAC protocol is a capability to perform concurrent transmission.

Media Access Control Protocol Classification

The Open System Interconnection (OSI) Reference Model, a seven-layer protocol architecture, of the International Organization for Standardization (ISO), is used widely in the telecommunications industry. The reader is referred to Ref. 9 for the definitions and the principles for defining the layers. In the LAN context, the MAC protocol is located at the bottom of the second, or data link, layer. By viewing the transmission channel as the server, MAC protocols can be categorized into fully informed, partially informed, and totally uninformed server models. A server that has full knowledge (i.e., is fully informed) of the system status schedules the participating stations for packet transmission to avoid packet collisions. Scheduling requires channel bandwidth and incurs additional delay. For example, a conflict-free, scheduled-access protocol (e.g., token passing, polling, or slotted access) is a fully informed server model. A *token* is a unique packet that contains only control information to give access rights to the token holder. In a polling scheme, a unique message is sent to each station in turn to inquire if the station has a packet to send. If the polling message is sent by a central scheduler, it is called *roll-call polling*; if it is sent by a neighboring station, it is called *hub polling*. In either case, time is needed to perform the polling function. Token passing or polling thus occupies bandwidth and incurs a delay. In a slotted access system, a master station issues empty slots, each of a duration long enough to hold one packet. The slot simply propagates along the channel. Stations observe the status of the slot and insert a packet into the slot if it is empty. CSMA/CD, which requires the stations to monitor the channel status but does not resolve collisions completely, is an example of a partially informed server model. Random access, in which a station transmits independent of any other stations, is an uninformed server model.

Collecting status information from and passing access rights to the stations expend energy. Thus, a fully informed server model incurs a penalty, normally in the form of loss of bandwidth and additional delay due to scheduling. This scheduling delay can be significant, particularly at light loads. The delay-throughput characteristics of the three types of multiple access models are depicted in Fig. 1. The vertical dashed lines indicate the maximum throughput attainable by each of the three server models. In a fully informed server model,

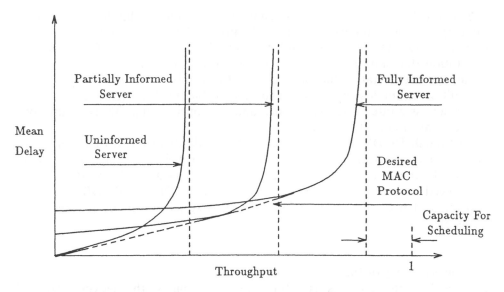

FIG. 1 Delay-throughput characteristics of different server models.

the participating stations are highly coordinated so that potential collisions are resolved prior to packet transmission. Such a server model attains a higher throughput range but suffers a larger scheduling delay, which is particularly significant at light loads. The label "capacity for scheduling" in Fig. 1 depicts the minimum bandwidth penalty sustained in order to achieve the fully informed status. The best MAC protocol would be one that exhibits a performance characteristic that follows the dashed envelope, which encaptures the desirable features of all three server models. For large networks ($\alpha \gg 1$), it is conjectured that a MAC protocol with a concurrent access capability would yield a delay-throughput characteristic approaching the dashed envelope in Fig. 1.

Standardized Local-Area Networks/Metropolitan-Area Networks

Depending on the manner in which channel access rights are exercised, one fully informed server model may incur a larger delay than another. For example, a token-passing scheme in a ring network would incur a token rotation time that is not in a slotted access scheme. There are now two protocol standards for high-speed LANs/MANs: the American National Standards Institute (ANSI) FDDI (fiber distribution data interface) (10) and the IEEE 802.6 DQDB (distributed queueing dual bus) (11). FDDI has a ring topology and uses a token-passing mechanism, while DQDB has a dual-bus topology and uses a slotted access scheme. Of particular importance is that a slotted access mechanism such as that used in DQDB supports concurrent transmissions, while a token-passing scheme such as that used in FDDI does not support concurrency. Therefore, compared to DQDB, FDDI incurs a bandwidth loss due to token rotation and due to the lack of concurrent transmission capability. This loss of channel capacity is significant particularly in lightly loaded or nonuniformly loaded large networks. On the other hand, the slotted access protocol of DQDB gives the station that started packet transmission at an earlier time an unfair access advantage over a station that started packet transmission at a later time. Thus, in addition to the loss of channel capacity (or bandwidth) incurred by a given media access strategy, fairness issues and appropriate algorithms to improve fairness have to be addressed.

Although DQDB and FDDI are MAN standards in terms of the network parameter α, the functions of very high speed LANs are similar to those in MANs. In the discussions below, attention is focused on the MAC layer of high-speed ring and bus LANs, which include FDDI and DQDB and other nonstandard networks. The performance of a system normally is assessed in terms of

1. the efficiency of the MAC protocol
2. the throughput ranges as a function of α, that is, the network size and speed
3. their ability to handle real-time and non–real-time traffic types
4. the delay-throughput characteristics
5. the fairness issues

The terms throughput, delay, and efficiency are defined above. The question of fairness is a matter of definition. We define fairness as the condition in which the MAC protocol offers the same degree of access rights to all participating network stations.

Unidirectional Propagation Networks

The most popular LAN topologies are the bus and the ring structures. In each case, the stations are connected to the bus or ring in a distributed and multidrop fashion. Here attention is focused on fully informed server models, which include MAC protocols using a scheduled access or a hybrid scheduled/random-access strategy. Scheduled access protocols are conflict-free and yield high throughput rates; in this sense, LANs with a scheduled access control mechanism are referred to as high-performance LANs. Most controlled-access protocols are variations of the following:

1. Slotted access for dual-bus architectures
2. In-band token passing for a single-ring topology
3. Out-of-band token passing for a dual-ring topology

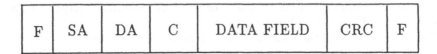

FIG. 2 Generic frame format (F = flag (frame delimiter); SA = source address; DA = destination address; C = control field; CRC = cyclic redundancy check).

These are network architectures with unidirectional signal propagation. Information transmissions are in a frame, or slot, format. Every frame has a header that contains, in addition to the control field, a source address (SA) and a destination address (DA). A generic frame format, which also contains a checksum field for error-detection purposes, is shown in Fig. 2. In the discussion below, we concentrate mainly on the dual-bus and the ring topologies.

The basic concept of a unidirectional propagation network is that each station repeats the frames it has received from its upstream neighbor to its downstream neighbor. If the destination address of the frame matches the station's MAC address, then the frame is copied into a local receive buffer. If no errors are detected, then the MAC establishes data transfer with the logical link control (LLC) (the top sublayer of the data-link layer) and the higher-level protocols.

The frame propagates along the transmission channel until one of two events takes place, depending on whether the network topology is a bus or a ring. If it is a bus, the frame simply will be absorbed at the end of the bus. If it is a ring, then the frame propagates around the ring to the transmitting station where it is removed. This mode of frame removal is referred to as *source removal*. There might be other variations in exercising frame removal (e.g., destination removal). Destination removal requires acknowledgment, which may be piggybacked onto other passing packets. An advantage of destination removal is that slots can be reused to conserve transmission capacity; otherwise, read slots will continue to occupy the transmission channel, which amounts to wasting channel capacity. On the other hand, destination removal can support only point-to-point communication, that is, there can be only one intended receiver. Source removal allows for point-to-multipoint communication and avoids the necessity for acknowledgment.

Dual-Bus Networks

A bus structure supports bidirectional and unidirectional signal propagation. With an optical fiber, present-day technology limits the signal propagation to unidirectional. To permit mutual communications among stations distributed along the bus, two buses with signals propagating in opposite directions are required. Fasnet is the earliest proposed dual-bus LAN (4), with a slotted MAC that can handle integrated voice and data services. The architecture of DQDB is based on the Fasnet dual-bus structure.

Fasnet and DQDB use a slotted access type of MAC protocol in which slots are passed along the bus unidirectionally. An active station transmits onto an empty slot when allowable. This mode of access control is not sensitive to the network size. In its basic form (e.g., in Fasnet), however, an upstream station always has an unfair access advantage over a downstream station in that the upstream station sees the slot passing by earlier. In DQDB, access rights are reserved in a reservation queue. A station that transmits first is tantamount to being an upstream station, that is, it will see the next available empty slot first, relative to another station that transmits at a later time. This unfair condition in unidirectional transmission networks is more prevalent under light and unbalanced loading.

Fasnet

The basic structure of Fasnet is shown in Fig. 3, in which stations are attached to both buses. In terms of signal transmissions, there is complete symmetry between the two buses. For each bus, the extremal stations are designated the HEAD and the END stations. The HEAD station on each bus has the task of issuing slots for packet transmission. For discussion purposes, it suffices to consider transmission on one of the buses.

For integrated voice and data services, Fasnet employs two types of slots, a voice type and a data type, and a cyclic service discipline with a cycle length equal to the time required for the preparation of a voice packet (12). Let T be the packet preparation time in μs, R be the LAN transmission speed in Mb/s, and P be the packet length in bits. Then the cycle length in slots is given by $N = RT/P$, where slot is a time interval long enough to transmit one complete packet, which is comprised of a header and an information field. The header

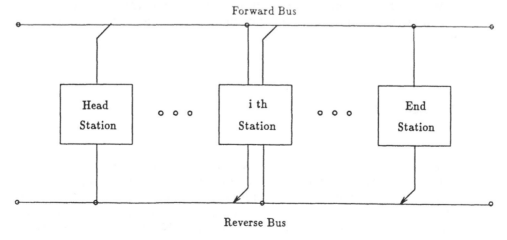

FIG. 3   Fasnet dual-bus topology.

carries the access control signals, consisting of a busy bit, a type bit, and an end (E) bit for the END station to echo status information to the HEAD station. The status information may be the termination of a service subcycle of a given priority.

The service cycle is partitioned into a voice subcycle and a data subcycle. The termination of a voice subcycle is signaled by the detection of an empty slot by the END station. For the purpose of discussion and performance analysis, it suffices to consider one of the buses as the forward bus that carries the packet transmissions and the other as the reverse bus for echoing the termination of a service subcycle. With respect to the forward bus, the extremal stations are designated as the HEAD and the END stations.

A station observing a passing slot has the right to transmit only if the busy bit indicates idle and the station's ready packet is of the correct type. At the slot level, Fasnet operates in an asynchronous mode. The cyclic service injects a synchronous-like service for voice. To implement the cyclic service discipline, the HEAD station keeps a timer that is set to a duration of T μs at the start of each service cycle. At the instant the first voice slot is issued, the timer starts to count down. When the timer expires (i.e., counts down to zero), the HEAD station reissues contiguous voice slots to start a new service cycle again. In this way, a voice service subcycle is enforced every T-μs interval. Consider, for example, if $T = 10$ ms, $R = 150$ Mb/s, and $P = 700$ bits, then the packet transmission time, or slot length, is $\tau = 4.667$ μs and the number of slots in a service cycle is $N = 2142.86$, which would be rounded to 2143 slots. In Fasnet, all slots are of equal length.

To start a service cycle, the HEAD station issues a train of idle voice slots. The voice service subcycle terminates when the END station detects an empty voice slot. The END station then informs the HEAD station of the end of voice subcycle by setting the E subfield in the header on the reverse channel. When the HEAD station detects a set E subfield, it issues contiguous data slots until the cycle timer expires, at which time it reissues a train of voice slots to start a new service cycle. This way of switching from a voice service subcycle to a data service subcycle incurs a switchover time, which represents a fraction of the network capacity that is not usable for information transfer. The propagation time between the last active voice station and the END station plus the propagation delay in the reverse channel equals the switchover time. At very high speeds and/or long end-to-end propagation delays, this switchover time can be a significant fraction of the available channel capacity.

The MAC protocol of DQDB differs from that of Fasnet in that there is no switchover penalty. Transmission rights are based on reservation. If packets are transmitted on one bus, called the *forward bus*, then reservations are made on the other bus, called the *reverse bus*. Packets for which a reservation has been made but not yet transmitted are enqueued locally. This enqueueing process is done by the individual stations on a distributed basis. Hence, we have the term *distributed queueing*. In its basic form, the packet-switching mode of DQDB is not suitable for real-time services. Instead, DQDB introduces an isochronous (circuit-switched) mode for voice service. This means that a fraction of the available channel capacity is dedicated to the individual voice users whether or not a voice user requires the channel capacity.

Distributed Queueing Dual Bus

The dual-bus architecture of DQDB, also known as QPSX (Queued Packet and Synchronous Exchange), shown in Fig. 4 is based on that of Fasnet. DQDB uses a distributed queueing MAC protocol for efficient channel access (13). To avoid the loss of channel capacity due to switchover from one level of priority service to another, the MAC of DQDB uses the reverse channel for stations to submit requests. The operation of the MAC protocol is based on two control bits, a busy bit and a REQ (request) bit, in the header of the slot. When set, the busy bit indicates that the slot already is occupied and the REQ bit indicates that a node in the network has a packet waiting for transmission. By counting the number of REQs it observes and the number of idle slots passing by, a node can determine the number of packets queued for transmission ahead of it. This counting procedure establishes a single ordered queue across the network for access to each bus. With this queued access, different levels of priority can be established by operating a number of queues, one for each priority level (14). Although priority always is given to packets in higher-level queues, nevertheless queueing does take place so that QPSX in its basic packet-switching mode is not suitable for servicing traffic with a critical timing requirement.

To accommodate synchronous service, it has been proposed that the MAC protocol of DQDB identify and dedicate some of the slots in a frame for isochronous service. The head end (frame generator) generates frames once every 125 μs. The frame is divided into a fixed number of units called *slots*. A slot may be used to transmit either isochronous (circuit-switched) or nonisochronous (packet-switched) traffic types. In keeping with the cell format of ATM, each slot is comprised of 53 octets. Each octet in a slot provides a 64-kilobits-per-second (kb/s) channel for the support of isochronous service. The frame format of DQDB is shown in Fig. 5. The first octet of every DQDB slot is the access control field (ACF), which leaves 52 octets of each slot available for isochronous transmission. By using multiple octets per frame, higher-rate circuits can be emulated. Slots that are not allocated to handle isochronous traffic are available for packet-switched communications. Thus, to support integrated real-time and non–real-time services, DQDB uses a hybrid emulated-circuit/packet-switching

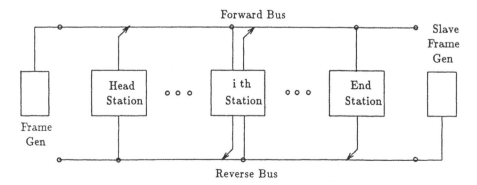

FIG. 4 Distributed queueing dual bus architecture.

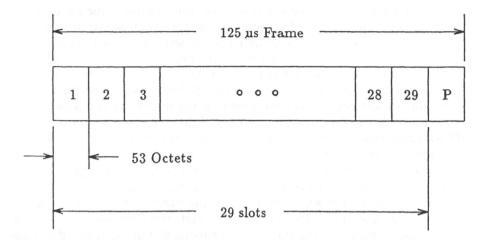

FIG. 5 Frame structure of distributed queueing dual bus.

operation. The boundary between isochronous (circuit-switched) and asynchronous (packet-switched) services is movable within a slot. If only $k < 52$ octets were used to handle synchronous service at any given time, then the other $(52 - k)$ octets in that slot would be wasted (i.e., could not be used for asynchronous service). The

As shown in Fig. 6, the MAC protocol of DQDB for asynchronous service can accommodate four priority requests. The distributed queueing protocol is

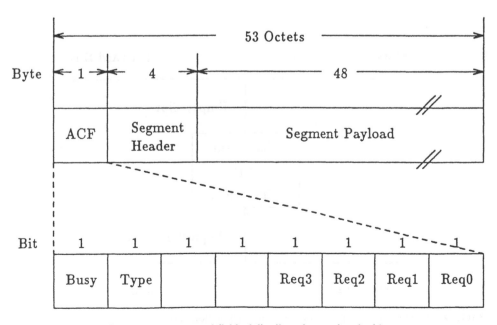

FIG. 6 Access control field of distributed queueing dual bus.

implemented using first-in, first-out (FIFO) transmit buffers, one for each priority level. For the purpose of discussion, it suffices to consider only one priority level. During a slot time, a station may observe a request from both a downstream station and locally. Each station places the requests it observes from the downstream stations and the local data segments to be transmitted into the FIFO transmit queue. When an empty slot passes by on the forward bus, the station services the transmit queue. If the next entry is a data packet, the busy bit is set and the packet is transmitted. Since every station maintains a FIFO transmit queue in this manner, a uniform FIFO transmit queue is established across the entire network.

The transmit queue is implemented with two counters: a request counter (RC) and a countdown (CD) counter. When the station has no data packets of its own in the queue, the RC keeps track of the number of unserved requests from downstream stations. This operation is performed by observing the forward and reverse buses for the status of the passing slot in the forward bus and the request bit on the reverse bus. If an empty slot is observed on the forward bus, the RC decrements its count by 1 if its content is nonzero. If a request is observed on the reverse bus, the RC increments its count by 1. This operation is depicted in Fig. 7.

When a station has a data packet of its own in the queue, it moves the RC value to the CD, which counts the number of requests that are ahead of the station's own data packet in the transmit queue, and the RC is cleared and used to count requests behind the station's data packet. The logical implementation of the FIFO transmit queue is shown in Fig. 8.

In DQDB, when a node has more than one data segment to transmit, it is not allowed to submit the next request until its current segment for which a

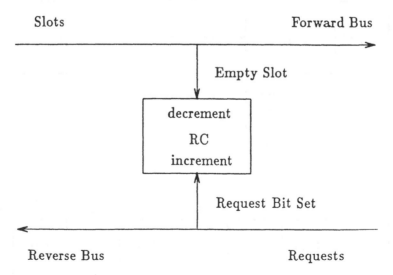

FIG. 7 Operation of request counter when the station has no data packet of its own for transmission.

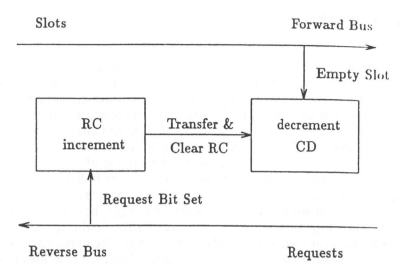

FIG. 8 Operation of request counter and countdown counter when the station has a data packet of its own for transmission.

request has already been submitted is transmitted. This restriction is imposed to ensure that no single large user would dominate the available bandwidth. The DQDB reservation mechanism degenerates as the propagation delay increases. The degree of degeneration depends on the network utilization and the length of the messages, as well as the propagation delay. Consider the situation of two active nodes 1 and 2 that are transmitting long messages (i.e., very high utilization). If the MAC protocol were fair, each node would receive half of the available bandwidth. Suppose we call the upstream node 1 and the downstream node 2, and let the two nodes be separated by an integer of D slots. If the message arrival times to the two nodes differ by an integer of Δ slots, the node with the earlier message arrival will capture a larger share of the available capacity. The severity of this imbalance in capacity distribution depends on the values of D and Δ (15). Basically, the node that transmits first has an unfair advantage over other nodes. Moreover, if the upstream node transmits first, this unfairness is more severe (15). The unfairness issue is particularly important in networks with a large value of α, the normalized propagation delay. An interesting algorithm to improve the fairness in sharing the available channel capacity in DQDB is described in Ref. 15.

A busy slot that has been read but has not yet been removed continues to occupy channel bandwidth. If the destination removes the read slot (i.e., by changing it from busy to empty), it can be reused. The IEEE 802.6 DQDB standard has not considered destination removal as an option. An alternate way to permit slot reuse is to implement erasure nodes that are located appropriately in the dual-bus architecture (16). The efficiency of slot reuse depends critically on the algorithm on which the erasure node operates and the use of the bandwidth balancing mechanism, for example, as described in Ref. 15.

Ring Networks

The ring topology with unidirectional signal propagation is particularly suited for fiber-optic-based implementation. Depending on whether the scheduling for channel access (e.g., token passing) is performed using an in-band signaling technique or an out-of-band signaling technique, there are in general two variations of ring topologies: the single ring (3,10,17), and the dual ring (18). In the dual-ring case (see Fig. 9), separate rings are used for the transmission of information packets and scheduling tokens. This mode of transmission can be mechanized using either a space-division or a wavelength-division multiplexing technique. In this sense, we may view the dual-ring topology as a single physical ring with the information packets and the scheduling tokens being transmitted in separate wavelengths. Both the single-ring and the dual-ring topologies can use an extra physical ring, with signals propagating in the opposite direction, as a standby to provide reliability and increased bandwidth. FDDI uses this twin-ring concept.

The performance of a network depends critically on the efficiency of the MAC protocol used. MAC protocols fall in one of three broad classes: the controlled-access, token-passing type; the slotted-access type; and the random-access CSMA type. As with the dual-bus networks, network efficiency and fairness among the network stations are important performance measures. A critical parameter that influences the performance of a ring network and the choice of MAC protocol is the normalized round-trip propagation delay α, defined in the section on known MAC protocols. For a value of α much larger than unity, the CSMA-type protocols exhibit severe performance degradation. The token rotation time in a token-passing-type protocol represents time segments (channel capacity) that cannot be used for information transmission. In addition, MAC protocols that do not have a concurrent transmission capability also incur a loss of channel capacity. The amount of wasted channel capacity due to token rotation and the lack of concurrent transmission capability increases with α, particularly under very light and/or unbalanced loads.

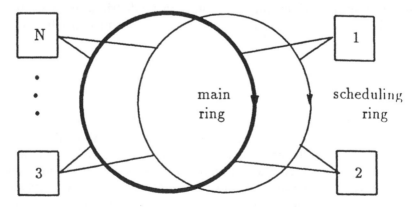

FIG. 9　Dual-ring topology.

Single-Ring Networks

The IEEE 802.5 token ring (3) and FDDI (10) are typical single-ring networks using in-band scheduling. At a transmission speed of 100 Mb/s, compared to the IEEE 802.5 token-ring network, FDDI is indeed high speed. While α in the IEEE 802.5 token-ring network is less than unity, α in FDDI is much larger than unity. The main difference is that, while there can be only one packet propagating in the IEEE 802.5 token ring, there can be multiple packets propagating in the FDDI ring simultaneously. The single-token protocol is not suitable for FDDI, as the capacity lost due to token rotation would be excessive. FDDI uses a multiple-token MAC protocol in which the token is released immediately following the completion of packet transmission.

The FDDI MAC uses a timed token rotation (TTR) protocol to control access to the transmission medium (19). Under the TTR protocol, each station, using a token rotation timer (TRT), keeps track of the time that has elapsed since a token was last received. Two classes of service are defined: synchronous and asynchronous. To provide a bounded response time for the ring, a threshold called the target token rotation time (TTRT) is introduced. During the initialization phase, stations bid to establish a value for the TTRT, which is set equal to the lowest value that is bid by any station. A station is allowed channel access for asynchronous service only when the time elapsed since a token was last received does not exceed the established TTRT. Multiple levels of priority for asynchronous frames may be provided within a station by specifying additional time thresholds for token rotation. Using the TTRT as a threshold, the TTR protocol establishes a guaranteed minimum response time for the ring since the time between the arrival of two successive tokens never will exceed twice the value of TTRT. For greater details on the timing requirements of the FDDI access protocol, the reader is referred to Ref. 20.

The original FDDI system, with multiple-priority token passing, is more suitable for asynchronous packet-switched transfer. A later version, FDDI-II, incorporates an isochronous service for the transmission of such real-time sources as voice and video. Similar to DQDB, the channel time is partitioned into frames of 125 μs, a frame length used by the public networks (e.g., the T1 frame). The transmission rate of FDDI-II is 100 Mb/s. Each FDDI-II frame is divided further into 16 wideband channels (WBCs), each composed of 96 octets (see Fig. 10). Each WBC may be assigned independently to handle either the packet-switched service or the circuit-switched service. FDDI-II is implemented using an additional sublayer, called the hybrid ring control (HRC), that is placed between the MAC sublayer and the physical layer. HRC multiplexes data between the asynchronous MAC and the isochronous MAC. Packet-switched data are interleaved with the isochronous data so that the asynchronous MAC is required to be able to transmit and accept data on a noncontinuous basis. As with DQDB, each octet in a WBC provides a 64-kb/s channel for the support of isochronous service. Thus, a single WBC theoretically can support up to 96 64-kb/s isochronous circuits, each of which is suitable for handling a voice connection. Isochronous service for higher-rate synchronous sources can be handled using multiple octets. An FDDI-II cycle has a 12-byte-cycle header to

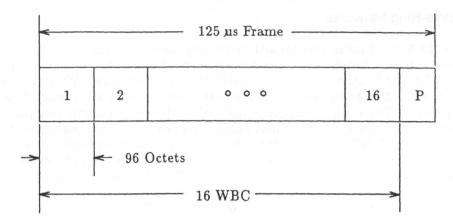

FIG. 10   Frame structure of fiber distribution data interface II (FDDI-II).

which the ACF is embedded so that not all 96 octets of each WBC are available for isochronous service.

Initially, the ring operates in the basic token-access mode, and switches to a hybrid mode of operation only after a station has negotiated for and won the right to be cycle master and has the synchronous bandwidth allocation required to support it (19). The cycle master then generates cycles at an 8-kHz rate (i.e., at a periodicity of 125 μs), and inserts the latency required to maintain an integral number of cycles synchronously on the ring.

FDDI-II offers four levels of priority, with the isochronous service given the highest level. The other levels are allocated depending on the token designation, that is, the tokens are distinguished in terms of priorities.

Dual-Ring Network

The dual-ring network described in Ref. 18 uses an auxiliary channel for out-of-band scheduling. The rationale for out-of-band scheduling is that the information and control signal transmissions can take place simultaneously. Of particular relevance is that an active station acquires the permit (token) to send while the packet transmitted by an upstream station is still passing through. Out-of-band token passing allows relatively simple implementation for scheduling channel access on a distributed basis. Distributed scheduling provides fair access to all stations and, at the same time, allows the superposition of a cyclic service mode on the basic asynchronous packet-switched operation to handle integrated synchronous and asynchronous services using pure packet switching. The service cycle length is the time interval required to prepare one synchronous packet. The cyclic service operation to emulate a synchronous service for handling real-time traffic is described in the next section.

Distributed scheduling also can be performed by in-band token passing on a single ring. The difference, in this case, lies in that the token transmission takes place immediately following the completion of packet transmission by the token-holding station and in the same transmission channel. Since distributed scheduling access can be performed on a dual-ring or a single ring, in the discussion here we use the term distributed scheduling ring (DSR) to represent the high-speed LAN with a distributed token-passing access control mechanism as described in the next section.

The Distributed Scheduling Ring Media Access Control

The DSR MAC protocol is implemented using a set of three tokens: SYNC, ASYNC, and NULL. The SYNC and ASYNC tokens are service tokens, while the NULL token is a nonservice token. The SYNC token provides synchronous service, the ASYNC token provides asynchronous service, and the NULL token is used to terminate a service cycle to enforce the cyclic service discipline. During any service cycle, one of the stations assumes the role of ring leader, with the task of issuing the three tokens to handle synchronous and asynchronous services during that service cycle as well as to terminate that service cycle. To enforce cyclic service, the ring leader keeps a timer that is set to a duration equal to the length of a service cycle less the mean time required for the termination action to take effect, as explained below. The term *distributed scheduling* comes from the fact that the leadership role changes dynamically among the stations to provide fair access to all the stations. Therefore, each of the stations connected to the ring has the capability of performing the role of ring leader (21). The cyclic service discipline operating in the jth service cycle is shown in Fig. 11. It is assumed that the capacity provided by one service cycle is sufficient to handle all synchronous service demand (otherwise the system would be lossy), and that during one service cycle, a synchronous station only has time to prepare one packet.

Suppose the system is operating at the instant jT in Fig. 11, at which time the leader's timer has counted down to zero and it is time to terminate the $(j - 1)$th service cycle. At this instant of time, the system is servicing asynchronous traffic and the ASYNC token is somewhere in the ring. The leader issues a NULL token to terminate the $(j - 1)$th cycle. The NULL token takes d seconds (s) to catch up to the ASYNC token-holding station. The ASYNC token-holding station completes transmitting its asynchronous packet; if transmission had already started, it kills the ASYNC token and assumes the leadership role by transmitting a synchronous packet, if any, and issuing a SYNC token to start the jth service cycle. Since the token-holding station may be anywhere in the ring relative to the ring leader, the NULL token propagation delay d is a random variable in the range $0 \leq d \leq \tau$, where τ is the ring propagation delay. On the assumption that the ASYNC token is equally likely to be held by any one of the stations, the mean value $\overline{d}$ equals $\tau/2$. With the leader's timer set equal to $(T - \overline{d})$, the service cycle length is a random variable with a mean value equal to T.

If the number of active synchronous stations is fewer than the number of slots in a cycle, the SYNC token will return to the ring leader before the timer expires. At this time, the ring leader issues an ASYNC token to start asynchro-

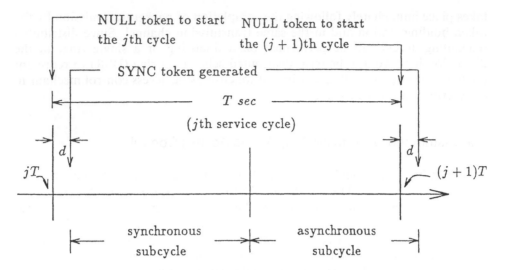

FIG. 11   Periodic service cycle structure.

nous service. On the other hand, if the number of active synchronous stations is greater than the number of slots in a cycle, there will be service interruption due to timer expiration. In any case, when the timer counts down to zero, the leader issues a NULL token to preempt the ongoing service (i.e., to terminate the current service cycle and to start a new service cycle as described above). Since the number of active synchronous users varies from cycle to cycle, the boundary between the synchronous and asynchronous subcycles is movable. The transmission rule is that, while in possession of a token of the correct type, a station transmits its quota of (asynchronous) packets and then releases the token downstream.

The token-passing mechanism can be formatted so that the asynchronous service supports multiple-priority service and that the SYNC token can permit an active synchronous station to transmit more than one packet while in possession of the SYNC token. For simplicity, we present performance results in a separate section for the scenario that each active synchronous station is allowed to transmit one packet while in possession of the SYNC token, and that there is only one priority level of asynchronous traffic.

Hybrid Access Protocols

Hybrid access protocols are attempts to use that fraction of the available capacity that otherwise would be wasted due to the transmission of control signals. Specifically, these are techniques that aim to drive the system to exhibit a performance characteristic close to that shown by the dashed envelope in Fig. 1.

Bhargava, Kurose, and Towsley (22) have studied a hybrid token/random-access protocol for ring networks, and Gerla, Wang, and Rodrigues (23) have proposed a hybrid scheme for a dual-bus network. In Ref. 21, Mark, Lee, and Mark proposed a combined token/random-access (CTRA) protocol in which a random-access mode is superimposed on the basic token-access mode as an add-on feature. In the CTRA protocol, the system always operates in the token-access mode; the random-access mode takes effect only during intervals when the token-access mode is ineffective (e.g., during token rotation times). Moreover, the add-on random-access mode of the CTRA protocol offers a concurrent access capability to make efficient use of the available capacity.

Hybrid Access Protocols for a Dual-Bus Network

As an attempt to improve the fairness in capacity sharing, Gerla, Wang, and Rodrigues proposed a hybrid slotted/random-access protocol in which, at light loads, all stations use the random-access mode. On detection of the onset of heavy traffic, a special "buzz" pattern is emitted on the bus to signal the switchover to the controlled access. The resultant network has been referred to as Buzz-net (23). If a collision takes place when the network is operating in the random-access mode, the system moves immediately to the controlled-access mode. This switchover involves the resolution of, and successful transmissions by, the collided stations while all others are being suspended in a hold state before the system can move to the controlled-access mode. It is shown in Ref. 23 that, under certain conditions, Buzz-net exhibits a higher channel utilization than Expressnet or Fasnet.

Hybrid Access Protocols for a Ring Network

In an attempt to capture part of the capacity wasted due to token rotation in a token-passing ring network, Bhargava, Kurose, and Towsley proposed a hybrid token-passing/random-access protocol in which an active network station senses the channel for idle conditions (22). If the channel is sensed idle, the station starts packet transmission using a random-access mode, but aborts its transmission when the channel is sensed busy from upstream. While such a hybrid access scheme potentially can capture some of the wasted channel capacity at light loads, it also can lead to many partial packets flowing around the ring, particularly at intermediate loading conditions.

Hybrid access schemes of the types described in the sections above do not offer a concurrent network access capability.

Combined Token/Random-Access (CTRA) Protocol for the Distributed Scheduling Ring Network

The combined token/random-access (CTRA) protocol described in Ref. 21 is not a hybrid access protocol; the random-access mode, as an add-on feature, backs off whenever the token-access mode needs the channel. That is, the system always operates in the token-access mode, and the random-access mode takes effect only when the channel appears to be idle. An active station that is not holding a token monitors the channel and, upon sensing the channel idle, transmits its asynchronous packet using the random-access mode, but must abort transmission when the ring is sensed busy from upstream. There will be partial packets flowing around the ring until they are removed by the transmitting stations or by some other mechanism. As with the hybrid protocols discussed above, certain rules must be exercised so that the ring is not occupied unnecessarily by partial packets.

As described in Ref. 21, each station is attached to the ring through a station-ring interface (SRI) in a manner shown in Fig. 12. Each SRI has two buffers: a packet-header buffer and a transit buffer. The header of the incoming upstream packet is buffered first to permit an interrogation of the source and destination addresses. If the packet is destined for the station in question, its contents are copied while being passed on to the next downstream station. On the other hand, if the source address matches the station's own address, then the packet is removed from the ring. The above functions are handled by the C/R (copy/remove) controller.

Consider a tagged station. If it has a packet ready for transmission, it transmits its packet by token access if it holds the token. Otherwise, it attempts random access if the upstream channel is sensed idle.

To describe the random-access operation, suppose the station is in an idle state; the switch S in Fig. 12 is positioned at 1, connecting the pointer and the

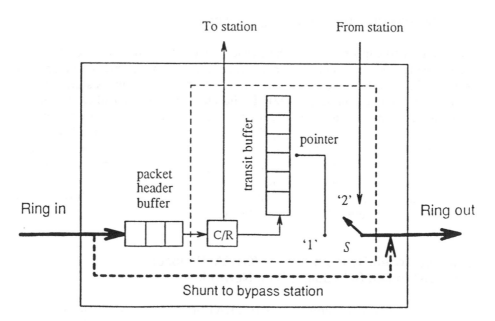

FIG. 12   Station-ring interface of combined token/random access.

outbound transmission medium of the ring. Assume that the content of the transit buffer initially is empty so that the pointer is pointing at the bottom of the transit buffer. If the station now has a packet ready for transmission and both the packet-header buffer and the transit buffer are empty, the station then initiates random-access transmission by repositioning the switch S to 2. However, if an incoming packet is detected from upstream, the station aborts its random-access transmission unless the source address of the incoming packet matches the station's own address, in which case it removes the incoming packet. Note that monitoring the contents of the packet-header buffer is equivalent to the operation of sensing the upstream channel.

Since the token is released immediately after the completion of packet transmission, the presence of any upstream traffic sensed by the token-holding station must have been transmitted via the random-access mode. As token access has priority over random access, the incoming random-accessed packet(s), if any, has to be stored temporarily at the SRI of the token-holding station to allow for packet transmission by token access. The transit buffer in the SRI provides this temporary storage. Upon completion of the token-access transmission, the SRI then transmits the transit traffic from the transit buffer. In this way, token access does not destroy incoming upstream packets. However, this transit buffering introduces a latency for the random-access traffic, which is tantamount to stretching the ring. To summarize, transit buffering has no effect on token-access transmission, but may increase the transmission delay of random-access traffic.

Capacity Consideration

In the sections above, it is stated that synchronous service can be handled using either an isochronous (circuit-switched) service (in FDDI-II or DQDB) or a cyclic service discipline (in DSR) to emulate a packet-switched synchronous service. In the isochronous service case, the capacity allocated to servicing real-time sources is fixed regardless of whether or not a particular real-time source is active. If the individual source is active, then the allocated capacity is used effectively; otherwise, the allocated capacity is wasted. In the emulated packet-switched synchronous service mode, a station occupies the channel capacity only when it has a packet or packets to send. The remaining capacity then is available for asynchronous service.

The hybrid isochronous/asynchronous service mode for handling service integration in FDDI-II or DQDB has a fixed boundary between synchronous and asynchronous services. The pure packet-switched operation of DSR, using a cyclic service discipline to emulate a synchronous service for real-time traffic, has a movable boundary. Intuitively, the movable boundary feature in DSR allows the asynchronous service mode to utilize the available bandwidth more efficiently.

To assess the capacity available for asynchronous service, we consider the integration of voice and data and assume that

1. there are N_v voice sources, each with a coding rate of 64 kb/s
2. the interarrival times of call-connection requests and connection holding times are independent and identically distributed (*i.i.d.*) and exponentially distributed with parameters λ_v and μ_v, respectively
3. each in-progress voice source with an accepted call-connection request generates voice packets only during the talkspurt intervals
4. the talkspurt and silent intervals are distributed exponentially so that a voice source can be modeled using a two-state Markov chain
5. the voice sources are *i.i.d.* so that the probability distribution of k out of N_v voice sources are active is binomial

Consider a hybrid isochronous/asynchronous integration scheme (e.g., FDDI-II or DQDB). Each active voice source occupies one octet of the channel capacity during any one use of the channel. Suppose at time t the number of isochronous slots allocated for voice service is S_t. Let L be the number of octets in a slot. If only $N_t \le S_t L$ octets are occupied by voice sources, then $W_t = S_t L - N_t$ octets are wasted. This is a consequence of the fixed boundary between isochronous and asynchronous services, so that data cannot cross the boundary to use the slots allocated to, but unoccupied by, voice. Let M be the total number of slots in a frame. The capacity available for data is then $Y(t) = M - S_t$ slots. Therefore, the capacity available for data packets is

$$C_p = 0.064 \, L(M - S_t) \qquad \text{Mb/s} \qquad (1)$$

Although the number of isochronous slots allocated for voice service can be changed over time, the boundary between isochronous and asynchronous services can be changed only on the basis of an integral number of slots. Therefore, if only part of a slot is occupied by a voice circuit, the other octets would be unoccupied and wasted.

For the emulated packet-switched synchronous service of DSR, we are interested in the average voice occupancy, that is, the average number of slots occupied by active voice sources per service cycle. Let

C_{total} = the total channel capacity in Mb/s.
C_s = the average channel capacity occupied by the voice sources (i.e., the channel occupancy of the emulated synchronous service) in Mb/s.
C_p = the average channel capacity available for asynchronous service in Mb/s.
T = the service cycle length in s.
τ = the round-trip propagation delay in s.
x_v = the voice packet length in s.
N_v = the maximum number of voice sources.
n = the total number of active voice sources (i.e., the number of voice packets generated during the service cycle interval T).
p = the speech activity factor.

The worst case is that the token rotation time equals the round-trip propagation delay τ. Then the channel occupancy of the emulated synchronous service is

$$C_s = C_{total} \left(\frac{\tau + \bar{n}x_v}{T} \right) \tag{2}$$

where $\bar{n}$ is the average number of active voice sources per service cycle. Since a voice source generates at most one voice packet per service cycle, $\bar{n}$ is also the average number of voice packets generated per service cycle. The channel capacity available for asynchronous service thus can be expressed as

$$C_p = C_{total} - C_s = C_{total} \left(1 - \frac{\tau + \bar{n}x_v}{T} \right). \tag{3}$$

The determination of C_s or C_p is tantamount to the determination of $\bar{n}$. Let P_n be the probability of n voice packets generated during a service cycle interval. We then have

$$\bar{n} = \sum_n nP_n. \tag{4}$$

Define $\rho_v = \lambda_v/\mu_v$ as the voice traffic intensity. Let p_k be the probability that k voice sources (call connections) are in progress during a given service cycle. Since each slot can be viewed as providing one voice channel and there is no waiting room for voice packets, our system is a pure loss system. The probability p_k is given by the Erlang formula of the first type:

$$p_k = \frac{\rho_v^k/k!}{\sum_{l=0}^{N_v} (\rho_v^l/l!)}. \tag{5}$$

Let $p_{n|k}$ be the probability that, given there are k calls, n are in the talkspurt state. On the assumption that the voice sources are independent of each other, we have

$$p_{n|k} = \binom{k}{n} p^n (1 - p)^{k-n}. \tag{6}$$

The distribution of the number of voice packets generated during a service cycle interval is given by

$$P_n = \sum_{k=n}^{N_v} p_{n|k} p_k$$

$$= \sum_{k=n}^{N_v} \binom{k}{n} p^n (1 - p)^{k-n} \frac{\rho_v^k}{k! \sum_{l=0}^{N_v} (\rho_v^l/l!)}, \tag{7}$$

where p is the speech activity factor. Substitution of Eq. (7) in Eq. (4) yields the expression for $\bar{n}$.

To compare the channel capacity available for packet-switched, or asynchronous, service between a hybrid isochronous/asynchronous method and a pure packet-switched method, we calculate the capacity C_p as a function of the input rate (the number of connection requests per unit time). Consider the case in which the LAN speed is 100 Mb/s, the speech activity factor is $p = 0.4$ and the average voice connection holding time is 3 min. While the isochronous service of DQDB and FDDI-II is not affected by the network size, the emulated synchronous-like service of DSR is. To calculate the round-trip propagation delay τ, a value of 5 μs/km is used. With a voice coding rate of 64 kb/s and a 48-byte (plus 5 bytes of header) packet length, the service cycle length is $T = 6$ milliseconds (ms). The voice packet length is then $x_v = 4.24$ μs. The channel capacity available for asynchronous service as a function of the connection requests λ for the three systems is plotted in Fig. 13. The maximum number of voice sources (connections), N_v, supported by SDR is made identical to the

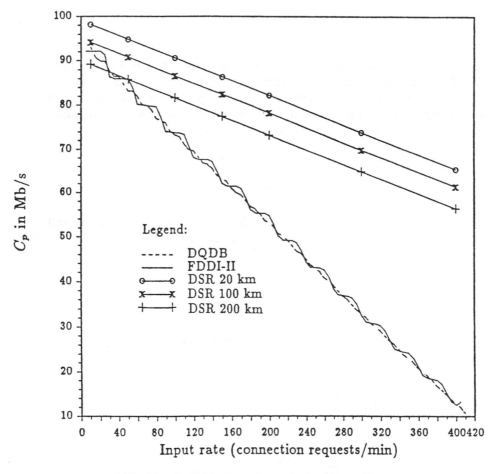

FIG. 13   Available channel capacity for data traffic.

maximum number of octets available in a DQDB frame, that is, 29 slots times 52 octets per slot.

Inspection of Fig. 13 shows that the capacity C_p curve of FDDI-II exhibits a stepwise decrement as a function of the input rate λ. The corresponding curve for DQDB only exhibits a similar behavior at small values of λ. This difference is attributable to the difference between the sizes of a DQDB slot and an FDDI-II wideband channel. It is also observed that DSR exhibits a dependency on the network size.

Based on this limited comparison, it is conjectured that an emulated synchronous-like service, which offers a movable boundary for service integration at the packet level, leaves on average more capacity for asynchronous service than a hybrid isochronous/asynchronous approach.

Performance of Asynchronous Service

In either the circuit-switched mode of FDDI-II or DQDB, or the priority cyclic service discipline of DSR, the synchronous service is not affected by the asynchronous service. Because of the state dependency associated with all three systems, it is difficult to analyze the queueing performance of asynchronous service exactly. In the literature, performance results have been presented mostly using approximate analysis, which is then validated by simulation.

Tangemann and Sauer have presented some performance results for FDDI and FDDI-II (24), while Zukerman has presented some performance results for DQDB (25). Simulation results are critically dependent on the simulation model and the parameters used. Unless one actually has performed the simulation, it is not reasonable to make performance comparisons based on simulation results obtained by others. At this writing, only extensive performance simulation results for the DSR are available (21). In this section, we present some of the delay-throughput numerical results for the DSR system using the CTRA protocol, and some results that demonstrate the fraction of channel capacity recaptured by the add-on random-access feature of CTRA.

The numerical results presented below pertain to the following sets of parameters:

- Transmission speed: 150 Mb/s
- Mean packet length: 3750 bits
- Number of stations: $N = 20, 100$
- The N stations are spaced equally around the ring.
- Poisson arrivals

The ring lengths of 5, 50, and 140 km correspond to the values of α, the normalized propagation delay, of 1, 10, and 28, respectively. It is noted that the use of a packet length of 3750 bits, rather than 53 bytes (424 bits), in the simulation is mainly for convenience. If the packet length were to be 53 bytes long, the values of α would have been much larger. In our simulation, we allow a packet quota of $k = 1$. If the packet size were 424 bits in length, we would let $k = 8$ to be comparable.

To assess the fraction of channel capacity that can be recovered by the add-on random-access component, we present simulation results for the mean access delay as a function of throughput. Here, *access delay* is defined as the delay experienced by a packet from the time it arrives at the source station until it is transmitted successfully by the source station (21).

Simulation results of mean access delay as a function of throughput are shown in Figs. 14 and 15, in which the labels have the following connotation:

- CTRA.$\alpha \equiv$ combined token/random access for a normalized propagation delay α.
- TA.$\alpha \equiv$ token access only for a normalized propagation delay α.

It is observed that, for the token-access-only cases, a large fraction of the transmission capacity is wasted due to token rotation and the lack of concurrent

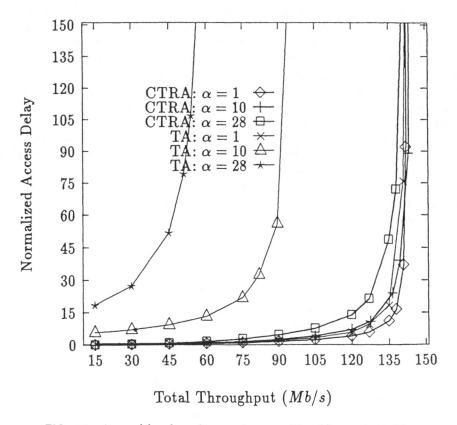

FIG. 14   Access delay-throughput performance ($N = 20$; $\alpha = 1, 10, 28$).

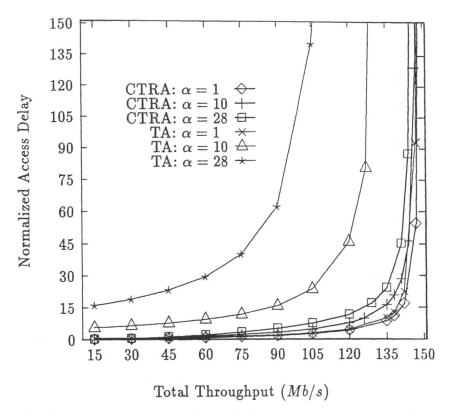

FIG. 15 Access delay-throughput performance ($N = 100$; $\alpha = 1, 10, 28$).

access. The fraction of traffic γ successfully transmitted via the random-access mode is shown in Figs. 16 and 17. It is observed that the fraction of capacity recovered by the random-access mode increases with the value of α. At $\alpha = 1$, virtually all of the traffic is transmitted via token access. At $\alpha = 28$, a large fraction of the traffic is transmitted via random access, especially at light loads.

Although it is not easy to compare system performance among the different high-speed networks, from the study of the DSR system using the CTRA protocol it is conjectured that the add-on random-access feature of the CTRA protocol offers a capability to recover the fraction of the channel capacity that otherwise would be wasted due to token rotation and the lack of concurrent transmission capability in the basic token-access protocol.

Conclusion

The international standards committees on high-speed LANs and MANs have been focusing attention on LAN speeds and cell sizes that are compatible with

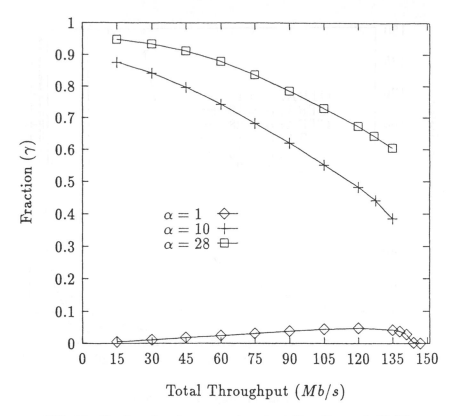

FIG. 16   Fraction of random-access throughput ($N = 20$; $\alpha = 1, 10, 28$).

those intended for Broadband Integrated Services Digital Network (BISDN). This means that the speed is about 150 Mb/s and the cell (slot) size is 53 octets. The intended use would be for the support of integrated real-time and non-real-time services.

Two standards with speeds of 100 Mb/s and above have been under evolution: the ANSI FDDI and the IEEE 802.6 DQDB. These two high-speed networks fit the definition of high-performance LANs. In addition, a DSR high-speed LAN, using a CTRA protocol, also fits the definition of high-performance LANs. It is believed that these networks will form the basis for supporting multimedia communications within local- and metropolitan-area coverages in the foreseeable future.

All three high-performance LANs considered in this article (FDDI, DQDB, and DSR) have origins from earlier networks, either in terms of the system architecture and/or the MAC. The single ring of FDDI has its origin in the IEEE 802.5 token ring (3), the dual-bus topology of DQDB is based on that of Fasnet (4), and the distributed scheduling of DSR is based on a concept developed in Welnet (8). Because of its topological relationship to DQDB, we have given a more detailed discussion of Fasnet compared to other earlier networks. The unidirectional propagation of Expressnet also offers high performance in terms of data throughput (26).

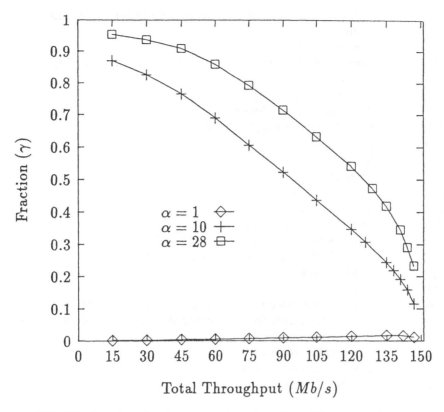

FIG. 17　Fraction of random-access throughput ($N = 100$; $\alpha = 1, 10, 28$).

There have been proposals of fiber-optic-based LANs at the Gb/s (gigabit per second) speed. Two such networks have been proposed by Yeh et al. (27) and Bergman and Eng (28). Because of the short time available for decision making, with the present-day technology, ultra-high-speed networks are restricted to using very lean MAC protocols.

Acknowledgments:　The author wishes to thank B. J. Lee and B. L. Mark for their contributions in obtaining the simulation results for the DSR system. This work has been supported in part by a grant from the Ontario Technology Fund and in part by the Natural Sciences and Engineering Research Council of Canada under Grant No. A7779.

List of Acronyms

ACF	access control field
ASYNC	asynchronous token for asynchronous service in DSR
ATM	asynchronous transfer mode

CSMA/CD	Carrier Sense Multiple Access with Collision Detection
CD	countdown
CTRA	combined token/random access
CTRA.α	combined token/random access for a normalized propagation delay of α slots
DQDB	distributed queueing dual bus
DSR	distributed scheduling ring
FDDI	fiber distribution data interface
FIFO	first-in, first-out
HRC	hybrid ring control
LAN	local-area network
LLC	logical link control
MAC	media access control
MAN	metropolitan-area network
NULL	null token for resetting the ring in DSR
QPSX	queued packet and synchronous exchange
RC	request counter
REQ	request
SYNC	synchronous token for synchronous service in DSR
TA.α	token access only for a normalized propagation delay of a slots
TRT	token rotation timer
TTR	timed token rotation
TTRT	target token rotation time
WBC	wideband channel

References

1. International Telegraph and Telephone Consultative Committee, CCITT Recommendation I.121 on the Broadband Aspects of ISDN. In: *Blue Book*, International Telecommunication Union, Geneva, 1989.

2. Metcalf, R. M., and Boggs, D. R., Ethernet: Distributed Packet Switching for Local Computer Networks, *Commun. ACM*, 19:395–404 (July 1976).

3. Institute of Electrical and Electronics Engineers, *Token Ring Access Method and Physical Layer Specification*, IEEE Std. 802.5, International Organization for Standardization Draft Proposal 8802/5, IEEE, Inc., New York, 1985.

4. Limb, J. O., and Flores, C., Description of Fasnet—A Unidirectional Local Area Communications Network, *Bell Sys. Tech. J.*, 61(7), Part 1:1413–1440 (September 1982).

5. Tobagi, F. A., Borgonovo, F., and Fratta, L., Expressnet: A High-Performance Integrated Services Local Area Network, *IEEE J. Sel. Areas Commun.*, SAC-1(5): 898–913 (November 1983).

6. Institute of Electrical and Electronics Engineers, *Token-Passing Bus Access Method and Physical Layer Specifications*, IEEE Std. 802.4, International Organization for Standardization Draft Proposal 8802/4, IEEE, Inc., New York, 1985.

7. Mark, J. W., Todd, T. D., and Field, J. A., Welnet: Architecture Design and Implementation, *J. Telecommun. Net.*, 3:225–237 (Fall 1982).

8. Mark, J. W., et al., Welnet: A High Performance Integrated Services Local Area Network, *Globcom '85*, 15.6.1–15.6.6 (December 2–5, 1985).

9. Zimmermann, H., OSI Reference Model – The ISO Model of Architecture for Open Systems Interconnection, *IEEE Trans. Commun.*, COM-28(4):425–432 (April 1980).

10. American National Standards Institute, *FDDI Token Ring Media Access Control*, Draft Proposal, ANSI Standard X3T9.5-1983, ANSI, New York, February 1986.

11. Institute of Electrical and Electronics Engineers, IEEE Std. 802.6, DQDB, Metropolitan Area Network, Draft Proposal D9, IEEE, New York, August 1989.

12. Mark, J. W., and Limb, J. O., Integrated Voice/Data Services on Fasnet, *Bell Sys. Tech. J.*, 63(2):307–336 (February 1984).

13. Newman, R. M., and Hullett, J. L., Distributed Queueing: A Fast and Efficient Packet Access Protocol for QPSX, *Proc. 8th Int. Conf. Comp. Comm.*, 294–299 (September 1986).

14. Newman, R. M., Budrikis, Z. L., and Hullett, J. L., The QPSX Man, *IEEE Commun.*, 26(4):20–28 (April 1988).

15. Hahne, E. L., Choudhury, A. K., and Maxemchuk, N. F., Improving the Fairness of Distributed-Queue-Dual-Bus Networks, *Proc. Infocom*, 175–184 (June 1990).

16. Rodrigues, M. A., Erasure Node: Performance Improvements for the IEEE 802.6 MAN, *Proc. Infocom*, 636–643 (June 1990).

17. Bux, W., Closs, F. N., Kummerle, K., Keller, H. J., and Mueller, H. R., Architecture and Design of a Reliable Token-Ring Network, *IEEE J. Sel. Areas Commun.*, SAC-1:756–765 (November 1983).

18. Mark, J. W., and Lee, B. J., A Dual-Ring LAN for Integrated Voice/Video/Data Services, *Proc. Infocom*, 850–857 (June 1990).

19. Ross, F. E., An Overview of FDDI: The Fiber Distributed Data Interface, *IEEE J. Sel. Areas Commun.*, 7(7):1043–1051 (September 1989).

20. Johnson, M. J., Proof that Timing Requirements of the FDDI Token Ring Protocol are Satisfied, *IEEE Trans. Commun.*, COM-35(6):620–625 (June 1987).

21. Mark, J. W., Lee, B. J., and Mark, B. L., A Combined Token/Random Access Protocol for High-Speed LAN/MAN Ring Networks, University of Waterloo, Computer Communications Networks Group (CCNG) Report, E-215, March 1992.

22. Bhargava, A., Kurose, J. F., and Towsley, D., A Hybrid Media Access Protocol for High-Speed Ring Networks, *IEEE J. Sel. Areas Commun.*, 6(6):924–933 (July 1988).

23. Gerla, M., Wang, G.-S., and Rodrigues, P., Buzz-net: A Hybrid Token/Random Access LAN, *IEEE J. Sel. Areas Commun.*, SAC-5(6):977–988 (July 1987).

24. Tangemann, M., and Sauer, K., Performance Analysis of the Timed Token Protocol of FDDI and FDDI-Il, *IEEE J. Sel. Areas Commun.*, 9(2):271–278 (February 1991).

25. Zukerman, M., Queueing Performance of QPSX. In: *Teletraffic Science, ITC-12* (M. Bonatti, ed.), Elsevier, New York, 1989, pp. 575–581.

26. Tobagi, F. A., and Fine, M., Performance of Unidirectional Broadcast Local Area Networks: Expressnet and Fasnet, *IEEE J. Sel. Areas Commun.*, SAC-11(5):913–926 (November 1983).

27. Yeh, C., et al., RATO-Net: A Random-Access Protocol for Unidirectional Ultra-High-Speed Optical Fiber Networks, *IEEE J. Lightwave Technology*, 8(1):78–88 (January 1990).

28. Bergman, L. A., and Eng, S. T., A Synchronous Fiber Optic Ring Local Area Network for Multigigabit/s Mixed-Traffic Communication, *IEEE J. Sel. Areas Commun.*, SAC-3(6):842–848 (November 1985).

JON W. MARK

Fast Packet Network: Data, Image, and Voice Signal Recovery

Introduction

The Integrated Services Digital Network (ISDN) has become a reality. Broadband ISDN and Asynchronous Transfer Mode (ATM) networks will include video signals. The future vision is the total integration of voice, video, and data in the same transmission medium and switching nodes; this integration simplifies the complexity of a network and in theory reduces the costs of implementation and maintenance of a telecommunications network.

One way to achieve this vision is the packetization of voice, video, and data using fiber optics for the transmission media and packet switching for the switching nodes (1,2). Packet networks customarily are used for data networks. The main reason is that the transmission of data is bursty and sometimes takes a long time. In order to avoid the full dedication of resources of a telecommunications network to long bursty data, packet switching is preferred. Packet switching, unlike circuit switching, does not dedicate a circuit for the period of a call. In case of congestion, data packets are stored in buffer memories for later transmission. Although this delay is acceptable for data transmission in many applications, data packet losses are intolerable.

Voice packets, on the other hand, cannot be stored in buffer memories and sent at a later time because telephone conversations are full duplex and in real time. Packet delays are limited to about 400 milliseconds (ms), similar to telecommunications standards for satellite communications. This means that if voice packets experience excessive delays due to switching node congestions, we have no choice but to drop those packets. Voice packets also can be dropped at the transmitter end when digital speech interpolation (DSI) is used to take advantage of silence in a telephone conversation to increase the capacity of the transmission medium. This implies that if voice signals are sampled at the Nyquist rate, packet losses result in the degradation of the quality of voice at the other end. This degradation, unlike data transmission, may be acceptable in many voice applications.

On the other hand, transmission of facsimile and video signals is a one-way communication system and delay is not as serious a problem as it is for voice. However, facsimile signals use the voiceband network and packet losses are possible due to a 400-ms delay limit. Although this problem could be solved by distinguishing between facsimile and voice packets, we still have the probability of packet losses due to buffer memory limitations. This problem is more realistic for video signals due to their bulk nature. In other words, although delay is not a serious problem for facsimile and video signals, packet losses are possible due to congestion and buffer memory limitations.

There are different techniques to remedy this problem. Automatic Repeat Request (or Query) (ARQ), forward error correcting (FEC) codes, and signal interpolation are potential methods. We briefly discuss these methods in this

453

article. Since we see more potential in signal processing as an alternative to ARQ and FEC, we bias our discussion to signal interpolation in the case of packet losses.

Before we get into the main theme of our discussion, it might be helpful to review the fundamental theorems of uniform and nonuniform sampling. These theorems are needed to understand the discussion of signal interpolation from lost samples.

Sampling Theorems

In this section, we discuss the uniform and nonuniform sampling theorems. The uniform sampling theorem is discussed briefly here as a reference for the reader. Our emphasis in this section is on nonuniform sampling theorem, which is necessary to our discussion of signal recovery from packet losses.

Let $x(t)$ be band-limited to W; the uniform sampling theorem can be stated formally as

Theorem 1. A band-limited signal $x(t)$ can be recovered uniquely from a set of uniform samples $\{x(nT), n = \ldots -2, -1, 0, 1, 2, \ldots\}$ if the sampling interval satisfies the Nyquist condition, that is, $T \geq 1/2W$. The interpolation is given by (3)

$$x(t) = \sum_{-\infty}^{+\infty} x(nT)\mathrm{sinc}\left(\frac{t}{T} - n\right), \tag{1}$$

where the sinc function in the equation above is the impulse response of an ideal low-pass filter (LPF) with a bandwidth of $1/2T$. Equation (1) implies that by using a linear time-invariant LPF, one can recover a band-limited signal from its ideal impulsive samples. One may wonder what happens if some of the samples are lost. Is it still possible to recover a signal from the remaining samples? We can answer this question after we study the theorems related to nonuniform sampling theory.

Theorem 2. If the band-limited signal $x(t)$ is sampled nonuniformly and if the sampling rate is above the Nyquist rate on the average, signal recovery is possible; this set of nonuniform samples uniquely determines $x(T)$ (4,5).

Proof: We prove this theorem by contradiction. Suppose signal recovery is not unique and we have at least two signals having bandwidths less than or equal to W. If we subtract the two signals, we get another signal that has a bandwidth of W or less. This difference signal has zero crossings at the nonuniform positions. According to our assumption, the average sampling rate is above the Nyquist rate, therefore the density of zero crossings also is above the Nyquist rate. This contradicts a well-established theorem that the average density of zero crossings of a signal band limited to W is less than or at most equal to $2W$ (6). Therefore, we conclude that the nonuniform sampling set uniquely determines $x(t)$.

For our application in this article, we need the following corollary, which is derived from the above theorem.

Corollary 2.1. Assume that the signal $x(t)$ is sampled uniformly and the sampling rate is more than the Nyquist rate. If some of the samples are lost and their positions are known, perfect signal recovery is possible from the remaining samples provided that the density of the remaining samples is above the Nyquist rate.

Some comments are in order. In our application, the positions of lost samples are known whenever the packets are dropped. In a communications system, this is called an *erasure channel*. The density of samples is calculated by counting the number of samples in an interval τ. The density is the limit of the ratio of the number of samples to τ when τ goes to infinity. Similarly to the uniform sampling interpolation (Theorem 1), we can find an interpolation for the nonuniform sampling case. The interpolation follows.

If we restrict the nonuniform positions (t_n) around nT (where $T = 1/2W$ at the Nyquist rate) such that (7)

$$|t_n - nT| \leq D < \frac{T}{4}, \qquad n = 0, \pm 1, \pm 2, \ldots, \tag{2}$$

then

$$x(t) = \sum_{n=-\infty}^{\infty} x(t_n)\Psi_n(t), \tag{3}$$

where

$$\Psi_n(t) = \frac{H(t)}{\dot{H}(t_n)(t - t_n)}, \qquad \Psi_n(t_n) = 1 \text{ and } \Psi_n(t_k) \neq 0, k \neq n, \tag{4}$$

and

$$H(t) = (t - t_0) \prod_{\substack{k=-\infty \\ k \neq 0}}^{\infty} \left(1 - \frac{t}{t_k}\right). \tag{5}$$

For proof of the convergence, see Refs. 8 and 9. $H(t)$ cannot be a band-limited function; $\Psi_n(t)$ is, however, a band-limited interpolation function. This function, unlike sinc functions, has its maximum at $t \neq t_n$. This might create a dynamic range problem when $t_n - t_{n-1}$ is large (10). If one replaces the sampling set $\{t_n\}$ by a finite set, say $\{t_n; 0 \leq n = N\}$, one derives the classical Lagrange interpolation of polynomials of degree less than or equal to N.

The above Lagrange interpolation is the most general interpolation for any sampling scheme. Indeed, most interpolating functions can be shown to be special cases of Eq. (3) (5). For instance, the well-known Shannon sampling theorem can be derived from the Lagrange interpolation by taking the sampling set $\{t_n\}$ to be $\{nT\}$. In this case, the product in Eq. (5) converges to $T/\pi \sin(\pi t/T)$. Equation (4) becomes

$$\Psi_n(t) = \frac{(-1)^n \sin\left(\frac{\pi t}{T}\right)}{\frac{\pi}{T}(t - nT)} = \text{sinc}\left(\frac{t}{T} - n\right).$$

The Lagrange interpolation is a linear process but is not a time-invariant system. Therefore, signal recovery from nonuniform samples, unlike uniform samples, cannot be achieved by such a simple linear time-invariant system as an LPF.

We discuss nonlinear and iterative methods for signal recovery, but, prior to that, we need to analyze the frequency spectrum of a set of uniform samples in which some of the samples are lost. This set is a special case of nonuniform samples. The spectral analysis gives an insight into how to devise a nonlinear and/or iterative method for signal recovery (11,12).

Spectral Analysis of a Set with Missing Samples

According to Corollary 2.1, a set of uniform samples, when some samples are lost, can determine uniquely a band-limited signal if the average density of the remaining samples satisfies the Nyquist rate. This means that, in general, if the sampling rate is a percent higher than the Nyquist rate, we can afford to lose up to $a/100/(1 + a/100)$ of the samples and still satisfy Corollary 2.1.

In case of sample losses, the remaining samples can be modeled as

$$x_s(t) = x(t)x_p(t), \tag{6}$$

where $x_p(t)$ is the comb function defined by

$$x_p(t) = \sum_k \delta(t - t_k); \qquad t_k = kT_1, k \neq \{i\} \tag{7}$$

where we have denoted the nonuniform positions t_k as uniform positions $t_k = kT_1$ except when the samples are lost; we denote the positions of the lost samples as $\{i\}$. The sampling rate $(1/T_1)$ is assumed to be a percent higher than the Nyquist rate $(1/T)$, so that

$$\frac{1}{T_1} = \frac{(1 + a/100)}{T}.$$

We assume that the percentage of lost samples is $b < [a/(1 + a/100)]$ giving an average sampling rate of

$$\frac{1}{T_2} = \frac{1}{T_1} - \frac{b/100}{T_1} = \frac{(1 - b/100)(1 + a/100)}{T} > \frac{1}{T}. \tag{8}$$

From the theory of generalized functions (13), we can write

$$\delta(t - t_k) = |\dot{g}(t)|\delta[g(t)], \tag{9}$$

provided that $g(t_k) = 0$ and $\dot{g}(t_k) \neq 0$, and that $g(t)$ has no zeros other than $\{t_k\}$. One possible $g(t)$ can be written as

$$g(t) = t - kT_2 - \theta(t), \tag{10}$$

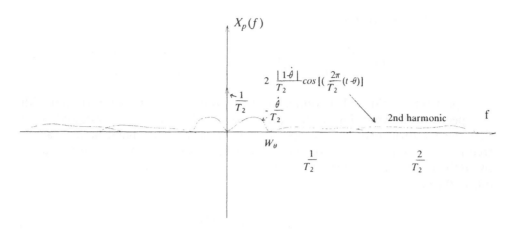

FIG. 1 Spectrum of nonuniform positions of $x_p(t)$.

where $\theta(t)$ is any function such that $g(t_k) = t_k - kT_2 - \theta(t_k) = 0$, that is, $\theta(t_k) = t_k - kT_2$ is the deviation of the nonuniform samples from the uniform position of an average sampling period* of T_2 defined in Eq. (8). Substituting Eqs. (9) and (10) in Eq. (7), we get

$$x_p(t) = [1 - \dot{\theta}(t)] \sum_{k=-\infty}^{\infty} \delta(\Phi(t) - kT_2) , \qquad (11)$$

where $\Phi(t) = t - \theta(t)$, and where we assume that $1 > |\dot{\theta}(t)|$. The Fourier expansion of Eq. (11) in terms of ϕ is

$$x_p(t) = \frac{[1 - \dot{\theta}(t)]}{T_2} \sum_{k=-\infty}^{\infty} e^{jk\frac{2\pi}{T_2}\Phi(t)}$$

$$= \frac{[1 - \dot{\theta}(t)]}{T_2} \left[1 + 2 \sum_{k=1}^{\infty} \cos\left(\frac{2\pi kt}{T_2} - \frac{2\pi k\theta(t)}{T_2}\right) \right] \qquad (12)$$

Equation (12) reveals that $x_p(t)$ has a direct current (DC) component [1 $- \dot{\theta}(t)]/T_2$ plus harmonics that resemble phase modulation (PM) signals. The index of modulation is $(2\pi k)/T_2$; the bandwidth is proportional to the index of modulation, the bandwidth of $\theta(t)$ and the maximum amplitude of $\theta(t)$, which in this case is related to $\theta(t_k) = t_k - kT_2$. The spectrum of $x_p(t)$ is shown in Fig. 1. The nonuniform samples can be expanded using Eqs. (6) and (12), that is,

---

*Since the average sampling rate of the nonuniform samples of $\theta(t_k)$ is $(1/T_2)$, we can determine uniquely a band-limited function $\theta(t)$ with a bandwidth of $(1/2T_2)$ or less.

$$x_s(t) = x(t)x_p(t)$$

$$= x(t) \frac{[1 - \dot{\theta}(t)]}{T_2} \left[1 + 2 \sum_{k=1}^{\infty} \cos\left(\frac{2\pi kt}{T_2} - \frac{2\pi k\theta(t)}{T_2} \right) \right]. \qquad (13)$$

The spectrum of Eq. (13) is obtained by convolving the spectrum of $x(t)$ with the spectrum shown in Fig. 1. Now, if we assume $(1/T_2) >> (1/T_2)$, and given that the bandwidth of $\theta(t)$ is less than $(1/2T_2)$, the PM signal at the carrier frequency $(1/T_2)$ is a narrowband PM and has a bandwidth of approximately twice the bandwidth of $\theta(t)$, that is, less than $(1/T_2)$. Therefore, low-pass filtering $x_p(t)$ (see Fig. 1) yields

$$x_{p_{lp}}(t) = \frac{[1 - \dot{\theta}(t)]}{T_2}. \qquad (14)$$

If the bandwidth of $\theta(t)$ is taken to be W_θ, then as long as $(1/T_2) - W_\theta - W > W + W_\theta$, there is no overlap between the narrowband PM signal and $X(f)*\mathcal{F}\left\{ \left(\frac{-\dot{\theta}(t)}{T_2} \right) \right\}$, where $\mathcal{F}$ stands for the Fourier transform. Thus, low-pass filtering $x_s(t)$ (with a bandwidth of $W + W_\theta$), we get

$$x_{s_{lp}}(t) = x(t) \frac{[1 - \dot{\theta}(t)]}{T_2}. \qquad (15)$$

Comparing Eq. (14) to Eq. (15), we can recover $x(t)$ by dividing $x_{s_{lp}}(t)$ by $x_{p_{lp}}(t)$, that is,

$$x(t) = \frac{x_{s_{lp}}(t)}{x_{p_{lp}}(t)}. \qquad (16)$$

The division is possible if $\dot{\theta}(t) \neq 0$ and hence the denominator of Eq. (15) is non-zero. In practice, we normally choose the bandwidth of the LPF for $x_{s_{lp}}(t)$ to be equal to $(1/2T)$. This assumption creates some minor distortion.

We thus developed a nonlinear technique for signal recovery from a set of nonuniform samples. The proposed method is shown in Fig. 2. The degraded signal (nonuniform samples of speech) is represented by $x_s(t)$. This subsequently is low-pass filtered, giving $x_{s_{lp}}(t)$. In parallel, $x_s(t)$ is hard limited, rectified, and low-pass filtered to yield $x_{p_{lp}}(t)$, which is a representation of the positions of the missing samples. Division of $x_{s_{lp}}(t)$ by $x_{p_{lp}}(t)$ yields $\hat{x}(t)$, an estimate of the original speech signal. The method shown in Fig. 2 is an approximation that works quite well despite its simplicity. We give some simulation results for speech signals using this method in the section on simulation and comparison of results (p. 472).

The above spectral analysis can be extended to image signals (14). The comb function for an image signal with missing pixels is denoted by $f_p(x,y)$ and defined as

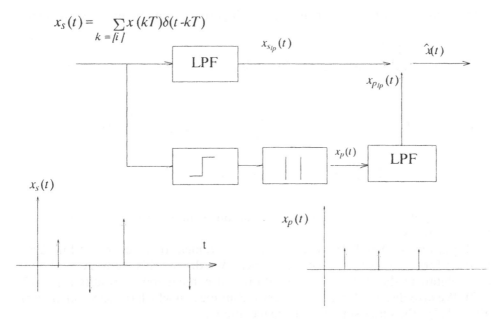

FIG. 2   Block diagram of the nonlinear reconstruction method (LPF = low-pass filter).

$$f_p(x,y) = \sum_m \sum_n \delta(x - x_{nm}, y - y_{nm}). \tag{17}$$

Using the same analysis as the one-dimensional case (14), the Fourier expansion of the above comb function is

$$f_p(x,y) = \frac{|J|}{T_1 T_2} \left\{ 1 + 2 \sum_{i=1}^{\infty} \cos \frac{2\pi i}{T_1} [x - \theta_1(x,y)] \right\}$$

$$\times \left\{ 1 + 2 \sum_{l=1}^{\infty} \cos \frac{2\pi l}{T_2} [y - \theta_2(x,y)] \right\} \tag{18}$$

where J is the Jacobian function defined as

$$|J| = \left| \left(1 - \frac{\partial \theta_1}{\partial x} \right)\left(1 - \frac{\partial \theta_2}{\partial y} \right) - \left(\frac{\partial \theta_1}{\partial y} \right)\left(\frac{\partial \theta_2}{\partial x} \right) \right|$$

and where $\theta_1(x,y)$ and $\theta_2(x,y)$ are any two-dimensional functions such that

$$\begin{cases} \theta_1(x_{nm}, y_{nm}) = x_{nm} - nT_1 \\ \theta_2(y_{nm}, y_{nm}) = y_{nm} - mT_2. \end{cases}$$

Equation (18) is equivalent to the sum of phase-modulated signals in two-dimensions. Therefore, the Fourier spectrum of $f_p(x,y)$ consists of a low-pass component $|J_1|/(T_1T_2)$ and band-pass or high-pass components around carrier frequencies $\left(\dfrac{2\pi n}{T_1}, \dfrac{2\pi m}{T_2}\right)$.

The image pixels are derived by multiplying Eq. (18) by the image $f(x,y)$, that is,

$$f_s(x,y) = f(x,y)f_p(x,y) = f(x,y)\ \frac{|J|}{T_1T_2} \times \left\{1 + 2\sum_{i=1}^{\infty} \cos \frac{2\pi i}{T_1} [x - \theta_1(x,y)]\right\}$$

$$\left\{1 + 2\sum_{l=1}^{\infty} \cos \frac{2\pi l}{T_2} [y - \theta_2(x,y)]\right\}. \tag{19}$$

Equation (19) can be interpreted as deviation of nonuniform points from uniform positions.

Equations (18) and (19) suggest that a two-dimensional version of Fig. 2 can be used for image recovery with lost pixels. We discuss this recovery method in more detail in the section on simulation and comparison of results (pp. 472–477). We also discuss iterative recovery techniques, in which the spectral analysis above helps the understanding of the technique.

Classical Techniques in Packet Recovery

So far we have analyzed the sampling theorems as related to packet loss and recovery. In this section, we briefly review the classical methods of FEC and ARQ as applied in a packet network. After this review, we discuss the signal processing approaches to packet recovery.

Forward Error Correcting Codes

In the forward error correcting code method, redundant bits (parity bits) are added to the packets at the transmitter side. At the receiver, the parity bits are used for error detection and correction. The main drawbacks of this method are the overhead and the complexity of the decoder. There are two kinds of FEC codes: block and convolutional.

In block codes, the packets are divided into fixed message blocks of k information bits. Each message block is mapped into an n-tuple codeword. The ratio of k/n is called the *code rate*. Since there is a set of 2^k codewords associated with the information blocks, any n-tuple sequence at the receiver that does not belong to the set is erroneous and may be corrected. All special cases of block codes (i.e., cyclic codes) are used to reduce the complexity of the decoder for error correction.

In convolutional codes, the n encoded message depends not only on the corresponding k input message block but also on the past m blocks. The (n,k,m)

convolutional code with the code rate $R = k/n$ is implemented by a k-input and n-output sequential circuit with m shift registers. For small m, the Viterbi algorithm is the best and the simplest decoder. For large m, sequential decoding is preferable.

The performance of an FEC code is measured by its code rate $R = k/n$, complexity of the decoder and the performance curve, that is, the plot of the probability of error versus the signal-to-noise ratio (SNR). In general, for a given code rate k/n, the longer the message size k (for block codes) and memory size m (for convolutional codes), the better the performance will be. The decoder, however, becomes more complex. For a fixed message size k, the smaller the code rate is, the better the performance of the code will be. The tradeoff is a higher transmission rate.

Automatic Repeat Request Systems

If a receiver node in a packet network can detect errors, it can ask the transmitter to repeat the packet; this system of detection and request for retransmission is called ARQ. This strategy is suitable particularly for a communications system with small transmission delay.

There are three types of ARQ: stop and wait, go-back-N, and selective repeat (15). In the stop-and-wait method, the transmitter does not send a packet unless it receives a positive acknowledgment (ACK). If the transmitter receives a negative acknowledgment (NAK) due to error detection at the receiver, the previous packet is sent again. In the go-back-N method, the transmitter does not wait for ACK to send the next packet, but rather it continues sending packets until a NAK is received. When a NAK is received, the transmitter backs up to the erroneous packet and retransmits that packet along with the next N − 1 succeeding packets that were sent during the round-trip delay of sending the packet and receiving a NAK. This scheme is more effective than the stop-and-wait method but requires more buffer and circuitry. This method becomes ineffective when the round-trip delay is large and the transmission rate is high. In the selective-repeat scheme, only the negatively acknowledged packets are retransmitted. This requires sequence headers for each packet and a large buffer at the receiver. However, the performance is better than with the other two schemes.

The performance of an ARQ system is measured by its throughput and reliability. The *throughput* is defined as the average information bit rate accepted by the receiver divided by the transmission rate at the transmitter. In general, the throughput of the select-repeat ARQ is better than the go-back-N scheme, which is better than the stop-and-wait ARQ (see Fig. 3). The reliability of any type of ARQ depends on the error-detection capability of the code used in conjunction with each packet and in general is very high; therefore, the reliability of all the ARQ schemes is the same.

In comparison to FEC codes, ARQ systems are simpler to implement and have a higher reliability since the probability of undetected error is much lower than the probability of a decoded error. However, as Fig. 3 demonstrates, the throughput of ARQ systems fails rapidly with higher channel error rates. On

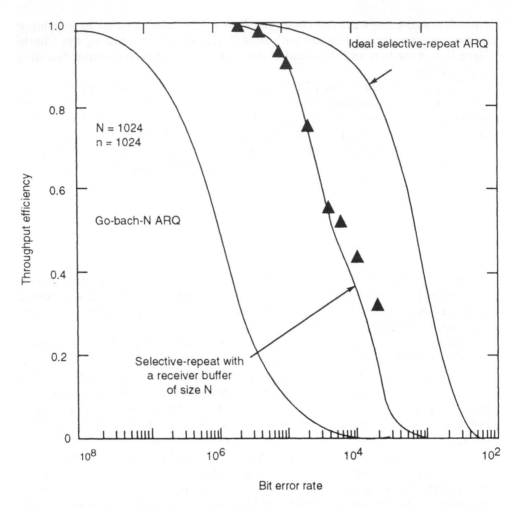

FIG. 3 Throughput efficiencies: the ideal selective-repeat automatic repeat request with infinite receiver buffer, the selective-repeat automatic repeat request with a receive buffer of size $N = 1024$ (the solid triangles represent simulation results), and the go-back-N automatic repeat request. (From Ref. 15.)

the other hand, FEC systems have a constant throughput equal to the code rate k/n irrespective of the channel error rates. For these reasons, ARQ is preferred in packet networks. However, if return channels are not available and/or the delay is too long, FEC and signal processing approaches (discussed in the next section) are the only choices. The overhead and the complexity of FEC, in general, are higher than that of the signal processing approaches.

Hybrid ARQ is a combination of FEC and ARQ schemes in which the drawbacks of both systems are reduced. In this scheme, errors are corrected first and then, in the case of error-detection, packets are retransmitted. This improves the throughput as well as the reliability; for more details, see Ref. 15.

Signal Processing for Recovering Lost Packets

An alternative to ARQ and FEC is signal interpolation. Instead of adding redundant bits, we oversample and, in the case of lost packets, we try to interpolate the missing samples from the remaining samples. According to Theorem 2, if the Nyquist rate is satisfied on the average, error-free recovery of lost packets is possible. An advantage of this technique might be simpler circuitry in comparison to that of classical methods. Another advantage is that, unlike data, voice and video signals do not require perfect recovery. This means that even at the Nyquist rate, the SNR can be improved for lost packets using interpolation. For the extension of this idea to data transmission, see Ref. 16. We discuss this improvement for speech and image signals in the simulation section.

Some simple techniques for lost speech packet recovery are discussed in Refs. 17 and 18. A summary of these methods is given in Ref. 19. We briefly discuss these methods here.

Substitution Techniques

The simplest way to handle a packet loss is to substitute zero amplitudes for lost samples. Obviously, this crude method does not perform too well and may be acceptable only for small probabilities of packet loss (rates less than 1%) (20,21). In addition, it does not work for any low-bit-rate coded signals (e.g., differential pulse code modulation [DPCM], adaptive DPCM [ADPCM], delta modulation [DM], adaptive DM [ADM]) because the recovered samples depend on previous ones. This means that in the case of packet losses, error will propagate to other packets after decoding.

Another method used in the case of a packet loss is to substitute the sample values of the previous packet (17,22). This scheme, however, introduces distortion due to discontinuities at packet edges (17). In order to reduce the subjective effects of discontinuities, one can taper the end samples of the substituted packet along with the neighboring packets.

A more sophisticated substitution method proposed in Ref. 17 is called *pattern matching*. Figures 4–6 show the method graphically. Figure 4 shows one lost packet. L is the number of samples per packet and M the template size that came just before the missing packet. The algorithm searches for the best match between the template and the window shown in Fig. 5; it then uses the L samples that follow the best match as a substitution packet. Figure 6 shows the reconstructed waveform.

Another waveform substitution is based on pitch detection for voiced segments. If the previous packet is unvoiced, we repeat the previous packet in the time slot for the missing packet. When the previous packet is voiced, if we can find an estimate of the period of the pitch then a missing packet can be reconstructed by repeating the pitch with the same period. This strategy makes no allowance for changes that might occur during missing packets. However, it has been proven to be the best waveform substitution method for PCM-type

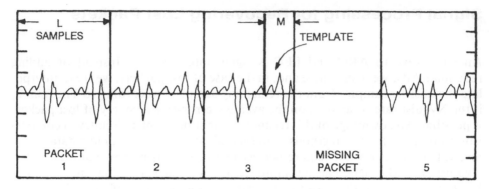

FIG. 4   Speech waveform divided into five packets, each with L samples. Packet 4 is missing, and the final M samples of Packet 3 comprise the template to be used in a search for a substitution packet. (From Ref. 17.)

samples (22). The above methods, in general, are not appropriate for differential types of PCM in which recovered samples are dependent on the previous samples.

Recently, another method has been suggested for waveform substitution. It consists of storing short-time energy information for different segments of a packet and the zero-crossing information of these segments in a previous packet (23). In case of a packet loss, a waveform is substituted based on the short-time energy information and the zero-crossing information stored in the previous packet. Since about 20–25% overhead is added to each packet, one would expect that this method should perform better than the other methods discussed above.

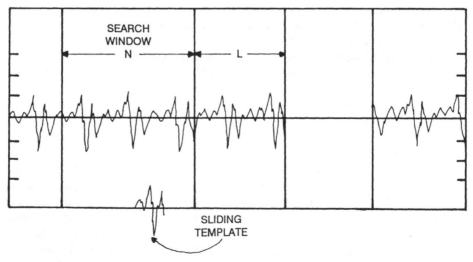

FIG. 5   The template slides along a search window containing N samples to find M samples that best match the template. (From Ref. 17.)

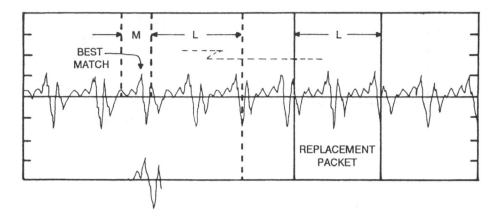

FIG. 6   The substitution packet contains the L samples immediately following the best match to the template. (From Ref. 17.)

Least Significant Bit Dropping Scheme

In the least significant bit (LSB) dropping scheme, the PCM codewords for each sample are divided into two segments: most significant bits (MSBs) and LSBs. Each segment is packetized separately. In case of network congestions, packets that consist of LSBs have a higher probability of being dropped. Under these circumstances, the receiver decodes samples despite LSB losses; this recovery is at the cost of more quantization error in case of packet losses (19). The SNR of the reconstructed waveform after dropping L LSBs of an M-bit quantizer PCM is

$$\frac{3}{A^2(2^{-2D} - 2^{-2M})},\tag{20}$$

where A is the maximum input amplitude and $D = M - L$. This procedure sometimes is called *embedded coding* and can be generalized to DPCM and ADPCM (24).

A typical DPCM is depicted in Fig. 7. The feedback loop in the encoder and decoder is a predictor approximating the input signal $x(k)$. DPCM, unlike PCM, is not an embedded code. This means that dropping L bits from M-bit DPCM codewords generates substantially more quantization noise than DPCM encoding with $D = M - L$ bits per sample. This is due to the correlation of samples at the encoder and decoder and propagation of errors. In order to make DPCM an embedded code, a modified DPCM is suggested (see Fig. 8) (25). This embedded DPCM has a slightly worse performance than the conventional M-bit DPCM. The same idea could be used for embedded ADPCM (24). Figure 9 shows the benefit of bit dropping on the average delay experienced in a packetized voice conversation (26). The above techniques are not applicable to data and video packets. The following methods are more general and are applicable to voice, fax, and video packets.

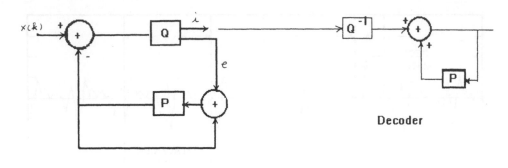

FIG. 7   Differential pulse code modulation.

Interpolation

Next, we discuss different interpolating functions for lost packets: linear (LI) interpolation, nonlinear (NL) interpolation, spline (SP) interpolation, Lagrange (LG) interpolation, and iterative (IT) interpolation. We consider sampling rates at the Nyquist rate and twice the Nyquist rate. Also, both PCM and ADPCM are considered.

Linear Interpolation

The simplest type of interpolation besides sample and hold is LI interpolation. For this technique to be effective, signal samples should be interleaved and deinterleaved at the transmitter and the receiver, respectively. One way to

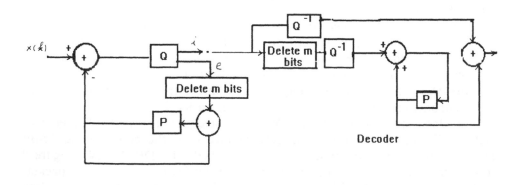

FIG. 8   Embedded differential pulse code modulation.

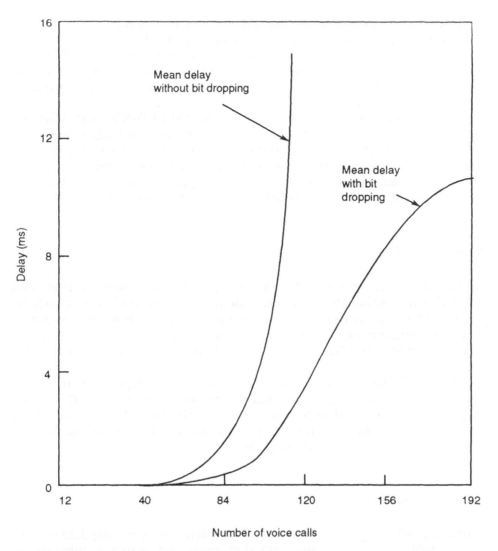

FIG. 9 The performance of bit dropping in adaptive differential pulse code modulation packetized speech.

achieve this is to partition odd and even samples into two adjacent packets. If we assume the probability of adjacent packet losses is independent of each other, the probability of losing two adjacent packets is very low. The samples of a lost packet then can be estimated by LI interpolation in the form

$$\hat{q}_P = \alpha q_P(n - 1) + \beta q_P(n + 1) \tag{21}$$

where $q_P(n)$ is the PCM signal for voice or video sources. α and β are constant parameters that can be optimized and depend on the autocorrelation function of the samples.* In other words, the odd-even interleaving reduces the problem of packet interpolation to sample interpolation (18). This reduction is at the expense of increased delay (twice the packet size) at the receiver end. For voice packets, this would translate into 32–64-ms delay, which is acceptable in telephone networks. If two adjacent packets are lost, we assume zero stuffing for the entire two blocks.

This odd-even LI interpolation can be extended to DPCM and ADPCM, but the difference signal for these cases has a lower autocorrelation and hence is not as predictable as a PCM signal. Reference 19 shows that embedded ADPCM (International Telegraph and Telephone Consultative Committee [CCITT] Recommendation G.721) and LSB dropping similar to that shown in Fig. 8 outperforms any LI interpolation technique for ADPCM-type coding.

Nonlinear Technique

We first convert the problem of packet interpolation into sample interpolation by interleaving. Although the odd-even interpolation discussed for LI interpolation could be used, we can envision a more sophisticated scheme over a span of n packets. Interleaving is a classical method to combat burst errors in error-correcting codes (15). The price of interleaving is a delay of n packet size at the decoder.

The circuit shown in Fig. 2 now can be used for interpolation of missing samples. In the section on simulation results, we show that the NL method outperforms the LI method for voice and still pictures. Although there have not been any experimental results, it would be safe to assume that the NL method also works well for video signals.

Spline Interpolation

Spline interpolation is a classical method for interpolating missing data in computer graphics (27). A common method in spline interpolation is cubic spline. Assume there are $n + 1$ samples and hence n intervals (the lost samples are ignored); to get a function that interpolates over these $n + 1$ samples, we use a third-order polynomial for each interval. For the interval i, the polynomial is in the form

$$f_i(x) = a_i x^3 + b_i x^2 + c_i x + d_i, \tag{22}$$

where $f_i(x_{i-1})$ and $f_i(x_i)$ are known.

---

*The optimum values are $\alpha_{opt} = \beta_{opt} = [R_{qq}(1)][1 + R_{qq}(2)]^{-1}$, where R is the autocorrelation function. If $R_{qq}(1) = R_{qq}(2) = 1$, we have $\alpha_{opt} = \beta_{opt} = 1/2$. These conditions are almost satisfied if the source signal is oversampled. Experimental results in Ref. 19 show that even at the Nyquist sampling rate we can assume $\alpha = \beta = 1/2$ without any noticeable degradation. Reference 18, however, shows that adaptive LI based on varying autocorrelation function does improve the performance for unvoiced speech but there is no significant improvement for voiced speech.

Since there are n intervals, there are $4n$ unknowns and $2n$ equations. In order to solve for these unknown parameters, we assume that the first and second derivatives of adjacent polynomials are equal and the second derivatives at the first and the last samples are zero. These assumptions also make the interpolated function "look" smooth. For computational purposes, the polynomial represented in Eq. (22), along with the assumptions for smoothness, can be written in terms of the second derivatives in the following form:

$$f_i(x) = \frac{\ddot{f}(x_{i-1})}{6(x_i - x_{i-1})} (x_i - x)^3 + \frac{\ddot{f}(x_i)}{6(x_i - x_{i-1})} (x - x_{i-1})^3$$

$$+ \left[\frac{f(x_{i-1})}{(x_i - x_{i-1})} - \frac{\ddot{f}(x_{i-1})(x_i - x_{i-1})}{6} \right](x_i - x)$$

$$+ \left[\frac{f(x_i)}{(x_i - x_{i-1})} - \frac{\ddot{f}(x_i)(x_i - x_{i-1})}{6} \right](x_i - x) \qquad (23)$$

The second derivatives can be found by using the assumption that the derivative of Eq. (23) is continuous at each sample point, that is,

$$\dot{f}_{i-1}(x_i) = \dot{f}_i(x_i)$$

The result is

$$[x_i - x_{i-1}]\ddot{f}(x_{i-1}) + 2(x_{i+1} - x_{i-1})\ddot{f}(x_i) + (x_{i+1} - x_i)\ddot{f}(x_{i+1})$$

$$= \frac{6}{(x_{i+1} - x_i)} [f(x_{i+1}) - f(x_i)] + \frac{6}{(x_i - x_{i-1})} [f(x_{i-1}) - f(x_i)] \qquad (24)$$

Since for a natural cubic spline the second derivatives at the end sample points are zero, we have $n - 2$ equations and $n - 2$ unknowns. In our application, n must be many times the number of samples in one packet to get acceptable results. However, because the system equations are tridiagonal, the solution can be obtained efficiently using the Gaussian algorithm for decomposition to invert the $(n - 2) \times (n - 2)$ coefficient matrix (28).

Simulation results as well as the computational complexity of the cubic spline are discussed in the sections on simulations and comparisons of results.

Lagrange Interpolation

We discussed Lagrange interpolation for a set of nonuniform samples in the material concerning Eqs. (3)–(5). We rewrite the equations as follows:

$$f_n(x) = \sum_{i=0}^{n} L_i(x)f(x_i) \qquad (25)$$

where

$$L_i(x) = \prod_{\substack{j=0 \\ j \neq i}}^{n} \frac{(x - x_j)}{(x_i - x_j)} = \frac{\pi_i(x)}{\pi_i(x_i)} \tag{26}$$

and where

$$\pi_i(x) = \prod_{\substack{j=0 \\ j \neq i}}^{n} (x - x_j); \qquad i = 0,1, \ldots, n \tag{27}$$

Computational complexity for LG interpolation is discussed in the sections concerning simulation.

Iterative Interpolation

For IT interpolation, we use the successive approximation proposed by the author and others (29–32). Each iteration improves the signal-to-distortion ratio and potentially can recover a signal from a set of pulses with missing points without any distortion after infinite iterations provided the Nyquist rate is satisfied on the average. To analyze a voice or video signal with missing samples, we can write such a signal as

$$z(t) = \sum_{k=-\infty}^{\infty} x(kT) \, \text{sinc} \left(\frac{t}{T} - k \right) - \sum_{k=\text{lost samples}} x(kT) \, \text{sinc} \left(\frac{t}{T} - k \right) \tag{28}$$

where $z(t)$ is the low-pass-filtered version of the samples. The first term on the right side of Eq. (28) is the signal $x(t)$ and the second term is the error signal after low-pass filtering the samples. Equation (28) can be written as

$$z(t) = x(t) + e(t), \tag{29}$$

where $e(t)$ is the error term. Now, if the error power is less than the signal power, the error power can be reduced with additional iterations (31). The iterations are depicted in Fig. 10. The SNR after i iterations is

$$\frac{S_i}{N_i} = i \cdot \frac{S_1}{N_1} \, \text{dB.} \tag{30}$$

Equation (30) shows that improvement is possible if $S_1/N_1 > 0$ dB.

A more rigorous but less intuitive representation for successive approximation is

$$x_{n+1}(t) = \frac{k_1}{k_2} (PS \, x - PS \, x_n) + x_n, \tag{31}$$

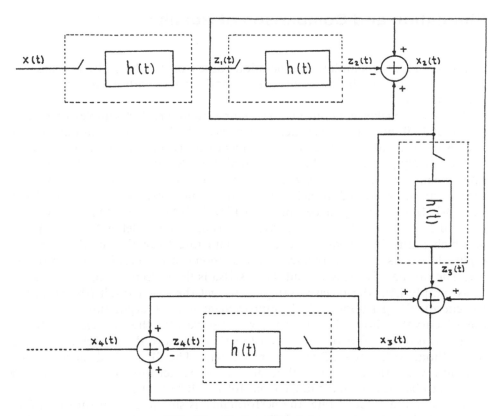

FIG. 10 The iterative technique.

where $x_0 = 0$ and

$$x(t) = \lim_{n \to \infty} x_n(t), \tag{32}$$

Equation (32) converges if the following inequalities are satisfied; that is, for any pair of band-limited square integrable signals $x(t)$ and $y(t)$, we have

$$\int_{-\infty}^{\infty} (Sx - Sy)(x - y)\, dt \geq k_1 \int_{-\infty}^{\infty} (x - y)^2\, dt \tag{33}$$

$$\int_{-\infty}^{\infty} (PSx - PSy)^2 \leq k_2 \int_{-\infty}^{\infty} (x - y)^2\, dt, \tag{34}$$

where P is the band-limiting operator and S is the sampling process (the lost samples are excluded). Simulation results and comparisons to other methods are discussed next.

Simulation and Comparison of Results

In this section, we simulate interpolations based on LI, NL, SP, IT, and LG techniques. The simulation results are for both images with missing pixels and voice packets.

For voice packet experiments, we assume there are 128 samples of speech signal (at 4 kilohertz [kHz]) in each packet. We take 125 packets, and the packet loss varies from 10% to 40%. We experiment with two different sampling rates: 8 kHz (the Nyquist rate) and 2–16 kHz (twice the Nyquist rate). At the Nyquist rate, ideally, we cannot afford losing any samples. However, the degraded speech can be smoothed in such a way that it is not noticeable to the ear. That is, a 4-kHz speech signal becomes 3.6 kHz if 10% of the samples are lost $(8 - (0.1 \times 8) = 7.2 = 2 \times 3.6\ kHz)$. This has no parallel in ARQ and FEC, in which parity bits always increase the data rate from that of the Nyquist. Figure 11 shows the performance of the different techniques in signal processing. The ordinate is the SNR and the abscissa is the percentage of lost packets. If we simply LPF the received samples, we get the worst result (shown in the LP curve). We get about 4.5-dB improvement if LI interpolation is used (see the LI curve). Above the LI curve, we observe that the IT method after 5 iterations is a few dB better than the LI method. Next to the IT technique is the SP method, with a few dB more improvement. The SP method, nevertheless, becomes worse than the IT method after 25% packet loss. The NL technique appears to be the best in the range from 10% to 40% packet loss.

At twice the Nyquist rate, the performance is similar to that shown in Fig. 11 except that the IT method definitely is better than the SP and the SNR improvements are about 20–30 dB better than that of the Nyquist rate. Figure 12 shows the performance of each method. This figure shows that the IT technique can outperform the NL technique if 10 or more iterations are used; this increases the cost of complexity and computation.

For facsimile images with missing pixels, we experimented with LI, NL, and IT interpolations. The results for a color image with 10% packet loss are shown in Fig. 13. Figure 13-*a* shows the degraded image with 10% pixel loss. The results of interpolations using NL and IT methods are shown in Figs. 13-*b* and 13-*c*, respectively. The performance of different techniques for black-and-white images is shown in Table 1. This table shows that, in general, NL and IT procedures are better than LI ones. A fair comparison of different techniques should take into consideration the complexity of interpolation at the receiver.

In the next sections, we use an illustrative example for speech packets to compare the computation required for the various interpolation methods. Note that all operations refer to floating-point calculations.

LI Method

To use the linear method, for each missing sample we need one addition and one multiplication. Therefore, for 125 packets of 128 samples each, if the packet

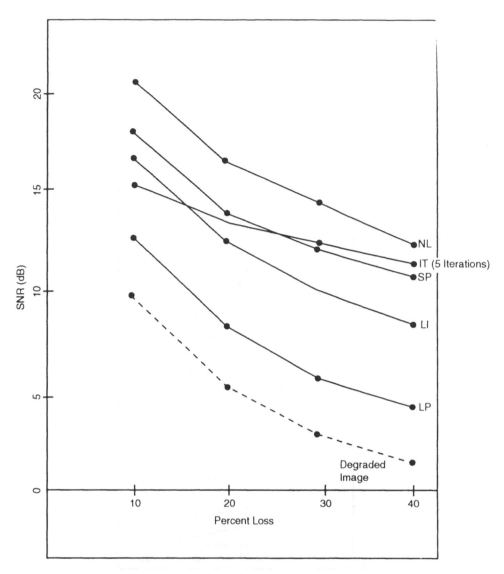

FIG. 11 Performance of the different interpolation techniques at the Nyquist rate (NL = nonlinear; IT = iterative; SP = spline; LI = linear; LP = low pass).

loss is 10% we need 1600 multiplications and 1600 additions. This method is by far the simplest method to implement.

NL Method

For the NL method, we need to LPF both the upper and lower branches (see Fig. 2). We use a 31-point LPF, requiring 31 multiplications and 30 additions for the upper branch. The lower branch requires only additions since the inputs

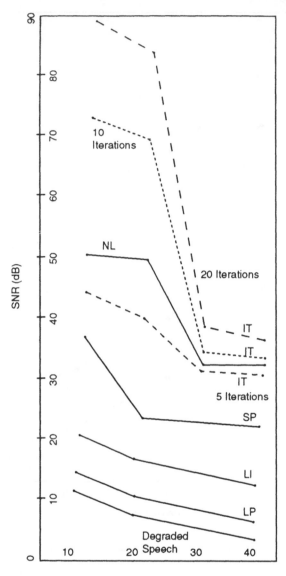

FIG. 12 Performance of different interpolation techniques at twice the Nyquist rate (see Fig. 11 for explanation of abbreviations).

to the LPF are all ones or zeros. Finally, we also require a single division (which we assume is equivalent computationally to a multiplication). Assuming a 10% packet loss rate and 125 packets (16,000 samples total), we have

$1600 \times 31 = 49,600$ multiplications
$1600 \times 60 = 96,000$ additions
1600 divisions

(a)

(b)

(c)

FIG. 13 Interpolation of facsimile image with missing pixels: *a*, degraded image after 10% pixel loss; *b*, recovered image using the nonlinear method; *c*, recovered image using the iterative method.

TABLE 1 The Signal-to-Noise Ratio of Different Image-Recovery Techniques (dB)

Percent of Lost Sample	Low-Pass Filter	Linear Interpolation	Nonlinear Method	Ten Iterations
10	19.0	35.0	44.8	51.1
20	14.2	32.6	39.5	40.5
30	11.0	30.4	32.6	30.3
40	8.6	27.8	28.2	22.6
50	6.7	23.9	24.4	16.5

Cubic Spline Interpolation (SP)

From Eqs. (23) and (24), since spline interpolation is done for each packet independently, there are 125 packets of 128 samples each ($n + 1 = 16,000$). We have $N - 1$ equations and $N - 1$ unknowns where $N = 128 - (0.1 \times 128)$, giving a total of $N = 116$. However, because the system equations are tridiagonal, the solution can be obtained efficiently using the Gaussian decomposition to invert the $N \times N$ coefficient matrix. The total number of calculations required are derived as follows

1. Set up 125 tridiagonal systems of N equations. This requires 8 additions and 6 multiplications for each equation. The total is 116,000 additions and 87,000 multiplications.
2. Solve the tridiagonal matrix via the Gaussian decomposition:
   a. Decomposition step = 29,000 multiplications and 14,500 additions
   b. Forward substitution = 14,500 multiplications and 14,500 additions
   c. Back substitution = 29,000 multiplications and 14,500 additions
3. Interpolate 125×13 points for 10% packet loss. Each point requires 18 multiplications and 12 additions; this yields 28,800 multiplications and 19,200 additions.

The total computation is 188,300 multiplications and 178,700 additions. This represents 3–4 times the requirement for the NL method. Also, the cubic spline method needs access to the entire data segment (128 samples) before any interpolation occurs, which may make it unsuitable for some real-time applications.

Iterative Method

Each iteration requires an LPF, a sampler, and an adder. We use the same LPF as the one used in the NL method. The sampler is a multiplexer that either transfers the samples or makes them zero. Therefore, the complexity is equivalent to an adder. If K is the number of iterations, for 10% packet loss the total number of additions is $48,000K$ for low-pass filtering plus $16,000K$ for sampling and $14,400K$ at the end, equaling $78,400K$; there are $49,600K$ multiplications. For five iterations, the results are 248,000 multiplications and 392,000 additions. This computation is 4–5 times that of the NL method.

Lagrange Method (LG)

In cases of Lagrange interpolation such as those given in Eqs. (26) and (27), for a 10% packet loss rate and $N = 128$ samples, for each packet we have the following:

1. For $\pi_i(x_i)$, we need $3 \times (0.9N)$ additions and $3 \times (0.9N)$ multiplications.
2. For $L_i(x)$, we need $3 \times (0.9N)$ additions and $6 \times (0.9N)$ multiplications for each lost data point.
3. For $f_n(x)$, we need $0.9N$ additions and $0.9N$ multiplications for each lost data point.

Since there are 125 packets, we need $125(3 \times 0.9N + 3 \times 0.1N \times 0.9N + 0.1N \times 0.9N) = 125(2.7N + 0.36N^2)$ multiplications and $125(3 \times 0.9N + 3 \times 0.2N \times 0.9N + 0.1N \times 0.9N) = 125(2.7N + 0.63N^2)$ additions. For $N = 128$, we have 780,480 multiplications and 1,333,440 additions. As seen, the computational complexity of the LG method becomes prohibitive as the number of samples increases. This represents about 16 times the requirement for the NL method. Besides, continuity of the signal and its derivative along the border lines is a problem in LG methods.

A comparison of the above methods shows that as far as computation is concerned, the order of increasing complexity is represented by the LI, NL, SP, IT, and LG methods, respectively. The LG method, like that of the SP, is also a "block" method. That is, we need all of the data samples in a packet available prior to performing the interpolation. In contrast, the LI and NL methods operate with a minimal delay and therefore are far more suited to real-time implementation. The IT method also can be implemented in real time. As far as the performance is concerned, the NL method is far better than the LI method and, in general, better than the other methods.

Conclusion

The integration of voice, data, and video signals has become a reality. This integration may create congestion at switching nodes and hence cause packet delays. For voice and video signals, packet loss is inevitable since excessive delay is not acceptable. In order to recover the lost packets, FEC, ARQ, and signal interpolation can be used. In this article, the emphasis has been on signal interpolation for voice and facsimile packets. We have shown the feasibility of various techniques to improve the SNR. The advantage of signal processing is the simplicity of implementation.

List of Acronyms

ACK   acknowledgment packet
ADM   adaptive delta modulation
ADPCM   adaptive differential pulse code modulation
ARQ   automatic repeat request (query)
ATM   Asynchronous Transfer Mode
DM   delta modulation
DPCM   differential pulse code modulation
DSI   digital speech interpolation
FEC   forward error correction
IT   iterative interpolation
LG   Lagrange
LSB   least significant bit
LI   linear interpolation
MSB   most significant bit
NAK   negative acknowledgment
NL   nonlinear interpolation
PCM   pulse code modulation
SNR   signal-to-noise ratio
SP   spline interpolation

References

1. Turner, L., Aoyama, T., Pearson, D., Anastassiou, D., and Minami, T., Packet Speech and Video, *IEEE J. Sel. Areas Commun.*, 7:629 (June 1989).
2. Marvasti, F., Wideband Packet Technology: Exploratory Directions and Issues, internal memorandum, AT&T Bell Laboratories, Document No. 55116-870408-01IM, 1987.
3. Jerri, A. J., The Shannon Sampling Theorem—Its Various Extensions and Applications: A Tutorial Review, *Proc. IEEE*, 65(11):1565–1596 (November 1977).
4. Marvasti, F., *A Unified Approach to Zero-Crossings and Nonuniform Sampling of Single and Multidimensional Signals and Systems*, Nonuniform Publication, Oak Park, IL, 1987.
5. Marvasti, F., Nonuniform Sampling. In: *Advanced Topics in Sampling Theory* (Robert Marks, ed.), Springer-Verlag, New York, 1993, pp. 121–156.
6. Titchmarsh, E. C., *The Theory of Functions*, 2d ed., University Press, London, 1939.
7. Kadec, M. I., The Exact Value of the Paley–Wiener Constant, *Soviet Math. Dokl.*, 5:559–561 (1964).
8. Higgins, J. R., *Completeness and Basis Properties of Sets of Special Functions*, Cambridge University Press, Cambridge, 1977.
9. Paley, R.E.A.C., and Wiener, N., *Fourier Transforms in the Complex Domain*, American Mathematical Society, New York, 1934.

10. Khurgin, Y. I., and Yakovlev, U. P., Progress in the Soviet Union on the Theory and the Applications of Band-Limited Functions, *Proc. IEEE*, 65(7):1005–1029 (July 1977).

11. Marvasti, F. A., Spectrum of Nonuniform Samples, *Electronic Letters*, 20:896–897 (1984).

12. Marvasti, F., Clarkson, P., Dokic, M., and Chuande, L., Reconstruction of Speech Signals from Lost Samples, *IEEE Trans. Signal Processing*, 2897–2902 (October 1992).

13. Rowe, H. E., *Signals and Noises in Communication Systems*, Van Nostrand, Princeton, NJ, 1965.

14. Marvasti, F., Chuande, L., and Adams, G., The Analysis and Recovery of Multidimensional Signals from Irregular Samples Using Nonlinear and Iterative Techniques, *Signal Processing J.* (forthcoming).

15. Lin, S., and Costello, D. J., Jr., *Error Control Coding: Fundamentals and Applications*, Prentice-Hall, Englewood Cliffs, NJ, 1983.

16. Marvasti, F., and Liu, C., Oversampling as an Alternative to Error Correction Codes in Digital Communication Systems, *Proc. ICC '92*, 344–347 (1992).

17. Goodman, D., Lokhart, G., Wassem, O. J., and Wong, W. C., Waveform Substitution Techniques for Recovering Missing Speech Segments in Packet Voice Communication, *IEEE Trans. ASSP*, ASSP-34:1440–1448 (December 1986).

18. Jayant, N. S., and Christensen, S. W., Effects of Packet Losses in Waveform Coded Speech and Improvements Due to an Odd-Even Sample-Interpolation Procedure, *IEEE Trans. Commun.*, COM-29:101–109 (February 1981).

19. Suzuki, J., and Taka, M., Missing Packet Recovery Techniques for Low Bit Rate Coded Speech, *J. Sel. Areas Commun.*, 7(5):707–717 (June 1989).

20. Gruber, J., and Strawcznski, L., Subjective Effects of Variable Delay and Speech Clipping in Dynamically Managed Voice Systems, *IEEE Trans. Commun.*, COM-33:961–962 (August 1985).

21. Gruber, J., and Le, N., Performance Requirements for Integrated Voice/Data Networks, *IEEE J. Sel. Areas Commun.*, SAC-1:981–1005 (December 1983).

22. Wasem, O. J., Goodman, D. J., Dovrak, C. A., and Page, H. G., The Effect of Waveform Substitution on the Quality of PCM Packet Communications, *IEEE Trans. ASSP*, ASSP-36(3):342–347 (March 1988).

23. Erdol, N., Ziluchian, A., and Castellucia, C., Recovery of Missing Speech Packets Using the Short-Time Energy and Zero-Crossing Measurement, Conference Proceedings BILKENT '90, Bilkent University, Ankara, Turkey, July 1990.

24. Goodman, D. J., Embedded DPCM for Variable Bit Rate Transmission, *IEEE Trans. Commun.*, COM-28:1040–1046 (July 1980).

25. Ching, Y. C., U.S. Patent 3 781 685, Differential Pulse Code Communications System Having Dual Quantization Schemes, issued December 25, 1973.

26. Decina, M., and Modena, G., CCITT Standards on Digital Speech Processing, *IEEE J. Sel. Areas Commun.*, 6(2):227–234 (February 1988).

27. Cheney, W., and Kincaid, D., *Numerical Mathematics and Computing*, 2d ed., Brook/Cole, Monterey, CA, 1985.

28. Schwarz, H. R., *Numerical Analysis*, John Wiley and Sons, New York, 1989.

29. Marvasti, F., and Analoui, M., Recovery of Signals from Nonuniform Samples Using Iterative Methods, IEEE Proc. of International Conference on Circuits and Systems, Oregon, July 1989.

30. Marvasti, F., Analoui, Mostafa, and Gamshadzahi, M., Recovery of Signals from Nonuniform Samples Using Iterative Methods, *IEEE Trans. ASSP*, 872–876 (April 1991).

31. Marvasti, F., Spectral Analysis of Random Sampling and Error Free Recovery by

an Iterative Method, *IECE Trans. Inst. Electron. Commun. Engs.*, Japan (Section E), E 69(2):79–82 (February 1986).

32. Wiley, R. G., Recovery of Band-Limited Signals from Unequally Spaced Samples, *IEEE Trans. Commun.*, COM-26(1):135–137 (January 1978).

FAROKH A. MARVASTI

The Federal Communications Commission of the United States

Authority of the Commission to Set Communications Policy

The Federal Communications Commission (FCC or commission) is an independent federal regulatory agency responsible directly to Congress. Established by the Communications Act of 1934 (47 U.S.C.A. §151) it is charged with regulating interstate and international communications by broadcasting (radio and television), cable, wireless cable, satellite operations, and common carriers or telephone companies, and private radio carriers. The commission also conducts equipment certification for electronic devices that emit radio frequencies. Its jurisdiction covers the 50 states and territories, the District of Columbia, and U.S. possessions. Its rules are codified in Title 47 of the Code of Federal Regulations, Parts 0.1 through 300.1.

Functions of the Commission

The FCC has the responsibility to adopt policies and rules affecting the public and a range of different communications and telecommunications companies including AT&T, the Bell Operating Companies, GTE, TCI, the broadcast television networks, and cable companies, to name just a few of its regulatees. These are the companies that keep the American people in touch with one another. The broadcast stations and cable systems provide news, information, and entertainment programming to the public. The telephone companies, through copper and fiber links connect people throughout the nation, and by satellite, throughout the world. The private radio companies provide needed communications for industries, local governments, and the public, including the frequencies used to support the public safety and health industries. Manufacturing companies produce electronic devices such as computers, cellular phones, and other equipment the public uses to process information.

The Commission, the Bureaus, and the Offices

The FCC (see Fig. 1) is directed by five commissioners who are appointed by the president and confirmed by the Senate for five-year terms. The president designates one of the commissioners to serve as chairperson. Three of the com-

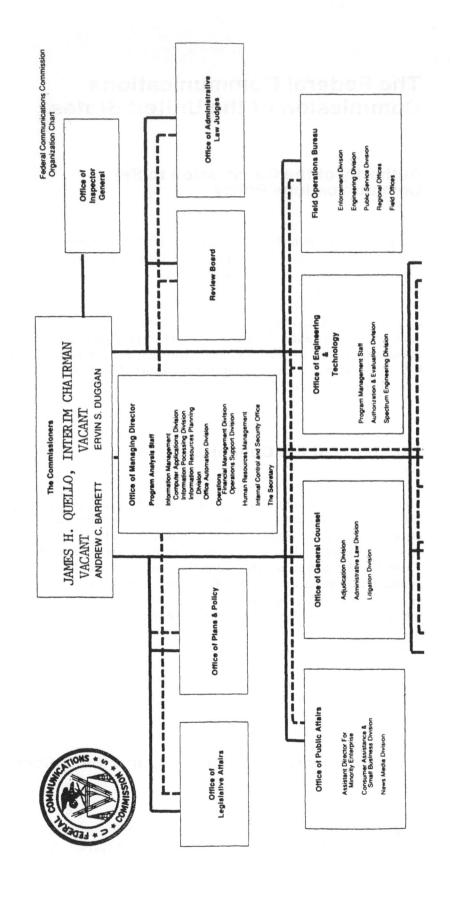

Federal Communications Commission
Organization Chart

The Commissioners

JAMES H. QUELLO, INTERIM CHAIRMAN
VACANT
VACANT
ANDREW C. BARRETT
ERVIN S. DUGGAN

Office of Inspector General

Office of Administrative Law Judges

Review Board

Field Operations Bureau
Enforcement Division
Engineering Division
Public Service Division
Regional Offices
Field Offices

Office of Managing Director

Program Analysis Staff

Information Management
Computer Applications Division
Information Processing Division
Information Resources Planning Division
Office Automation Division
Operations
Financial Management Division
Operations Support Division
Human Resources Management
Internal Control and Security Office
The Secretary

Office of Engineering & Technology
Program Management Staff
Authorization & Evaluation Division
Spectrum Engineering Division

Office of General Counsel
Adjudication Division
Administrative Law Division
Litigation Division

Office of Plans & Policy

Office of Legislative Affairs

Office of Public Affairs
Assistant Director For Minority Enterprise
Consumer Assistance & Small Business Division
News Media Division

482

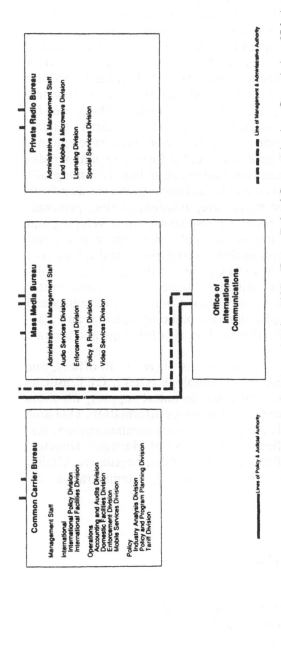

FIG. 1 Federal Communications Commission organization chart (October 1990). (From: *Federal Communications Commission, 57th Annual Report/Fiscal Year 1991,* U.S. Government Printing Office, Washington, DC, 1992.)

missioners are members of the president's political party (majority party) and the other two are members of the other political party or an independent party (minority party). One of the commissioners from the majority party performs the role of chairperson. The other commissioners also perform different roles to assist in the orderly functioning of the commission. As the chief executive officer of the commission, the chairperson delegates management and administrative responsibility to the managing director. Certain other functions are delegated to staff units, bureaus, and to committees of the commissioners.

The four bureaus conducting principal FCC operations are the Mass Media, Common Carrier, Private Radio, and Field Operations Bureaus. The Mass Media Bureau regulates AM (amplitude modulation), FM (frequency modulation), and television broadcast stations and related facilities; administers and enforces cable TV rules; and licenses private microwave radio facilities used by wireless cable systems or educational institutions. The Common Carrier Bureau regulates wire and radio communications common carriers such as telephone, telegraph, and satellite companies. The Private Radio Bureau regulates radio stations serving the communications needs of businesses, individuals, nonprofit organizations, and state and local governments, including such uses as private land mobile, private operational fixed microwave, aviation, marine, personal, amateur, and disaster. These three bureaus are responsible for developing and implementing regulatory programs, processing applications for licenses and other filings, analyzing complaints, conducting investigations, and taking part in FCC hearings. The fourth bureau, the Field Operations Bureau, detects violations of radio regulations, monitors transmissions, inspects stations, investigates complaints of radio interference, and issues violation notices. It examines and licenses radio operators; processes applications for painting, lighting, and placing antenna towers; and furnishes direction-finding aids for ships and aircraft in distress.

In addition to the 5 commissioners and 4 bureaus, there are 10 offices, one of which is the Office of Engineering and Technology (OET). OET uses its expertise to provide the FCC with scientific and technical support, monitors scientific and technological developments, and analyzes information. One area in which OET currently is involved is the PCS (personal communications services) docket. The remaining nine offices are the offices of Managing Director, Small Business Activities, Plans and Policy, Public Affairs, Legislative Affairs, Inspector General, International Communications (OIC), Administrative Law Judges, and General Counsel.

How the Commission Works

The commissioners hold regular open and closed agenda meetings and special meetings. They also may act between meetings by "circulation," a procedure in which a document is submitted to each commissioner for consideration and official action.

The chairperson presides over all FCC meetings, coordinates and organizes the work of the commission, and represents the agency in legislative matters and in relations with other government departments and agencies. If the chairperson is absent or the office is vacant, the commission designates one of its commissioners to act temporarily as chairperson.

The commission holds at least one open agenda meeting each month. Members of the public are permitted to watch these meetings to see the government in action. These meetings allow the commission staff to bring items forward for commission decision making. The majority of the items at these meetings involve major policy decisions that affect the future of the communications industry. In addition, the commission conducts closed meetings that members of the public are not allowed to attend. However, a record is maintained of the proceedings. Closed meetings generally involve adjudicatory proceedings in which two or more parties are vying for a commission license or proceedings in which there is a petition to deny a commission license. Finally, routine items often are handled through the commissioner circulation voting process or on delegated authority to one of the commission's bureaus or offices. Through this process, items are given to each commissioner's office with opportunity for edits or rewrites.

In each type of proceeding, the commissioners or commission staff discuss the legal and/or policy implications involved in the case. Under the Sunshine in Government Act, however, no more than two commissioners are allowed to discuss any issues at the same time, unless public notice of the meeting is given. Despite this rule, the commissioners do manage to get input individually from one another before making major decisions. The commissioners are assisted by legal and other professional staff members, who can hold discussions on various voting items. Because of this system, commissioners are required to rely on staff to keep them informed and represent their views with representatives of the other commission offices or bureaus.

The Commission on the International Front

Staff from the FCC participate with staff from the telecommunications regulatory authorities of other countries in the formation of international standards governing technical and operational characteristics of radio systems, interference matters, and worldwide frequency allocations. For example, the World Administrative Radio Conference convened in Spain from February 3, 1992 to March 3, 1992 to address frequency allocations in certain parts of the radio spectrum.

The OIC mentioned above coordinates commission international policy activities and ensures that the commission's international policies are uniform and consistent. The OIC, in conjunction with international staff from the appropriate commission bureau or office, assumes the principal representational role for commission activities in international forums, serves as the focal point for international activities, and advises the commission on international matters. The OIC provides coordination among the bureaus in the development of international telecommunications policy, representation of this policy, and participation in international conferences.

The Commission and the Economy

The combined assets of the entities regulated by the commission total several hundred billion dollars. The financial viability of many of these companies has a significant impact on the overall economic health of the United States.

FCC rules also have an impact on companies not directly under its jurisdiction. For example, financial interest and syndication rules have an impact on Hollywood studios. In addition, the FCC's regulation of equipment affects many computer and manufacturing companies whose devices have radio-frequency emissions. Under Part 15 of its rules, the commission is responsible for establishing and enforcing emission standards for intentional and unintentional radiators that operate without an individual license. Part 18 of the commission's rules covers devices of this nature that operate as industrial, scientific, or medical equipment. Under these two sections of its rules, the commission affects the equipment manufactured in the United States and the equipment imported from other nations.

Many companies regulated by the FCC contribute to the positive trade balance of the United States in the services sector. Further, advanced communications infrastructure creates a positive atmosphere for foreign nations wishing to conduct business in the United States. Those companies doing business in the United States know that this country attempts to stay at the forefront of modern communications technology.

Recent Actions

The FCC is at the cutting edge of new technology. For example, in recent years the commission modified its telephone cable crossownership rules to permit local telephone companies to offer video dial-tone services. In addition, the FCC modified its radio station ownership rules to allow a single licensee to own up to 36 stations (18 AM, 18 FM), up from the previous 24 stations (12 AM, 12 FM). In 1994, the limit will jump to 40 stations (20 AM, 20 FM). The new rules also allow broadcasters that have reached the national limit to hold a noncontrolling interest in up to three stations owned by minority or "small-business" entities. Following this decision, the commission formed the Office of Small Business Activities and a Small Business Advisory Committee to address other issues affecting small- and minority-business enterprises. In addition, the commission is currently in the process of making decisions concerning the technical standards for high-definition television, digital audio broadcasting, and PCS. PCS is a new wireless communications service that is expected to grow to 23 million subscribers by the end of 1994 and 68 million subscribers by the year 2002. The FCC recently allocated 220 megahertz (MHz) of a 2-gigahertz (GHz) spectrum for such emerging technologies as PCS and adopted a transition framework designed to prevent disruption to incumbent 2-GHz fixed-microwave licensees. In addition, the FCC tentatively awarded pioneer's preferences to three applicants for new personal communications services. Finally, the commission has formulated new cable rate regulations pursuant to the Cable Act of 1992.

The commission's decisions in these areas bring such companies as COMSAT (Communications Satellite Corporation), the Bell Operating Companies, interexchange carriers like AT&T and MCI, cable and entertainment companies like Time Warner, and large manufacturing and service conglomerates like General Electric to the FCC. The new services offered by these types of major corporations has an impact on all facets of our daily lives—such as radio station programs, telephone calls, cable television services, and satellite services.

A Commission for the Future

Although the commission accomplished much in 1992, it has even more to do in terms of improving the competitive posture of the United States in the telecommunications field. As American companies increasingly face global competitive challenges, this agency must endeavor to minimize regulating that hampers their international competitiveness. The FCC increasingly is addressing the international ramifications of regulatory issues.

Over the years, the commission has improved its responsiveness to public demands for action and industry demand for regulatory review. There is still much room for improvement. Resources are the key to making this a better agency. The FCC must update its equipment in terms of modern technology and increase its staff levels in order to serve the future needs of a dynamic communications industry.

Freedom of Information Act

The FCC is in full compliance with the Freedom of Information Reform Act of 1986 (Public Law No. 99-570, Sections 1801–1804). FCC rules specify the required contents of requests for records not routinely available, and set time limits for acting on requests for records and applications to be used in reviewing adverse decisions. In addition, agency rules require disclosure of the names and titles of persons responsible for denying an inspection request and provide for deletion of exempted portions of a record and disclosure of the remainder. Readers interested in information on the FCC's rules and/or rulings should address their inquiries to

Office of Associate Managing Director
Public Service Division
1919 M Street, N.W., Room 254
Washington, DC 20554
(202) 632-7260

Acknowledgment: Lisa McArthur, a student at the University of California at Los Angeles School of Law, assisted in the preparation of this article.

Bibliography

The Consumer Assistance and Small Business Division, Office of Public Affairs, Federal Communications Commission, *Information Seekers Guide*, April 1989.

Federal Communications Commission, *57th Annual Report/Fiscal Year 1991*, U.S. Government Printing Office, Washington, DC, 1992.

BYRON F. MARCHANT